Global Optimization in Action

Nonconvex Optimization and Its Applications

Volume 6

The titles published in this series are listed at the end of this volume.

Global Optimization in Action

Continuous and Lipschitz Optimization: Algorithms, Implementations and Applications

by

János D. Pintér

Pintér Consulting Services,
Dalhousie University

KLUWER ACADEMIC PUBLISHERS

DORDRECHT / BOSTON / LONDON

A C.I.P. Catalogue record for this book is available from the Library of Congress.

ISBN 0-7923-3757-3

Published by Kluwer Academic Publishers,
P.O. Box 17, 3300 AA Dordrecht, The Netherlands.

Kluwer Academic Publishers incorporates
the publishing programmes of
D. Reidel, Martinus Nijhoff, Dr W. Junk and MTP Press.

Sold and distributed in the U.S.A. and Canada
by Kluwer Academic Publishers,
101 Philip Drive, Norwell, MA 02061, U.S.A.

In all other countries, sold and distributed
by Kluwer Academic Publishers Group,
P.O. Box 322, 3300 AH Dordrecht, The Netherlands.

Printed on acid-free paper

Printed in the Netherlands

Dedicated

to the memory of my Father

and

to my Mother

— both far, yet close ...

Contents

List of Figures

List of Tables

Preface

Long, windy prefaces are boring...
Besides, it's far better to see the arguments for yourself.
(John Casti, *Searching for Certainty*,
William Morrow and Co., New York, 1990.)

This (admittedly not too short) preface serves to put the subject of global optimization—and, within that scope, the present volume—into perspective, especially for the novice in this exciting field.

Making decisions is an everyday activity, for all of us. Many of these decisions are simple and barely consequential, while many others are serious and responsible. The prior limitations influencing these decisions and the ranking of their consequences cannot always be simply expressed in quantitative terms. Nevertheless, proper quantification frequently assists the decision-making process, with an attempt toward objectivity. This endeavour is especially characteristic of scientific and many other professional investigations and studies.

In the sciences, engineering and economics, decision problems are frequently modelled as optimizing the value of a (primary) objective (criterion, performance, loss etc.) function, under stated feasibility constraints to be met by all 'acceptable' decisions. To reflect this paradigm, the classical theory of constrained optimization, including also the body of mathematical programming, investigates problems of the generic form

$$\min f(x) \quad \text{subject to} \quad x \in D \subset \mathbb{R}^n. \tag{OPT}$$

The function f symbolizes the objectives (goals) in the decision problem, and D denotes the set of all a priori admissible decisions (n-vectors) x.

Note that, most typically, it is also assumed that the formulation (OPT) will lead to a singular optimal solution, corresponding to the unique, unambiguously 'best' decision. Uniextremality is frequently postulated or implied by the mathematical problem structure (for example, by the strict convexity of the objective function f, and the convexity of the feasible set D). This generic optimization model form corresponds to the situation in which one supposedly has a sufficiently close initial 'guess' of the feasible (sub)region where the unknown optimal solution is located. Hence, the global optimality of the solution directly follows, having found the single local—and thus also global—optimum of the criterion function f on D.

Although many practically important optimization models naturally belong to this category, there is also a broad class of problems in which the property of uniextremality cannot be simply postulated or verified. Consider, for example, a system of nonlinear equations in which only rather vague ranges can be provided with respect to the location of the solution(s): then it may well happen that the system does not have a solution in the region specified, or it has a number of 'suboptimal' solutions (with corresponding non-zero residuals), besides one or several 'true' (global) solutions. In other cases—for example, in studying complex engineering or environmental systems—the system performance (objective) function f and/or some constraints defining the feasible region D may be determined only by running some computationally intensive numerical model, algorithm, performing a set of experiments or Monte Carlo simulation cycles, etc. Therefore the performance of each particular decision alternative needs to be evaluated by activating a large number of 'oracle' or 'black box' operations. One may refer, for example, to the calibration of descriptive physical, chemical, biological, ecological, etc. system models, the optimized 'tuning' of industrial or laboratory instruments and processes, and so on.

Frequently, in these or similar circumstances the underlying optimization problem does not possess the (verified) uniextremality property. In addition, the number of 'candidate' local optima is typically also unknown, and some of the locally optimal solutions may be very 'far'—in terms of the criterion function f—from the best possible decision. Therefore, in such cases, it is quite natural to search for the global solution.

Although finding globally best quantitative decisions must have been a valid issue for a long time, it has not been systematically studied until very recently. It should be emphasized in this context that—because of the a priori set limitations of their underlying problem model—standard local scope optimization methods are (on their own) inherently non-applicable for solving multiextremal problems, in the 'true' global sense. Therefore such problems have been thought basically intractable until not too long ago. To illustrate this point, the reader may observe that many prominent textbooks in

- operations research/mathematical programming (consult, for example, Himmelblau, 1972a; Simmons, 1975; Avriel, 1976; Moder and El Maghraby, 1978; Hestenes, 1980; Bachem, Grötschel and Korte, 1983; Dennis and Schnabel, 1983; Hillier and Lieberman, 1986; Winston, 1994)
- algebraic modelling and optimization systems and software (Murtagh and Saunders, 1983, 1987; Lasdon and Waren, 1986; Fourer, Gay and Kernighan, 1987, 1993; Brooke, Kendrick and Meeraus, 1988; Schrage, 1990, 1991)
- statistical decision theory (Berger, 1985; French, 1986)
- numerical analysis (Press, Teukolsky, Vetterling and Flannery, 1992)
- computational mathematics (Wolfram, 1991)

either completely omit or restrict the discussion of multiextremal optimization issues to at most a few pages and remarks. At the same time—driven by apparent necessity—practitioners in diverse fields of applied sciences and engineering have been widely using rudimentary tools, and (semi)heuristic approaches to solve 'their own' global optimization problems. Although the application of such methods may have been quite successful in many cases, their theoretical foundations, robustness, reliability and efficiency have raised questions which have not been systematically addressed until recently.

The emerging field of global optimization deals with the theory, implementation and application of models and strategies to solve (potentially) multiextremal decision problems. In light of the discussion above, it is probably not surprising that there exist quite a few early theoretical results and numerical attempts related to the subject: see, for instance, Brooks (1958), Moore (1962, 1966), Kushner (1964), Tuy (1964), McMurty and Fu (1966), Chichinadze (1967, 1969), Danilin and Piyavskii (1967), Mockus (1967, 1972), Piyavskii (1967, 1972), Vaysbord and Yudin (1968), Zhidkov and Schedrin (1968), Clough (1969), Neimark and Strongin (1969), Bremmerman (1970), de Haan (1970), Danilin (1971), Evtushenko (1971), Martos (1971, 1975), Sukharev (1971), Branin (1972), Forgó (1972), Ivanov (1972), Shubert (1972), Ueing (1972), Hartman (1973), Strongin (1973), Majthay and Whinston (1974). However—at least to my knowledge—the first collection of mathematical programming papers devoted exclusively to global optimization in the English literature was published only in the mid to late seventies (Dixon and Szegö, 1975, 1978).

Following these pioneering works, multiextremal optimization has gradually become an increasingly popular subject of research. A number of promising approaches have been suggested for solving certain global optimization problem-classes. Surveys and monographs have appeared. See, for example, Strongin (1978), Wilde (1978), Zielinski and Neumann (1983), Evtushenko (1985), Fedorov (1985), Žilinskas (1986), Pardalos and Rosen (1987), van Laarhoven and Aarts (1987), Forgó (1988), Ratschek and Rokne (1988), Mockus (1989), Rinnooy Kan and Timmer (1989), Törn and Žilinskas (1989), Diener (1990), Horst and Tuy (1990, 1993), Zhigljavsky (1991), Zhigljavsky and Žilinskas (1991), Floudas and Pardalos (1992), Hansen (1992), Pardalos (1994), Horst and Pardalos (1995). The increasing maturity of the field is indicated also by the publication of the *Journal of Global Optimization* since 1991, by Kluwer Academic Publishers. In 1993, this publisher also launched a series titled 'Nonconvex Optimization and Its Applications,' in which the present volume appears.

The wide variety of uniextremal optimization strategies (for example, for solving problems in linear programming, classical game theory, convex quadratic programming, general convex minimization under linear constraints, nonlinear convex programming) is essentially based on assumptions concerning the underlying problem structure. Similarly, there exists a broad range of global optimization problem-types such as, for example, concave minimization under linear or general convex constraints, (indefinite) quadratic programming, reverse convex programming, differential convex (D.C.) programming, general fractional programming, geometric programming, Lipschitz-continuous, or continuous global optimization, etc. The underlying specific theory and the solution approaches to such diverse global optimization problems—as can be expected—also vary to a considerable extent.

The present volume offers a comprehensive treatment of a broad class of deterministic, direct (gradient-free) optimization algorithms. These methods are based on adaptive, sequential partition and search in the feasible set D. The complete, 'exhaustive' search feature of such procedures guarantees their theoretical convergence to the global solution(s). In addition, a variety of fairly efficient numerical implementations can be realized on personal computers, workstations or modern mainframes, including parallel computers. The methodology discussed here can primarily be recommended for solving Lipschitzian and continuous global optimization problems, in which the lack of further analytical information realistically excludes the possibility of applying more specifically 'tailored' approaches. Besides the deterministic methodology suggested to solve these problems, a shorter discussion will also be devoted to adaptive stochastic algorithms which also enjoy wide popularity, especially in solving difficult 'black box' global optimization problems with a (merely) continuous objective function. In fact, (artificial) stochastic elements will be included also in the discussed implementation of partition algorithms. Moreover, stochastic phenomena and processes play an essential role in several application areas discussed in the present volume.

In accordance with these objectives, Part 1 of this book presents a short introduction to the subject of multiextremal optimization. The aim of this part is to review some of the most important model-types and corresponding solution approaches. Part 2 contains the principal theoretical core of the book, providing a detailed treatise of the general convergence properties of adaptive deterministic partition algorithms. Some further theoretical results and important aspects of efficient implementations of partition strategies are treated in Part 3. Finally, Part 4 discusses several existing and prospective applications of continuous and Lipschitz global optimization: many of these applications originate from the field of environmental systems modelling and management.

As one can see from this brief overview, there are many branches of global optimization which will not be treated here in much detail. With respect to these areas, the interested reader is referred to the rapidly growing literature of the field. Similarly, although the present book contains a fairly extensive list of references, it is certainly far from being complete. Many additional references—typically dealing with more specific problem-classes—can be found, for example, in the monographs and surveys cited.

Apart from Part 1, the book intends to be self-contained. Parts 2 to 4 are based to a large extent on the author's research. In a number of cases—

particularly with respect to a range of applications—the material draws on work accomplished in cooperation with other researchers (please see the Acknowledgements).

It is assumed that the reader is familiar with the elements of mathematical analysis and probability theory. In addition, some familiarity with numerical optimization algorithms is obviously desirable, and an Operations Research, Science, Economics and/or Engineering background is certainly an asset. In terms of the applications discussed, some experience in modelling engineering-environmental systems can be useful, but is not required. The prospective circle of readers may, hence, encompass a fairly broad range, including

- operations research, management science, engineering or computer science professors, lecturers, and students
- researchers active in the field of optimization
- researchers and other professionals who want to apply global optimization in diverse fields of science, engineering and economics.

As the book is devoted to a rapidly changing field, it certainly does not mean to summarize something 'finished'. Constructive comments and suggestions from all readers will be sincerely appreciated.

Acknowledgements

Since graduating from the University of Sciences (Eötvös University), Budapest, in 1973, I have been interested in many of the subjects treated in this book. Without an attempt to put together a sort of 'complete' list, I would like to acknowledge at least a few of my professors and other outstanding researchers who certainly motivated and shaped my professional interests. In alphabetical order these include:

Blair T. Bower, Ekaterina V. Bulinskaya, Rolf A. Deininger, Ferenc Forgó, Boris V. Gnedenko, Pierre Hansen, Reiner Horst, Brigitte Jaumard, Emil Klafszky, László Béla Kovács, Antal Majthay, András Prékopa, Charles R. Scherer, László Somlyódy, Roman G. Strongin and András Szöllősi-Nagy.

Next, I would especially like to thank my co-authors for their valuable contributions to a number of joint research projects and papers discussed and referred to in this book. Again, in alphabetical order, these include the following individuals:

Carlijn Bak-Ijsberg, Pál Benedek, Johan G. Boon, János C. Fodor, Roger M. Cooke, Tibor Csendes, Attila Darázs, Willem J. De Lange, Mort Fels, Russell J. Finley, Sjur D. Flåm, Eligius M. T. Hendrix, David S. Lycon, Dirk J. Meeuwig, Jay W. Meeuwig, Géza Pesti, Mysore Satish, László Somlyódy, János Szabó, Dorien ten Hulscher, Diederik T. Van der Molen, Johan Van Zetten and Leo Voogt.

The support of the following institutions—where I was working on subjects discussed here—is also gratefully acknowledged:

Computing Center for Universities, Budapest
University of Economics, Budapest
Eötvös University of Sciences, Budapest
Moscow State University (Lomonosov University)
Water Resources Research Centre, Budapest
Delft Technical University
National Institute for Inland Water Management and Waste Water Treatment, Lelystad
Dalhousie University, Halifax
Atlantic Industrial Research Institute, Halifax
The Technical University of Nova Scotia, Halifax.

At this point, I also wish to acknowledge the support received from the Hungarian Scientific Research Fund (OTKA Grant No. 2879/1990), Lahey Computer Systems, Inc., and Wolfram Research, Inc.

This book does not necessarily represent in any way the views, opinions and positions of the persons or institutions named above. The author assumes responsibility for all possible errors and omissions. Again, all constructive comments and suggestions, which may lead to improvements of the material covered and the presentation, are welcome.

It is a pleasure to express my gratitude for the help and support received during the preparation of this book.

I am grateful to John Martindale and Kluwer Academic Publishers for initiating this project and for providing assistance.

Quite a bit of the finalization was done during several hot weekends of a consulting mission which took me to Indonesia for several months. Thanks are due to Pak Ugan and his crew at Crawford Lodge in Bogor (Java), and to the staff of the Swarna Dwipa Hotel in Palembang (Sumatra), for providing a relatively quiet island in a per definitionem bustling world.

Ray Côté, Tibor Csendes, János Fülöp, Eldon Gunn, Eligius Hendrix, Reiner Horst, Don Jones, Walter Stortelder, and Tamás Terlaky—some of them years ago, some recently—have put considerable effort into reading materials discussed in this book and/or parts of the manuscript, and have provided a large number of constructive remarks and suggestions. As a matter of course, similar appreciation goes also to all of my distinguished co-authors listed above.

Leslie Stockhausen was kind not only to typeset the text, but also to provide numerous comments and improve the style of presentation. Again, whatever errors remain are mine.

Last, but very far from least, I would like to thank Susan and Joci for their invaluable support.

J.P. May 1995

PART ONE

Global Optimization: A Brief Review

At the start, great research areas look rugged and mountainous; but it may be still possible to find a square yard to work in. The first job is to find the mountain.

...

Authors sometimes misrepresent their intentions, and politicians are famous for doing so.

...

Disagreement is the liveliest form of flattery...and the only useful form...

(George Watson, *Writing a Thesis.* Longman, London, 1987)

Chapter 1.1

General Problem Statement and Special Model Forms

1.1.1 Introduction

Global optimization is aimed at finding the best solution of constrained optimization problems which (may) also have various local optima. The objective of Part 1 (Chapters 1.1 and 1.2) is to provide a relatively short and informal survey of the spectrum of models and methods in global optimization, with a few concise references to applications, when appropriate.

As mentioned in the Preface, there exist a number of textbooks which also provide more or less detailed surveys, related to certain areas or 'all' of global optimization. Among the more recent works in the English literature, one can mention, for example, Archetti and Schoen (1984), Horst (1984b), Evtushenko (1985), Pardalos and Rosen (1987), van Laarhoven and Aarts (1987), Forgó (1988), Ratschek and Rokne (1988), Mockus (1989), Törn and Žilinskas (1989), Horst (1990), Horst and Tuy (1990, 1993), Zhigljavsky (1991), Floudas and Pardalos (1992), and Hansen (1992). Some of these surveys—found, for instance, in Horst and Tuy (1990, 1993), Zhigljavsky (1991), and especially in chapters of the recent handbook edited by Horst and Pardalos (1995)—are quite extensive, and the interested reader is referred to them for additional details on the general subject of global optimization. For these reasons, only some essential, specific references will be made here (in Part 1), since the monographs and surveys mentioned above—and also the subsequent parts of this book—cite numerous further works.

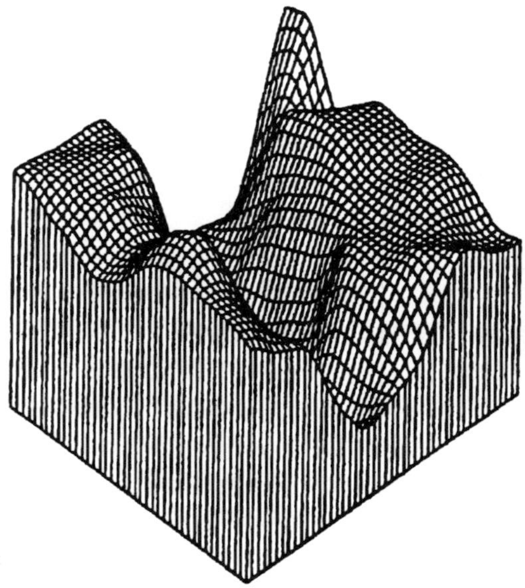

Figure 1.1.1
A multiextremal function in two variables

We shall consider the following general global optimization problem (GOP) statement: given a bounded, robust set D in the real n-space $\mathbb{R}^n$ and a continuous function $f : D \to \mathbb{R}$, find

$$\text{global min } f(x), \quad \text{subject to the constraint } x \in D. \tag{1.1.1}$$

For illustration, see Fig. 1.1.1, which displays a problem-instance of (1.1.1); in this example D is simply a two-dimensional interval (rectangle). In the figure one can observe several local minima of the objective function, and can immediately visualize the potential difficulty of the general problem statement (1.1.1).

Note that since, by definition, a robust set is the closure of its nonempty interior, D is obviously compact. Therefore—by the Weierstrass theorem of classical analysis—problem (1.1.1) has solution(s). The existence of the global solution could be guaranteed under somewhat more general analytical conditions than stated above, but the problem statement presented will be sufficiently general to cover all specific GOP instances treated in this book.

In (1.1.1), D is called the feasible set, and f is the objective function. Note that, obviously, global maximization problems can also be investigated in this framework, since $\max f(x) = -\min(-f(x))$ on arbitrary D. For brevity, explicit notational reference to the global minimum will be omitted in the remainder of this book: hence—instead of (1.1.1)—we shall simply write

$$\min f(x), \qquad x \in D. \tag{1.1.2}$$

At the same time, we shall keep in mind that the classical theory of optimization, as well as the 'standard' techniques of modern mathematical programming, are aimed at finding 'only' local optima. By contrast, we shall deal with global approaches to solve diverse specifications of (1.1.2).

Denote by $X^\star$ the set of solutions to the GOP (1.1.1); frequently, $X^\star$ consists only of a single point $x^\star$, but this is not necessarily the case in general. In fact, it is easy to visualize instances of (1.1.1) in which $X^\star$ is finite, countable, non-countable, or is a subset of D which has a positive volume. For the sake of algorithmic tractability, we shall suppose throughout this book that $X^\star$ is, at most, countable.

Introducing the notation $f^\star = f(x^\star)$, the more precise meaning of the problem statement (1.1.1) (or (1.1.2)) is the following:

$$\text{'find all elements of } X^\star \text{ and the function value } f^\star\text{'.} \tag{1.1.3}$$

In numerous important cases, it is very difficult to solve the GOP (1.1.2) exactly: hence, the requirement (1.1.3) may be too 'ambitious'. Consequently, one has to accept diverse numerical approximations, for example, in the following sense:

find an $x \in D$ which satisfies the relation

$$\rho(x, X^\star) := \min_{x^\star \in X^\star} \rho(x, x^\star) \leq \delta, \tag{1.1.4}$$

in which ρ is a given distance function in $\mathbb{R}^n$ (typically defined by the norm introduced), and $\delta > 0$ is a fixed tolerance parameter; or

find an $x \in D$ which satisfies the relation

$$f(x) \leq f^\star + \varepsilon, \tag{1.1.5}$$

in which $\varepsilon > 0$ is another fixed tolerance parameter.

There exists a considerable variety of specifications of the GOP (1.1.2). These may substantially differ in their underlying (additional) analytical assumptions, related to the structure of D and f. Exploiting the special structure of the GOP, corresponding solution strategies can then be suggested. In the remainder of Part 1 we shall briefly review some important special model forms and solution approaches.

1.1.2 Concave Minimization

Assume that in the problem statement (1.1.2), f is a concave function, and D is a convex set. That is, we shall consider

$$\min f(x), \qquad x \in D, \tag{1.1.6}$$

in which, for arbitrary $x \in D$, $y \in D$, and $0 \leq \lambda \leq 1$,

$$\begin{aligned} &\lambda x + (1-\lambda)y \in D, \quad \text{and} \\ &f(\lambda x + (1-\lambda)y) \geq \lambda f(x) + (1-\lambda)f(y). \end{aligned} \tag{1.1.7}$$

The concave minimization problem (1.1.6)–(1.1.7) is one of the most frequently studied paradigms in global optimization. Its importance stems, for instance, from a range of engineering and economic applications in which concavity of f models so-called 'economies of scale'. Note that the convexity of D can significantly contribute to establishing favourable structural properties, and to assisting the numerical solution procedure: for these reasons, it is often an essential prerequisite, even in certain GOPs.

The most important distinctive analytical feature of the concave minimization problem (CMP) is expressed by the following assertion.

Theorem 1.1.1. *Given the concave minimization problem* (1.1.6)–(1.1.7), *the global minimum of f is attained at a certain extreme point of D.* ■

The proof of this statement is elementary, and can be found in many standard textbooks; consult, for example, Rockafellar (1970), Martos (1975), or Forgó (1978, 1988).

In the special case where D is a polytope, the extreme points are its vertices: therefore, the solution(s) of the CMP can be found among these.

This simple observation already sheds light on some of the inherent difficulties that can be expected facing (general) GOPs, since—even in this rather special, 'simple' GOP—the number of local minima can be an exponential function of the problem dimensionality (Kalantari, 1986). To illustrate this point, consider the trivial CMP of minimizing $-\|x\|$ ($\|\cdot\|$ is the Euclidean norm function in $I\!R^n$) on an interval $[a, b] \subset I\!R^n$: it is not difficult to see that all 2^n vertices of $[a, b]$ are local solutions of this problem.

Several specific global optimization problems that can be easily represented as CMPs will be listed below: for additional details, see, for example, Horst (1984b), Pardalos and Rosen (1986), Horst (1990), Horst and Tuy (1990), or Benson (1995). Note that the subsequent CMP representations do not indicate per se that these problems should necessarily be viewed and/or solved as a CMP; rather, they show the theoretical connection between these problem-classes, and also indicate—from diverse angles—the complexity of GOPs.

General Discrete Programming (Combinatorial Optimization)

Consider the discrete (integer) programming problem (DPP) of minimizing a twice continuously differentiable function $f \in C^2$ on a given subset D of the binary vectors in $I\!R^n$:

$$\begin{aligned} &\min f(x) \\ &x \in D \\ &D = D_0 \cap B^n. \end{aligned} \tag{1.1.8}$$

In (1.1.8), B^n is the set of all vectors $x = (x_1, \ldots, x_n)$ with the property $x_i = 0$ or 1, $i = 1 \ldots, n$; furthermore, D_0 is an appropriate—but otherwise arbitrary—robust convex subset of $I\!R^n$ such that D_0 contains (exactly) the feasible points from B^n. (For instance, D_0 can be the convex hull of the feasible subset of B^n.)

Note first of all that combinatorial optimization problems (of minimizing an objective function $f \in C^2$ on a finite—possibly very large—set of discrete alternatives) can be, at least in principle, directly reduced to the form (1.1.8). For this, it is sufficient to enumerate these alternatives as integers (or integer vectors), and then take their binary vector representation. Needless to say, in many cases there are much more insightful problem representations than this 'automatic' procedure, but the present discussion is concerned only with the existing theoretical connection between the general GOP and combinatorial optimization paradigms. (On the subject of combinatorial optimization, consult, for example, Kovács, 1969; Lawler and Wood, 1969; Taha, 1975; Glover, 1978; Lawler, Lenstra, Rinnooy Kan and Schmoys, 1985; Ibaraki, 1987.)

Observe also that the feasible set D in (1.1.8) is not robust (its interior is empty): hence, the DPP would not formally belong to the class of GOPs encompassed by the problem statement (1.1.1). It is not difficult to show, however, that it can be properly transformed into an equivalent CMP. Namely, if $e = (1, \ldots, 1) \in \mathbb{R}^n$, and $E = \{x \in \mathbb{R}^n : 0 \leq x_i \leq 1,\ i = 1, \ldots, n\}$ is the unit interval in $\mathbb{R}^n$, then—for sufficiently large multiplier M—the problem

$$\begin{aligned} &\min f(x) + Mx^T(e - x) \\ &x \in D_0 \cap E \end{aligned} \tag{1.1.9}$$

(x^T is the transpose of x) is a CMP, which—in terms of optimal solutions and optimum value—is equivalent to (1.1.8). (For details, consult, for example, the classical paper of Raghavachari, 1969; or Horst and Tuy, 1990, and further references therein.)

Concave Complementarity Problem

Given a convex set $D \subset \mathbb{R}^n$, and two concave vector functions $g : \mathbb{R}^n \to \mathbb{R}^I$, $h : \mathbb{R}^n \to \mathbb{R}^I$, find an x (or all such x) in D which satisfies the conditions

$$g(x) \geq 0, \qquad h(x) \geq 0, \qquad g(x)^T h(x) = 0, \tag{1.1.10}$$

or show that such x does not exist. (In (1.1.10), the inequalities $g(x) \geq 0$ and $h(x) \geq 0$ are meant for all component functions of g and h, and $g(x)^T h(x)$ is their dot product value at argument x.)

The concave complementarity problem (CCP) plays a central role in many applications, since it is closely related to fixed point problems, systems equilibria, and variational inequalities: see, for example, Garcia and Zangwill (1981), Murty (1988), or Pang (1995). Note that an important special case of (1.1.10) is the linear complementarity problem (LCP), in which both g and h are linear (affine) functions and D is a polyhedron.

Let us denote the component functions of the mappings g and h by $g_i(x)$, $h_i(x)$, $i = 1, \ldots, I$, respectively. Then it is not difficult to see that the CCP (1.1.10) is equivalent to the CMP

$$\min\ f(x), \qquad f(x) := \sum_{i=1}^{I} \min\{g_i(x), h_i(x)\}, \tag{1.1.11}$$
$$g(x) \geq 0, \qquad h(x) \geq 0, \qquad x \in D,$$

in the following sense. If the solution set of (1.1.10) is nonempty, then it is identical with the solution set of (1.1.11), and the corresponding optimum value in (1.1.11) equals 0; the opposite case is equivalent to the case of $\min\ f(x) > 0$ in (1.1.11).

For a straightforward proof, special cases, and additional discussions, consult, for example, Horst and Tuy (19[illegible] [illegible]nd Pang (1995), with further references.

Without going into further detail, note that other important classes of mathematical programming problems can also be reformulated as CMPs. These include, for example, the bilinear programming problem (BPP):

$$\begin{aligned} &\min p^T x + x^T C y + q^T y && (\text{given } p \in \mathbb{R}^n, q \in \mathbb{R}^m, C \in \mathbb{R}^{n \cdot m}), \quad (1.1.12)\\ &x \in X,\ y \in Y && (X \subset \mathbb{R}^n,\ Y \subset \mathbb{R}^m \text{ are polytopes}); \end{aligned}$$

and the linear maximin problem (LMMP):

$$\begin{aligned} &\max_x \min_y\ c^T x + d^T y && (\text{given } c \in \mathbb{R}^n,\ d \in \mathbb{R}^m), \quad (1.1.13)\\ &Ax + By \leq b && (\text{given } A \in \mathbb{R}^{k \cdot n},\ B \in \mathbb{R}^{k \cdot m},\ b \in \mathbb{R}^k),\\ &x \in \mathbb{R}^n, x \geq 0, &&\\ &y \in \mathbb{R}^m, y \geq 0. && \end{aligned}$$

For recent reviews including the discussion of the BPP and LMMP—with extensive reference lists—consult Floudas and Visweswaran (1995), and Du (1995), respectively.

As mentioned previously, the equivalent transformations of diverse problems into CMPs provide valuable theoretical insight into the connections among these problems and, hence, allow for a unified treatment of the related computational complexity issues. Moreover, they make possible the adaptation of general CMP solution principles (to be reviewed later, in Chapter 1.2) to specific problem-instances.

1.1.3 Differential Convex (D.C.) Programming

Definition 1.1.1. Let the function $f(x)$ be defined on a convex set $C \subset \mathbb{R}^n$. If $f(x)$ can be decomposed as the difference of two convex functions

$$f(x) = p(x) - q(x), \qquad p(x), q(x) \text{ are convex on } C, \tag{1.1.14}$$

then it is called a differential convex (d.c.) function; (1.1.14) is (one of) its d.c. decomposition(s).

Note that functions representable in a d.c. form were discussed decades ago by Alexandrov (1950) and Landis (1951).

Definition 1.1.2. The set $D \subset \mathbb{R}^n$ is called d.c. if it is the difference of two convex sets, that is,

$$D = P - Q, \qquad P \text{ and } Q \text{ are convex subsets of } \mathbb{R}^n. \tag{1.1.15}$$

Given a convex set $C \subset \mathbb{R}^n$, and the finite collection of d.c. functions f, g_j, $j = 1, \ldots, J$,

$$\begin{aligned} &\min f(x) \\ &g_j(x) \leq 0, \qquad j = 1, \ldots, J, \\ &x \in C \end{aligned} \tag{1.1.16}$$

is called a d.c. programming problem.

The class of d.c. programming problems is very general, and hence it forms one of the most comprehensive modelling paradigms in global optimization; see, for example, Singer (1980), Thach and Tuy (1987), Tuy (1987a,b), Thach (1988, 1993). Below we shall cite several important results to illustrate this point: for proofs and additional detailed discussion, consult Horst and Tuy (1990), Tuy (1995); both contain also an extensive list of related further references.

Theorem 1.1.2. *Let D be an arbitrary closed set in $\mathbb{R}^n$; then there exists a d.c. set $DCS \subset \mathbb{R}^{n+1}$, such that its projection onto $\mathbb{R}^n$ is identical to D.* ■

Theorem 1.1.3. *Let f, g_j, $j = 1, \ldots, J$, be d.c. functions. Then*

$$\begin{aligned} &\max\{0, f(x)\}, \quad \min\{0, f(x)\}, \quad |f(x)| \\ &\max_{j=1,\ldots,J} g_j, \quad \min_{j=1,\ldots,J} g_j, \quad f(x)g_j(x), \quad j = 1, \ldots, J \end{aligned} \tag{1.1.17}$$

are also d.c. functions. ■

Theorem 1.1.4. *If $f(x)$ is locally d.c.—that is, for every $x_0 \in \mathbb{R}^n$ there exists a neighbourhood $N(x_0) \subset \mathbb{R}^n$ of x_0 and convex functions $p_N(x)$, $q_N(x)$, such that*

$$f(x) = p_N(x) - q_N(x), \qquad x \in N(x_0), \tag{1.1.18}$$

then f is d.c. on $\mathbb{R}^n$. ■

Corollary 1.1.1. *Let f be twice continuously differentiable; then f is d.c.* ■

Corollary 1.1.2. *The general (possibly indefinite) quadratic function $c^T x + x^T Q x$ ($x, c \in \mathbb{R}^n$, $Q \in \mathbb{R}^{n \cdot n}$) is d.c.* ■

Theorem 1.1.5. *A real valued continuous function f defined on a convex, compact subset D of $\mathbb{R}^n$ is the limit of a sequence of d.c. functions $\{f_k\}$ on D; the convergence $\{f_k\} \underset{k}{\to} f$ is uniform in D.* ■

Corollary 1.1.3. *The global optimization problem, defined by a convex feasible set $D \subset \mathbb{R}^n$ and continuous function f on D, can be uniformly approximated by a sequence of d.c. problems defined in $\mathbb{R}^n$.* ■

Corollary 1.1.4. *The GOP* (1.1.1) *can be converted into a corresponding d.c problem in $\mathbb{R}^{n+1}$.* ■

The last statement shows, in particular, the very general structure incorporated by the d.c. model of global optimization problems.

It should be observed at this point, however, that the d.c. structure may not be apparent or easy to determine (in a practically viable manner) in all cases of relevance—although it is known in a number of structured GOPs. Note additionally that, even if the d.c. decomposition structure is known or can be elicited, then it may be difficult to use it in a computationally tractable form, and to construct/apply an appropriately efficient algorithmic solution procedure. Nevertheless, the known analytical results permit the application of a very general view to a broad class of GOPs; consequently, generic solution methodologies in d.c. programming can, in principle, be directly brought to many—more specific—GOP classes.

In light of the general results cited above, it is not surprising that many important special problem-classes in global optimization can be interpreted as d.c. problems. We shall conclude the present brief discussion of d.c. problems by citing several examples.

Canonical Form: Reverse Convex Programming

Note first (Horst and Tuy, 1990) that the general d.c. problem (1.1.16)

$$\begin{aligned} &\min \ f(x) \\ &g_j(x) \le 0, \qquad j = 1, \ldots, J, \\ &x \in C, \end{aligned}$$

in which—by supposition—C is given by a finite system of convex inequalities, can be directly reduced to a so-called canonical form, by introducing two additional real variables u and v. In other words, one can define three convex functions C_1, C_2 and C_3 (all in $I\!R^{n+2}$) such that the problem statement

$$\begin{aligned} &\min C_1(x, u, v) \\ &C_2(x, u, v) \le 0 \\ &C_3(x, u, v) \ge 0 \end{aligned} \tag{1.1.19}$$

is equivalent to (1.1.16). The canonical form (1.1.19) is also called a reverse convex program (Hillestad and Jacobsen, 1980); observe that it differs from the standard (canonical) convex programming model only by the additional reverse convex constraint $C_3(x, u, v) \ge 0$.

Concave Programming

Since the CMP—evidently (by definition)—is a d.c. problem, all examples mentioned above as its specific cases belong also to the d.c. category.

As an immediate extension—and an important prototype model in many applications in diverse fields—let us consider a general single-facility location problem. Locational decisions often can be modelled in the generic form

$$\begin{aligned} &\min\Big\{\sum_{j=1}^{J} p_j(x) - \sum_{j=1}^{J} q_j(x)\Big\}, \\ &x \in C, \end{aligned} \tag{1.1.20}$$

in which p_j and q_j are convex functions, and C is a convex set in $\mathbb{R}^n$. Most typically, $n = 2$ or 3 (reflecting actual physical location), and C is often a given simplex or a rectangle in n dimensions; in addition, the functions p_j, q_j, $j = 1, \ldots, J$, express, respectively, the 'attraction' or 'repulsion' of the facility located at x, in relation to a given set of 'customers' $j = 1, \ldots, J$. Obviously, (1.1.20) is a d.c. problem; furthermore, it will still have a d.c. structure, if C is defined as an arbitrary bounded, robust set.

General (Indefinite) Quadratic Programming

In light of Corollary 1.1.2, the general quadratic programming problem under quadratic constraints

$$\begin{aligned} &\min c_0^T x + x^T Q_0 x \qquad\qquad (1.1.21)\\ &c_i^T x_i + x^T Q_i x \le 0, \qquad i = 1, \ldots, I,\\ &x \ge 0 \end{aligned}$$

(x, $c_i \in \mathbb{R}^n$, $Q_i \in \mathbb{R}^{n \cdot n}$, $i = 0, 1, \ldots, I$) is also a d.c. problem.

Note, for instance, that design centering, tolerance design and related problems also belong to the d.c. class, since these are most typically modelled by a specialized variant of the problem statement (1.1.21).

Multiplicative Programming

Consider two collections of convex functions $f_j(x)$, $j = 1, \ldots, J$, and $g_k(x)$, $k = 1, \ldots, K$. The convex multiplicative programming problem (CMPP) is stated as

$$\begin{aligned} &\min \prod_{j=1}^{J} f_j(x) \qquad\qquad (1.1.22)\\ &g_k(x) \le 0, \qquad k = 1, \ldots, K,\\ &x \in \mathbb{R}^n. \end{aligned}$$

Note that, for instance, some important applications of multiple objective decision making—in which the individual criterion functions f_j are simply multiplied to arrive at an aggregated indicator value—can be written in the form (1.1.22): consult, for example, Konno and Kuno (1995), and further references therein. Considering the assertion of Theorem 1.1.3, specifically related to products of d.c. functions, it can be immediately seen that

(1.1.22) is a d.c. problem. In fact, the same statement remains valid, if—more generally—all functions f_j and g_k are d.c.

Fractional Programming

Let the continuous functions f, h, and g_j, $j = 1,\ldots,J$, be defined on the robust, bounded set $C \subset I\!R^n$. Assuming that $h(x) > 0$ on C, define $q(x) = f(x)/h(x)$. The fractional programming problem (FPP) is then stated as

$$\begin{aligned} &\min q(x) \\ &g_j(x) \le 0, \qquad j = 1,\ldots,J, \\ &x \in C. \end{aligned} \tag{1.1.23}$$

For a detailed treatment of this problem-class—including a number of applications and an extensive list of references—see Schaible (1995). From the very general representation result expressed by Corollary 1.1.4 it follows that problem (1.1.23) also belongs to the d.c. category. More constructively, if f, h, and g_j, $j = 1,\ldots,J$, are all d.c. functions with known d.c. representation, then (1.1.23) can be reduced to a parametric set of (canonical) d.c. programs.

1.1.4 Lipschitz Global Optimization

Consider the following specification of the general GOP (1.1.1):

$$\begin{aligned} &\min\ f(x) \\ &x \in D = \{x \in D_0 : f_j(x) \le 0,\ j = 1,\ldots,J\}. \end{aligned} \tag{1.1.24}$$

In (1.1.24), $D_0 \subset I\!R^n$ represents a 'simple' explicit constraint set: frequently, D_0 is a finite n-dimensional interval or simplex, or $D_0 = I\!R^n$. Furthermore, the functions $f_0 := f$, f_j, $j = 1,\ldots,J$, are Lipschitz-continuous on D_0; that is, they satisfy the relations

$$|f_j(x_1) - f_j(x_2)| \le L_j \|x_1 - x_2\| \tag{1.1.25}$$

for an arbitrary pair of points x_1, x_2 from D_0, and for some suitable constants $L_j = L_j(f_j, D_0) > 0$, $j = 0,1,\ldots,J$. Most frequently, the Euclidean (l_2) norm is used in (1.1.25), but other norms could also be applied.

In accordance with the statement of the GOP (1.1.1), we shall assume throughout the discussion that D is bounded and robust. This requirement is evidently satisfied, for example, if D_0 itself is bounded and robust, and the intersection of the sets D_0 and $\{x \in I\!R^n : f_j(x) \leq 0,\ j = 1, \ldots, J\}$ has a nonempty interior. Recall that—by the Weierstrass theorem—the postulated properties guarantee the existence of the global solution set $X^\star \subset D$.

The Lipschitzian global optimization problem (LGOP) class (1.1.24)–(1.1.25) is a principal subject of this book: therefore, the present discussion will be confined to a few key observations, summarized below.

The basic LGO model stated above is motivated by a number of practical applications, in which the available analytical information (expressed by the functions f_j, $j = 0, 1, \ldots, J$) is either very limited, or is not easy (or convenient) to use. For instance, the function values may be provided by some sort of 'black box' or 'oracle' (running a computer program, or accomplishing a sequence of system simulation cycles, to arrive at deterministic/stochastic performance estimates). That is, for any given collection of function arguments, corresponding function values can be generated, but not much more is supposed to be known. The only essential analytical assumption related to the functions f_j, $j = 0, 1, \ldots, J$, in the model (1.1.24) is that their 'slope' (variability) is bounded, with respect to the system input (decision) variables x. This assumption is frequently supported by some basic information and (scientific, engineering) knowledge related to the system in question. Since numerous cases will be discussed later, here we refer only to the following examples:

- general (Lipschitzian) nonlinear approximation, including systems of nonlinear equations and inequalities
- parameterization of numerical models in statistics
- calibration of (arbitrary complex) nonlinear system models
- 'black box' engineering systems design
- optimization of complex hierarchical systems (related, for example, to configuration modelling/design, facility location, mass service systems, etc.).

Let us note that exceptions—system models which do not satisfy the basic Lipschitz assumption (1.1.25)—naturally do occur: in that case, however, there is simply no general way to assure the computability of guaranteed optimum estimates (on the basis of the sample points and function values, as shown below).

We shall review the most important analytical properties of Lipschitz functions. Observe first of all that the assumption (1.1.25) is sufficient to establish valid numerical approximations of the global optimum. If $x \in D$ is an arbitrary feasible point, and an upper bound $\overline{r}$ of the diameter of D is available:

$$\overline{r} \geq \max \|x_1 - x_2\|, \qquad x_1, x_2 \in D, \tag{1.1.26}$$

then, by (1.1.25), follows the estimate

$$f^\star \geq f(x) - L\overline{r}, \qquad x \in D. \tag{1.1.27}$$

Furthermore, for any given $\varepsilon > 0$ one can—in principle—construct a suitably fine grid G on D, such that the minimal function value attained on the points of G gives an ε-estimate of the optimum, in the sense of (1.1.5). To guarantee this, it is sufficient to construct G such that for any $x \in D$ there exists $x_{gr} = x_{gr}(x) \in G$ for which

$$\|x - x_{gr}(x)\| \leq \varepsilon / L, \tag{1.1.28}$$

and then to take the best sample point $x^\star_{gr} : f(x^\star_{gr}) = \min f(x_{gr})$, $x_{gr} \in G$.

Without going into details unnecessary at this point, note additionally that—on the basis of the Lipschitz property of f_j, $j = 1, \ldots, J$, and a suitable grid of sample points on D_0—analogous bounds to (1.1.27) can be provided with respect to the constraint functions. These bounds may then assist with respect to deciding upon the feasibility or infeasibility of points in subsets of D_0.

It can also be seen that—instead of the 'exact' (minimal suitable) constants $L_j = L_j(f_j, D_0) > 0$—their arbitrary overestimate can be used to validate the above remarks. This observation is from a practical point of view very important, since the 'exact' Lipschitz-constant of functions is often unknown (while overestimates can be provided in a great many cases).

The Lipschitz assumption is not only sufficient but also necessary to establish the simple deterministic bound estimation procedures outlined above. For instance, it is easy to see that in the absence of the property (1.1.25)—without a guaranteed (over)estimate of the Lipschitz-constant—on the sole basis of sample points and corresponding function values, one cannot provide a lower bound of f, after any finite number of function evaluations on D (f could even be an analytical function, with a priori unboundable 'jumps'). Of course, on a bounded robust set D, one can always construct a grid which will guarantee the arbitrarily close approximation of $X^\star$, in the sense of (1.1.4)—the important point to make is that one would not, however, be able to establish meaningful optimum estimates, as required by (1.1.5).

The above remarks suggest that the Lipschitz assumption is 'minimal', in order to arrive at 'guaranteed' (deterministic) and computable optimum estimates, on the basis of a finite number of sample points and corresponding function evaluations. In the absence of a criterion like (1.1.25), either only probabilistic optimum estimates can be provided, using appropriate sampling techniques, or—theoretically—an everywhere dense set of sample points is necessary to approximate the global solution(s): both types of strategies will be discussed later.

The following propositions show that the class of Lipschitz functions is also very general.

Theorem 1.1.6. *Let D be a bounded, robust set, and f a convex function on D. Then f is Lipschitz-continuous on D.*

The proof of this statement can be found, for example, in Rockafellar (1970). ■

Theorem 1.1.7. *Let D be a bounded, robust set, and f a smooth (continuously differentiable) function on an open convex set D_0, which contains D. Then f is Lipschitz-continuous on D.*

This statement will be proved later (in Part 2). ■

Theorem 1.1.8. *Let D be a bounded, robust set, g_j, $j = 1, \ldots, J$, an arbitrary collection of Lipschitz functions on D, and h a univariate Lipschitz function defined on an arbitrary set which contains the range of g_1. Then the following functions are also Lipschitz-continuous on D:*

$$\sum_{j=1}^{J} c_j g_j(x) \quad (c_j,\ j = 1, \ldots, J, \text{ are arbitrary real constants}) \tag{1.1.29}$$

$$\min\{g_j(x),\quad j = 1, \ldots, J\}$$

$$\max\{g_j(x),\quad j = 1, \ldots, J\}$$

$$g_i(x) g_j(x),\quad i, j = 1, \ldots, J$$

$$h\{g_1(x)\}.$$

The proof of these assertions is elementary, since it is simply based on the definition of Lipschitz-continuity. ■

Note that Theorems 1.1.6–1.1.8 are certainly not exhaustive, in the sense of covering all possibilities to define Lipschitz functions. Nevertheless, it follows immediately that all problem-classes discussed above are LGOPs.

Corollary 1.1.5. *Every d.c. function defined on a bounded, robust set D is Lipschitzian.*

This assertion is a straightforward consequence of Theorems 1.1.6 and 1.1.8. ■

Since the class of d.c. problems incorporates all the GOPs mentioned in Sections 1.1.2 and 1.1.3, the LGOP (1.1.24)–(1.1.25) indeed encompasses all of the more special problem-types discussed above—except the GOP (1.1.1), considered with some (merely) continuous objective function-class.

To mention another—more concrete, and practically also important—class of problems, note that, by Theorems 1.1.7 and 1.1.8, all given polynomial functions and their appropriate Lipschitzian composites (1.1.29) can be part of the definition of LGOPs.

Due to its practical significance, Lipschitz optimization has been an important field of research within the general subject of global optimization. In addition to the topics which will be discussed subsequently in detail, let us refer here to the recent review of Hansen and Jaumard (1995), which contains a fairly comprehensive summary of research related to the subject.

As shown above, the Lipschitzian model incorporates a very broad class of GOPs, and—of course—a similar remark applies to the (most) general GOP (1.1.1): hence, the methodology presented later in this book is, in principle, applicable to all such problems. Needless to say, this does not mean that all global optimization models encompassed should be solved (solely) by algorithms based on the continuous or the Lipschitzian model. For instance, when a proper d.c. decompositional structure is known, then it is natural to exploit that structure in an algorithmic solution procedure. If even more specific descriptors also apply—for example, we have a concave minimization problem which is defined by a piecewise linear objective function, many linear and just a few 'hard' nonlinear constraints—then, apparently, this information can be very valuable in 'tailoring' some generic solution procedure. Note in this context that even in case of 'black box' systems, concurrent developments in global optimization and other areas of numerical mathematics may provide alternative and, in fact, synergistic approaches. One can refer, for instance, to interval arithmetic based approaches (see, for example, Ratschek and Rokne, 1988, 1995; Hansen, 1992), or to advances in computational (automatic) differentiation (consult, for example, the survey of Griewank, 1994).

As the reader may realize even on the basis of this very brief introduction to global optimization models, the field of global optimization poses—and, for a long time to come will remain to pose—a sufficient theoretical and numerical challenge for the relevant repertoire of mathematical programming and numerical mathematics. With this 'truism' in mind—and, more constructively, with respect to the specific problem-types discussed above—in the next chapter we shall highlight several important solution approaches that can be applied in solving global optimization problems.

Chapter 1.2

Solution Approaches

The brief discussion of some problem types in global optimization, presented in Chapter 1.1, points towards the need of choosing suitable strategies to solve diverse instances of the generic GOP statement (1.1.1). Indeed, there exists a broad variety of global optimization methods. In existing monographs and surveys, these are classified following several different organizing principles: consult, for example, Dixon and Szegö (1975, 1978), Strongin (1978), Žilinskas (1986), Pardalos and Rosen (1987), Forgó (1988), Rinnooy Kan and Timmer (1989), Törn and Žilinskas (1989), Horst (1990), Horst and Tuy (1990, 1993), Zhigljavsky (1991), Zhigljavsky and Žilinskas (1991), Horst and Pardalos (1995).

In this chapter a 'methodology oriented', as opposed to 'problem-oriented', classification will be presented. This is motivated by the fact that several approaches are sufficiently general to tackle diverse (specific) GOP instances. With respect to each solution method, we shall mention problem-types which can be solved advantageously by that approach. This discussion will be preceded by a few remarks pertinent to the relation of global and local optimization strategies, and to the (expectable) complexity of the general class of GOPs.

1.2.1 Global versus Local Optimization

By its very definition, the GOP statement (1.1.1)—and all of its special variants discussed above—fundamentally calls for a 'complete' (exhaustive) investigation of the feasible set D, in order to guarantee the finding (or, in practice, often the numerical approximation) of the globally optimal solution(s). One can immediately see that this objective differs form the aim of the main body of 'classical' continuous optimization theory and mathematical programming: in the latter—given a sufficiently close 'guess' of the optimal solution vector—(only) local search is performed, in order to arrive at a refined (local) solution. The acceptability of the solution found in this way crucially depends on the problem structure. In classical optimization theory the postulated structure typically guarantees the global optimality of a locally optimal solution.

Some general analytical characterizations of optimization problems whose locally optimal solutions are—at the same time—globally optimal, are reviewed, for example, by Hiriart-Urruty (1995), with additional references. This category of problems includes the well-known convex programming problem (CPP):

$$\begin{aligned} &\min f_0(x) \\ &x \in D = \{x \in I\!R^n : f_j(x) \leq 0,\ j = 1, ..., J\}. \end{aligned} \tag{1.2.1}$$

In (1.2.1), D is a bounded, convex set and f_0 is a quasi-convex function, that is, for arbitrary x_1, x_2 from D and $0 \leq \lambda \leq 1$, we have

$$f_0(\lambda x_1 + (1-\lambda)x_2) \leq \max\{f_0(x_1), f_0(x_2)\}. \tag{1.2.2}$$

(Quasi)convexity is an essential concept in classical optimization studies. It is easy to verify that quasi-convexity of the functions f_j, $j = 1, \ldots, J$, is necessary and sufficient for assuring convexity of the set D. Note also that convex functions—that is, functions which multiplied by -1 become concave—are evidently also quasi-convex; thus so are also linear functions.

The local optimization paradigm is, obviously, of great importance, and has found a great many successful applications (for instance, think only of linear programming). In many practically important cases, however, the conditions that would assure the global optimality of local solutions simply cannot be postulated or verified: all these problems—at least in principle—call for an appropriate global optimization approach.

Clearly—in general terms—the GOP class is (much) more difficult than its local counterpart. Every proper numerical global optimization procedure could also serve as a local optimization method—although its efficiency may be (and is typically) hindered by the global scope required. Frequently, a global optimization method—perhaps following the guaranteed 'localization' of the global solution(s)—includes also some local (convergent) search procedure.

These general statements on local versus global optimization are, of course, not 'judgemental' in any reasonable sense. A convex programming model in, say, 300 nonlinear variables can be computationally very difficult, similarly to a huge linear program. The solution of such problems certainly needs solid underlying theory, skillful algorithm construction, organization and implementation; the computational demand (for example, number of function evaluations needed during the solution process, and/or necessary runtime) can be very substantial, etc. The point to be emphasized is that the same proportion of difficulties may easily be encountered when solving, say, a 10-variable (somewhat) complicated system of nonlinear equations in the precise (global) sense, on a 'vague' search region. To illustrate this, see, for example, the numerical experiment—solving a circuit design model—described by Ratschek and Rokne (1993), or some of the applications discussed later in Part 4. One can also observe at this point that, by definition, most of the commercially available solvers are not meant to solve continuous global optimization problems: although in many cases they are sufficiently well-constructed to produce 'quite good' results, the global solution scope is not warranted.

There are a number of theoretical results which confirm this notion of expectable complexity of GOPs. We shall not go into detail here: for related discussions, see, for instance, Khachiyan (1979), Dewdney (1989), Pardalos and Vavasis (1991), Wasilkowski (1992), Ralston and Reilly (1993, pp. 1394–1401), Vavasis (1995), with further references. To illustrate informally the (expectable) complexity of global optimization problems, one may keep in mind the simple example discussed in Section 1.1.2: namely, that even the 'relatively simple' problem of globally minimizing a concave function on a box region could involve an exponential increase of computational demand, as a function of the problem dimensionality n. This point will be illustrated by further discussions and examples in Part 2.

1.2.2 'Globalized' Local Optimization Strategies, Combined with Grid Search or Random Search

Consider again the general GOP

$$\min f(x), \qquad x \in D \subset \mathbb{R}^n, \tag{1.2.3}$$

as stated by (1.1.1).

Since there is a well-developed theoretical basis for applying various local optimization methods (when appropriate), and there exist many efficient implementations, it is a natural and straightforward approach to apply a—possibly quite rudimentary—initial global search, and then refine the search results by 'standard' local optimization tools.

To analyse such strategies in some detail, we shall introduce the following concept.

Definition 1.2.1. The basin (region of attraction) of a local—not necessarily global—minimum point $x_l^\star \in D$ is the set $D_l^\star \subseteq D$, which has the following property: initiating local search from an arbitrary point of $D_l^\star$, this search finds (that is, terminates at) $x_l^\star$.

One can immediately visualize this definition as corresponding to 'valleys' on a hilly landscape (recall Fig. 1.1.1). In a strict theoretical sense, every proper nonlinear local optimizer should lead to the same $x_l^\star$, when started from $D_l^\star$. Although for various numerical reasons this may not be the case, for the purpose of the present exposition we shall assume that an 'ideal' local optimizer LOPT applied in $D_l^\star$ will always terminate exactly at $x_l^\star$; furthermore, that these regions of attraction completely cover D; and that LOPT never leads out from D, if initiated from a feasible point. (These idealizations model many practically important GOPs, but certainly there are exceptions.) Hence, the remaining issue is 'only' to determine the 'valleys' $D_l^\star$.

Additionally—to simplify the presentation—we shall suppose that a unique globally optimal solution $x^\star$ exists. Finally, for obvious numerical reasons, we shall suppose that its basin $D^\star$ is robust (and thus has positive volume in $\mathbb{R}^n$).

Grid search

Uniform grid search (UGS)—mentioned briefly above, and to be discussed later—may obviously serve as a 'folklore' deterministic approach to accomplish a global scope sampling of the objective function (and the constraints, when present). Whenever this sampling leads to at least one point from the basin $D^\star$ of the global optimum, we need only to 'polish' this approximation, applying LOPT. Hence, global convergence is trivially guaranteed by the fact that the grid can be made arbitrarily dense: recall in this context the discussion which leads to (1.1.28).

The essential problem is, of course, that we typically do not know where basin $D^\star$ would be within D, and what is its relative (to D) size—otherwise we would gladly use this information. Therefore, only the generation of a sufficiently fine grid would guarantee the success of the global search phase: this, in turn, can be extremely demanding in a computational sense, since the number of uniformly taken sample points will increase as an exponential function of the dimensionality n of D. For this reason, in practice often only a 'small' grid sample is taken, and then LOPT is launched from the best sample point(s); but again, only the global scope grid sampling effort can guarantee the finding of $D^\star$. Evidently, more 'intelligent' adaptive deterministic sampling procedures to solve GOPs are of considerable interest—the investigation of such strategies will be the principal subject of Part 2.

Random Search

A similar basic philosophy to that of grid search is realized by an initial random search (RS) on the feasible region D. That is—in the simplest case—taking a number of randomly generated, uniformly distributed global sample points $\{x_r\}$, $r = 1, 2, 3, \ldots, R$, LOPT can be launched from all or (by common practice) from the best point(s) found.

This method is again extremely simple to apply—at least in principle. Furthermore, theoretical global convergence can be established easily, under 'minimal' analytical assumptions. (For several variants of such global convergence results, see, for example, Gaviano, 1975; Devroye, 1978; Solis and Wets, 1981; Pintér, 1984a; Patel, Smith and Zabinsky, 1988; Zabinsky and Smith, 1992; or the survey of Boender and Romeijn, 1995.) To show an example, one can easily prove the following assertion.

Theorem 1.2.1. *Under the analytical conditions stated with respect to the GOP (1.1.1), for arbitrarily small $\varepsilon > 0$ there is a sample size $R = R(\varepsilon)$ such that uniform RS plus LOPT will find $x^\star$, with probability $1 - \varepsilon$.*

Denoting by $p = \operatorname{vol}(D^\star)/\operatorname{vol}(D) > 0$ the relative volume of the basin of

$x^\star$, the elementary proof immediately follows from the fact that the probability of randomly 'hitting' $D^\star$ by at least one of the sample points approaches 1—and then LOPT will find $x^\star$. Taking the random sample $x_1, \ldots, x_R$, and denoting by $P\{\cdot\}$ the probability of an arbitrary random event $\{\cdot\}$, one can write

$$P\{(x_1 \in D^\star) \cup \ldots \cup (x_R \in D^\star)\} = 1 - (1-p)^R \underset{R\to\infty}{\longrightarrow} 1. \tag{1.2.4}$$

Specifically, for any $0 < \varepsilon < 1$, one can choose a minimal $R = R(\varepsilon)$ which satisfies the inequality

$$(1-p)^R < \varepsilon. \tag{1.2.5}$$

That is, any integer $R > (\log \varepsilon)/\log(1-p)$ determines a suitable sample size, in the sense of the assertion. ∎

One can observe that—in the case of multiple optima—the combined strategy RS+LOPT may typically find several of them. Launching LOPT from all RS points, the number of minima found will be between 1 and R, since some LOPT results may be identical. To reduce the repetition of the—as a rule, computationally expensive—LOPT phases, ingenious schemes for clustering the sample points have been suggested (Törn, 1978; Boender, Rinnooy Kan, Stougie and Timmer, 1982; Rinnooy Kan and Timmer, 1987a,b; Törn and Viitanen, 1992, 1994).

It is also evident—in complete analogy to (1.2.2)—that RS per se (without LOPT) can guarantee global convergence, in the sense of the relation

$$\mathrm{P}\{\min_{r=1,\ldots,R} \|x^\star - x_r\| \leq \delta\} \underset{R\to\infty}{\longrightarrow} 1 \tag{1.2.6}$$

for arbitrarily small $\delta > 0$; furthermore that, in fact, only its application within the combined search strategy RS+LOPT makes it sure to happen. This remark points towards numerical difficulties in applying this strategy in an 'unsophisticated' fashion. Namely, since we do not know the location and size of $D^\star$, theoretically the RS sampling effort should be 'infinite', to assure global convergence. More practically, by (1.2.5) one can see that the RS sampling effort needed to find at least one point of $D^\star$ with a high probability rapidly tends to infinity, if the relative basin size p of $x^\star$ becomes small and/or ε approaches 0.

Another computationally important point to mention here is that, in many constrained GOPs, sampling on the—possibly complicated (nonconvex, disjoint, etc.)—set D may not be trivial to realize in an efficient manner.

Bearing these caveats in mind, random search—with a large variety of adaptive refinements—nevertheless enjoys considerable popularity in global optimization. Indeed, it is one of the few global optimization approaches that can be applied under very mild 'black box' type assumptions, even if the number of variables is not very small (instead of, say, $n = 1$ to 5, it certainly can be a few tens, even in a personal computer environment). Another salient point is that the application of random sampling makes possible the inclusion of other probabilistic concepts and techniques in the algorithms. These, in turn, can lead to statistical estimation of the problem structure, Bayesian stopping criteria, and optimum value estimates applying extreme order statistics: see, for instance, Zielinski (1981), Boender and Rinnooy Kan (1987, 1991), Betrò (1984, 1991), Betrò and Schoen (1987, 1992).

For additional discussions of random search algorithms and related stochastic optimization approaches—including genetic algorithms, evolution strategies, simulated annealing, etc.—see, for instance, Archetti, Betrò and Steffe (1975), Born and Bellman (1982), Timmer (1984), Bohachevsky, Johnson and Stein (1986), Van Laarhoven and Aarts (1987), Aarts and Korst (1988), Michalewicz (1992), Romeijn and Smith (1994), Feo and Resende (1995). The surveys provided in Devroye (1979), Törn and Žilinskas (1989), Schoen (1991), Zhigljavsky (1991), and Boender and Romeijn (1995) contain numerous further references on the subject. A general framework for stochastically combining optimization strategies, and the subject of estimating problem characteristics using extreme order statistics, will be discussed in Part 3.

1.2.3 Sequential Improvement of Local Minima

Given the theoretical strength and demonstrated numerical effectiveness of local optimization strategies, a heuristically attractive concept is to generate an improving sequence of local optima.

Well-known approaches belonging to this class are, for instance, the so-called deflation technique (Goldstein and Price, 1971), tunnelling (Vilkov, Zhidkov and Schedrin, 1975; Levy and Gomez, 1980, 1985; Levy and Montalvo, 1985), and the filled function method (Ge and Qin, 1987). For illustration, the tunnelling algorithm will be discussed. To simplify the exposition, we shall suppose that the objective function does not possess multiple local minima with the same function value.

The algorithm consists of two phases, which can be iteratively repeated. In Phase 1 of iteration cycle $k = 1$, LOPT is launched from some point x_1^0 of D, leading to a (generally speaking, local) minimizer $x_1^\star$. In general, after Phase 1 of cycle k, the minimizer $x_k^\star$ is available. The purpose of Phase 2 ('tunnelling') is then to find a point $x_k^t \neq x_k^\star \in D$ such that $f(x_k^t) = f(x_k^\star)$. If this is possible, then setting $x_{k+1}^0 := x_k^t$, $k := k+1$, Phase 1 of the next iteration cycle follows.

The crucial issue in realizing this straightforward idea is to construct an appropriate (auxiliary) 'tunnelling' function. In the simplest case discussed here, at main iteration cycle k, this function can be defined as

$$T_k(x) = \frac{f(x) - f(x_k^\star)}{||x - x_k^\star||^{a_k}}, \qquad x \neq x_k^\star; \tag{1.2.7}$$

$a_k > 0$ is a suitable procedure parameter. (It is possible to choose just one $a = a_k$, for all indices k.) The objective of Phase 2 in iteration k is then to find a zero of the function T_k on D. Obviously, all points x for which $f(x) > f(x_k^\star)$ are eliminated as a possible solution of

$$T_k(x) = 0, \qquad x \in D; \tag{1.2.8}$$

furthermore, $x_k^\star$ itself is not a solution. Therefore if a point x_k^t of D is found for which $T_k(x_k^t) = 0$—or, in fact, if $T_k(x_k^t) \leq 0$—then such an x_k^t can serve as the starting point in Phase 1 of the next main iteration cycle. The iterative tunnelling procedure is terminated, if—in Phase 2 of a certain iteration k—the equation $T_k(x) = 0$ has no solution: in this case $x_k^\star$ is the global optimum estimate.

Apparently, the auxiliary tunnelling functions replace the initial GOP with a sequence of nonlinear equations (inequalities). According to numerical experience, in some cases the iterative improvement of the optimum estimates can be fast, while, in solving other problems, finding a solution to (1.2.8)—or verifying that it does not have a solution—can be just as difficult as solving the initial GOP itself. There is no rigorously established convergence theory associated with these methods, except for certain classes of more generally valid approaches (to be discussed later) which can then be directly applied also to the original function $f(x)$. Note additionally that the proper choice of parameter a (or $\{a_k\}$) is an open issue; and a general, well-established stopping rule is also absent, lacking a usable underlying (specific) theoretical convergence result. These remarks also apply, mutatis mutandis, to the other methods mentioned above as belonging to this category.

1.2.4 Enumeration of all Minima

One can observe that the determination of all minima—or, under proper analytical (smoothness) conditions, of all stationary points—of f in D will certainly also lead to finding the global optimum. There exists a class of (local and global) optimization methods which would essentially follow such a seemingly 'naive' approach, with theoretically elegant results, and various implementations which work well, at least on small-dimensional GOPs. In the following brief discussion of two related types of such methods, we shall assume that the objective function $f(x)$, $x \in D$, is twice continuously differentiable; $g(x)$ will denote the gradient of f, and $H(x)$ is its Hessian matrix.

Path-Following (Trajectory) Methods

In one of the earliest papers on global optimization methodology, Branin (1972) considers the differential equation (d.e.)

$$\frac{dg(x(t))}{dt} = mg(x(t)), \tag{1.2.9}$$

where $m \in [-1, 1]$ is a procedure parameter. The solution of (1.2.9) is

$$g(x(t)) = g(x(0)) \exp(mt). \tag{1.2.10}$$

Now, if $m = -1$, then the points x satisfying (1.2.10) will tend to a stationary point x_s (in which $g(x_s) = 0$), when t increases. Conversely, when $m = 1$, then starting from a stationary point, the points $g(x(t))$ will leave that x_s. Consequently, adaptively changing m lets $g(x(t))$ move along a trajectory which depends on $x(0)$ and the choice of m.

Note furthermore that if H is non-singular, then expressing $x(t)$ in (1.2.9) leads to the continuous Newton equation

$$\frac{d(x(t))}{dt} = mH^{-1}(x(t))g(x(t)). \tag{1.2.11}$$

(As is known, (1.2.11) serves as the basis of many nonlinear optimization algorithms.) Starting now from a feasible point x^0, Branin suggests a numerical approximation method which follows the trajectory $x(t)$—solving (1.2.11)—until a stationary point x_s is attained. Then—using an extrapolation technique—x_s is passed, and the path (1.2.11) is followed further.

One can observe that since, by (1.2.10), $g(x(t))$ is proportional to the fixed $g(x(0))$, for a given starting point $x(0)$ the trajectory points are a subset of the set

$$C(x(0)) = \{x \in D : g(x) = \lambda g(x(0)),\ \lambda \in I\!R^1\}. \tag{1.2.12}$$

The set $C(x(0))$ contains all stationary points ($\lambda = 0$), and thus all global and local minimizers can, in principle, be found following the trajectory defined by (1.2.10). Note that in global optimization problems, $C(x(0))$ may frequently consist of several components, and hence the trajectory should be started only once in each of them. Diener (1986, 1987b) suggests a method of appropriately connecting the components of $C(x(0))$, and discusses the application of path-following methods to finding all solutions in systems of nonlinear equations.

Trajectory methods and their variants in global optimization have been the subject of extensive theoretical analysis and numerical studies since Branin's seminal paper: see, for example, Gomulka (1975, 1978), Hardy (1975), Treccani (1975), Incerti, Parisi and Zirilli (1979), Jongen, Jonker and Twilt (1979, 1983, 1986), Garcia and Gould (1980), Garcia and Zangwill (1981), Griewank (1981), Twilt (1981), Aluffi-Pentini, Parisi and Zirilli (1985), Sturua and Zavriev (1991), Kaplinskij and Propoj (1993). Guddat, Guerra Vazquez and Jongen (1990) and the review of Diener (1995) contain detailed expositions and further references.

Homotopy (Parameter Continuation) Methods

The definition of $C(x(0))$ in (1.2.12) can be viewed also in the following way. Notice that for the parameter value $\lambda = 1$, $x = x(0)$ is an element of $C(x(0))$; and for $\lambda = 0$ the elements of $C(x(0))$ are stationary points. Hence, starting a trajectory from $x(0)$, one can attempt to trace $x(\lambda)$, as λ decreases from 1 to 0. This basic idea—namely, 'find first the solution to a simple problem, and then transform it, by parameterization, into a related complicated one'—leads to the concept of homotopies, that is, parametric transformations between (general) mappings. Given the appropriate homotopy structure, simplicial approximation techniques can be applied to compute the path $x(\lambda)$, and this approach can be extended to search for all stationary points of f.

Without going into further detail on these strategies (which are not within the immediate scope of the present book), one can refer, for instance, to Forster (1980), Garcia and Gould (1980), Robinson (1980), Allgower and Georg (1983, 1990), and Talman and Van der Laan (1987). Forster (1995) provides a lucid and detailed review of this field, with an extensive list of references.

1.2.5 Relaxation (Successive Outer Approximation) Strategies

Consider again the GOP

$$\min f(x), \qquad x \in D, \tag{1.2.13}$$

under the standard assumptions stated earlier (f continuous, D bounded and robust). Specifically, we shall assume that D may have a complicated structure, as being determined, for example, by a finite number of Lipschitz inequalities:

$$D = \{x \in I\!R^n : g_j(x) \leq 0,\ j = 1, \ldots, J\}. \tag{1.2.14}$$

The application of relaxation strategies is a very general concept in numerical optimization. The essential idea is to replace the original problem by a sequence of relaxed subproblems which are (by supposition, much) easier to solve than the initial problem. These subproblems are then successively adapted, by adding constraints to them in such a manner that they will become increasingly more precise approximations of the initial GOP.

We shall outline an algorithmic procedure which realizes this idea. Let k denote the main iteration index of the procedure. Let us assume that at $k = 1$, a compact set D_1 is available which includes D, and is 'easy to handle': for example, D_1 can be a sufficiently large interval or simplex region in $I\!R^n$.

At each iteration $k = 1, 2, 3, \ldots$—by construction—$D \subset D_k$, and one has to solve the relaxed subproblem

$$\min f(x), \qquad x \in D_k. \tag{1.2.15}$$

If the solution $x_k^\star$ of (1.2.15) belongs to the original feasible set D, then, obviously, we have a globally optimal solution $x^\star = x_k^\star$, and the procedure terminates. In the opposite case, an additional constraint function c_k is introduced which satisfies the requirements

$$c_k(x_k^\star) > 0, \quad \text{and} \quad D \subset C_k := \{x \in I\!R^n : c_k(x) \leq 0\}. \qquad (1.2.16)$$

Setting $k := k+1$, and defining

$$D_{k+1} = D_k \cap C_k, \qquad (1.2.17)$$

the procedure continues at the next main iteration cycle.

Evidently, by (1.2.16)–(1.2.17) the new constraint c_k separates $x_k^\star$ from the resulting relaxation D_{k+1}; at the same time, D_{k+1} will still contain the feasible set D.

Without going into detail, note that numerous special cases of the general relaxation strategy were studied and implemented before applications emerged in global optimization: see, for instance, the classical cutting plane method (Kelley, 1960), or the supporting hyperplane method (Veinott, 1967). The general relaxation approach, in various contexts, is discussed in most standard operations research monographs; consult, for example, Daellenbach, George and McNickle (1983), Hillier and Lieberman (1986), Winston (1994).

The theoretical convergence of the relaxation strategy outlined above can be established by the following general result (Horst, Thoai and Tuy, 1989).

Theorem 1.2.2. *Suppose that the following analytical requirements are satisfied.*

- *The functions c_k are continuous, for $k = 1, 2, 3, \ldots$.*
- *For each convergent subsequence of $\{x_k^\star,\ k \in N\}$, and $x^\star = \lim_{k\in N} x_k^\star$, $\lim_{k\in N} c_k(x_k^\star) = c(x^\star)$, with an appropriate limit function $c(x)$.*
- *The relation $c(x^\star) = 0$ implies that $x^\star \in D$.*

Then every such accumulation point $x^\star$ is a globally optimal solution of (1.2.13). ∎

An important point that deserves mentioning here is that, in the classical applications of relaxation strategies (related to convex programming), the additional constraint functions c_k are linear. In global optimization, however, it can be desirable to have more general—for example, convex, concave, d.c., or Lipschitzian—cuts which possibly could follow the curvature of D and the subproblems D_k more flexibly. The general convergence result stated above makes this possible.

Of course, another issue of crucial importance is the efficient solution of the embedded subproblems (1.2.15): within each main iteration cycle k, this embedded procedure still may be, theoretically, infinite. In certain 'well-structured' global optimization problems, relaxation strategies can be successfully applied. Examples include the concave minimization problem (1.2.1), see, for example, Hoffman (1981); and the canonical d.c. problem (1.2.16), see Tuy (1987), Thoai (1988). Observe also at this point that—according to the general relaxation strategy outlined above—at each iteration a new constraint is added to the definition of the current feasible region. This phenomenon has been a major stumbling block in applying relaxation methods. For this reason, efficient constraint dropping strategies—which allow one to ignore redundant constraints without losing points of D (or just optimal points)—have been of considerable interest: consult, for example, the pioneering work of Eaves and Zangwill (1971), Kiwiel (1985) (on non-differentiable convex optimization), and Horst, Thoai and Tuy (1989) (in the present general context). To enhance the numerical efficiency of the relaxation strategy, the latter is frequently combined with branch-and-bound methodology, to be discussed briefly below.

1.2.6 Branch-and-Bound Strategies

Branch-and-bound (B&B) is again a very general algorithm concept that has been playing a prominent role in optimization theory in solving 'complicated' problems. In fact, B&B is one of the principal ideas applied in the theoretical framework, discussed subsequently in Part 2. Therefore, the present exposition will be devoted only to its general principles, while also mentioning several applications of B&B methodology in global optimization. Other applications in the general field of operations research are discussed, for example, in the OR monographs mentioned above.

Given the—possibly more specific than (1.1.1)—optimization problem

$$\min f(x), \qquad x \in D, \tag{1.2.18}$$

the essence of B&B is to decompose this problem into 'parallel' (simultaneously existing) subproblems that may or will become gradually easier to handle. Hence, the B&B procedure can be visualized by a corresponding, adaptively generated search tree: the initial problem (1.2.18) is located at the root, and the branches contain hierarchically constructed subproblems. This construction is governed by a search strategy that will determine the order of solving the subproblems. The sequential decomposition and solution of subproblems continues until it can be verified that either the given branch cannot yield a better solution than the incumbent solution (found already by the B&B procedure) or the subproblem branching process is completed, that is, all 'elementary feasible solutions' on the given branch have been examined. Note here that B&B was originally developed for solving integer programming problems: consult, for example, Mitten (1970), Garfinkel and Nemhauser (1972). Later, B&B has been successfully applied to other 'hard' optimization problems: these include, for example, diverse specifications of the general concave minimization problem (1.2.1), and the combinatorial optimization problem (1.2.3); for a survey, see, for example, Pardalos and Rosen (1987). Note further that, in the—continuous, general—global optimization context, appropriate modifications of the original (discrete) B&B concept were necessary: in particular, the case of 'completed' branch enumeration had to be replaced by a suitable convergence theory (Benson, 1982; Horst, 1986, 1988a; Pintér, 1986a, 1988; Horst and Tuy, 1987).

Let us now formally describe an algorithm framework which corresponds to the generic B&B concept outlined.

Prototype Branch-and-Bound Algorithm

Step 0. (Initialization) Set the iteration counter $k = 1$. Start with a—possibly relaxed—feasible set D_1, $D \subset D_1$. Denote the collection of active subsets by $\mathcal{D}_1 = \{D_1\}$.

Step 1. (Bounding) At iteration k, for each active subset D_i, $i = i(k) \in \mathcal{D}_k$, determine a lower bound $l(D_i)$ and an upper bound $u(D_i)$ which enclose the best possible function value, $\min f(D_i)$, on D_i:

$$l(D_i) \leq \min f(D_i) \leq u(D_i). \tag{1.2.19}$$

Observe that the bounding operations lead also to the obvious optimum estimates

$$l_k := \min_{D_i \in \mathcal{D}_k} l(D_i), \qquad (1.2.20)$$
$$u_k := \min_{D_i \in \mathcal{D}_k} u(D_i).$$

Two numerical criteria can be included at this point. First, if

$$u_k - l_k \leq \varepsilon \qquad (\varepsilon > 0 \text{ is a given tolerance}),$$

then—since typically these estimates are established by certain sample points on the partition subsets—an ε-optimal solution (namely, the feasible solution $x_k^\star$ corresponding to u_k) is available.

Second, if, for some i, u_k is smaller than l_i, than the corresponding subset D_i can be fathomed (eliminated from the further analysis, since it cannot lead to a better solution than the one already guaranteed). Following the possible fathoming operations (reducing the collection of active subsets), Step 2 follows.

Step 2. (Selection) Choose one or several subsets from the remaining collection of subsets. Frequently the natural choice is to select that subset D_t for which we have

$$l_k = l(D_t). \qquad (1.2.21)$$

Step 3. (Partition/branching) Partition D_t into a finite number of subsets, add them to the collection of currently active sets. Setting $k := k + 1$ (and denoting by $\mathcal{D}_{k+1}$ the set of active subsets), return to Step 1 of the next main iteration cycle.

As can be seen from this algorithm-scheme, in solving (1.2.18) by B&B strategies, lower bounding operations play a central role as they serve as a basis of branching and, possibly, 'pruning' the decision tree. These lower bounding operations are typically established by minorant functions of f, which are required to be increasingly more accurate, in the following sense. Let l_i be a minorant of f on a given subset D_i of D, that is,

$$l_i(x) \leq f(x) \quad \text{for all } x \in D_i. \qquad (1.2.22)$$

Now, if D_j is an arbitrary subset of D_i—possibly leading to subproblems branched from D_i—then it is required that the bounding operation improves in the sense

$$l_i(x) \leq l_j(x) \leq f(x), \quad \text{for all } x \in D_j, \quad D_j \subset D_i. \qquad (1.2.23)$$

Furthermore, if $x \in D$ is the limit point of a subset sequence $\{D_i\}$, $i = i(k)$, $k \in N$, then the corresponding minorant function values have to converge (from below) to the 'true' function value

$$\lim_{k \in N} l(D_i) = f(x), \quad i = i(k). \tag{1.2.24}$$

It can be expected that these conditions, in essence, guarantee the convergence of the B&B approach outlined: a detailed exposition with convergence proofs (related to specific problem-instances) follows in Part 2.

Summing up the aforesaid, in implementing the B&B scheme important points are to be specified: these are related to proper sampling, bounding, selection, and partition refinement. In particular, the minorant functions $l(D_i)$, $i = i(k)$, $k \in N$, referred to should be relatively 'easy' (computationally not too expensive) to determine, and—at the same time—ideally they should provide 'tight' bounds. These requirements are, of course, somewhat contradictory in the general sense (since added meaningful effort should lead to more precise bounding operations). Again, this point will be treated in detail in Parts 2 and 3; therefore, only two important examples of bounding methods (not to be discussed in detail later) will be mentioned briefly below.

Convex Underestimating Procedures

A frequently applied technique is to provide (lower) bounds of f, via suitable convex minorant functions. The 'tightest' such minorant is called the convex envelope of f on D. Let us define the epigraph of f by

$$\text{epi}(f) = \{(x, r) \in D \times \mathbb{R} : f(x) \le r\}. \tag{1.2.25}$$

Then the convex envelope can be visualized as the convex hull of the (lower) boundary points of epi(f).

Convex envelopes have been extensively investigated in the context of global optimization: see, for example, Falk and Soland (1969), Rockafellar (1970), Horst (1976, 1980, 1984a), McCormick (1976, 1983), Al-Khayyal and Falk (1986), Kalantari and Rosen (1987). Additional examples—within the Lipschitzian framework—will be discussed later in Part 2.

Methods Based on Interval Arithmetic Operations

Let $I(\mathbb{R})$ denote the set of finite real intervals. The basic arithmetic operations $\circ = \{+, -, *, /\}$ can be naturally extended to operations between intervals as follows. For arbitrary A and B from $I(\mathbb{R})$, define

$$A \circ B = \{a \circ b : a \in A, \, b \in B\} \qquad (0 \notin B \text{ in } A/B). \tag{1.2.26}$$

Resulting intervals which correspond to these rules are easy to provide (Moore, 1962). The definition (1.2.26) can be extended to handle general classes of functions, including, for example, polynomials; additionally, it can be extended to multivariate functions. Interval arithmetic is then a natural way to provide bounds of f, and is especially straightforward to apply if D is an interval in $\mathbb{R}^n$. Global optimization based on interval arithmetic operations is a promising research area, and has been subject to extensive studies: consult, for example, the monographs by Moore (1966, 1979, 1988), Ratschek and Rokne (1984, 1988), Hansen (1992), and the survey of Ratschek and Rokne (1995).

1.2.7 Concluding Remarks

The discussion in the previous chapter and this one clearly indicates the broad diversity of global optimization models and solution strategies (although this review is not 'complete' in either direction). The conceptual tools and techniques outlined can be adapted, combined and applied to analyse and solve a large variety of global optimization problems. Such adjustments and combinations can be of particular relevance in cases where the underlying problem can be decomposed into subproblems that can be efficiently handled by applying different 'tailored' solution methods. Pointers towards this challenging but very promising direction are provided, for example, by Tind (1990), Hansen, Jaumard and Lu (1991b), Flippo and Rinnooy Kan (1993), Horst and Thoai (1993, 1994), and Thach and Konno (1993).

In the main body of the present volume which follows, the exposition will be focused on the solution of (general) continuous and Lipschitzian GOPs. With this objective in mind, broad classes of adaptive deterministic and stochastic algorithms will be analysed: these investigations will lead to a comprehensive underlying theory, convergence analysis, implementations, and a diversity of applications.

PART TWO

Partition Strategies in Global Optimization: The Continuous and the Lipschitzian Case

No claim is true or false because it can be shown to be an archetype or a stereotype.

...

'Scientific' is not the synonym for truth, since many assertions made by scientists since Aristotle have been shown to be false, and are none the less scientific for that.

(George Watson, *Writing a Thesis.* Longman, London, 1987)

Sometimes people will argue that because a theory is not proven fact, one theory is as good as another ...

(Bernard J. Nebel and Richard T. Wright, *Environmental Science – The Way the World Works.* Prentice-Hall, Englewood Cliffs, 1991.)

To get to the fruit on a tree, you have to go out on a limb.

(Shirley MacLaine, *Out on a Limb.* Bantam Books, New York, 1983.)

Chapter 2.1

An Introduction to Partition Algorithms

2.1.1 Problem Statement and Basic Assumptions

Consider the general continuous global optimization problem (CGOP) stated as

$$\min f(\boldsymbol{x}) \quad \text{s.t.} \quad \boldsymbol{x} \in D. \tag{2.1.1}$$

With respect to this GOP, we shall postulate the following basic assumptions:

- the feasible region D is a subset of the n-dimensional real Euclidean space $\mathbb{R}^n$.
- D is bounded, that is, there exist finite real n-vectors a and b such that $D \subset [a, b]$ ($[a, b]$ is called an n-dimensional interval, rectangle, or box);
- D is robust ('body-like'), that is, $\text{int}(D) \neq \emptyset$, and D is the closure of its interior;
- the objective function f is continuous and (possibly) multiextremal on D.

The above conditions directly imply the following classical result.

Theorem 2.1.1. (Weierstrass) *As D is bounded and closed, and f is continuous, the solution set of the CGOP* (2.1.1) *is nonempty. (In other words, f attains its minimum on D.)*

Note that it is easy to see, by using tools from elementary analysis, that the above requirements are 'nearly minimal'. For instance, dropping the

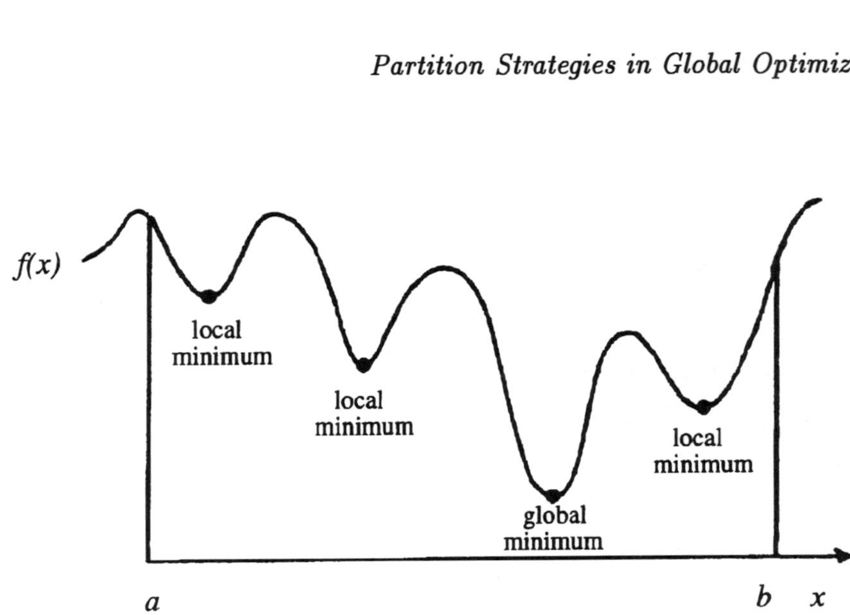

Figure 2.1.1
A multiextremal function with several local optima
$(D = [a, b] \subset \mathbb{R}^1)$

boundedness or closedness assumptions on D, or the continuity of f, one can construct simple examples of optimization problems (2.1.1), which do not have a well-defined, globally optimal solution.

We shall denote by $X^\star \subset D$ the set of global minimizers in (2.1.1); for an arbitrary optimal solution $x^\star \in X^\star$, $z^\star = f(x^\star)$ denotes the optimum value.

The main objective of Part 2 is to introduce a general class of optimization strategies which are aimed at the solution of problem (2.1.1), via 'exhaustive' partition and search of the set D. This approach is motivated by the fact that, in general, the traditional local search methods cannot guarantee any sort of convergence to the set $X^\star$, as they would typically approximate the deepest point of the 'valley' from which the local scope search algorithm was initiated; see Fig. 2.1.1.

In other words, if f has multiple local minima in D, then the set X^a of limit points of the search (sample) point sequence, generated by a local search algorithm, may or may not contain points of $X^\star$.

Although the Weierstrass theorem ensures the existence of the optimal solution(s) to (2.1.1), for numerical reasons additional problem specifications are necessary. First, we shall postulate the finiteness of $X^\star$. In relation to this assumption, note that it is easy to construct optimization problems which have an infinite number of global solutions. One may think, for example, of a one-dimensional function $f(x)$, $x \in [a, b]$, which attains its minimum on a subinterval of $[a, b]$. However, in most practical contexts it is quite natural to suppose the finiteness of $X^{\star}$. In fact, many well-posed global optimization problems have a singular global solution $x^{\star}$—although the theoretical convergence results presented later in Part 2 remain valid if $X^{\star}$ is (at most) countable.

Evidently, both the robustness of D and finiteness of $X^{\star}$ are important for the numerical implementability of convergent search algorithms. The following example shows that it is reasonable to postulate also an additional property of the objective function f, whenever such assumption is justified.

Example 2.1.1. Suppose that a finite sequence of search points and corresponding function values

$$x_i \in D, \; z_i = f(x_i), \qquad i = 0, 1, \ldots, k,$$

have been generated by a certain algorithm on the finite interval $D = [a, b] \subset \mathbb{R}^1$. Now, if f is merely continuous, then for arbitrary real number r it is easy to define another continuous function $g(x)$ which has the following properties:

$$g(x_i) = z_i, \quad i = 0, 1, \ldots, k, \quad \text{and} \quad g(x_{k+1}) = r. \tag{2.1.2}$$

For example, g can be a $(k+1)$st degree polynomial which is determined by the set of equations (2.1.2). But this implies that, after observing an arbitrary finite number of sample points x_i and corresponding function values z_i, one cannot 'guess' the function value at a new sample point. Specifically, it is not possible to provide a guaranteed (deterministic) lower bound of $z^{\star} = f(x^{\star})$.

This simple example indicates that—whenever possible—a stronger quantitative property should be postulated, instead of the mere (uniform) continuity of f on the set D.

Definition 2.1.1. The function f is Lipschitz-continuous (Lipschitzian, or simply Lipschitz) on the set D, if there exists a constant $L > 0$, such that for all pairs $x, y \in D$ we have

$$|f(x) - f(y)| \leq L\|x - y\|. \tag{2.1.3}$$

In (2.1.3), $\|\cdot\|$ denotes the Euclidean norm, although other norms could also be used. Note that, obviously, $L = L(D, f)$, that is, the Lipschitz-constant L depends not only on f, but also on the set D: this fact, however, is typically suppressed by the notation.

Remark 2.1.1. It is easy to see that, for bounded set D and Lipschitzian f, the optimum value can be estimated on the basis of a finite sample sequence $\{x_i, z_i\}$. For instance, if we know the function value z_0 at x_0, then, for arbitrary $x \in D$, (2.1.3) implies the inequality

$$f(x) \geq z_0 - L\|x - x_0\| \geq z_0 - Lr \quad \text{for all } r \geq \overline{r}(D) = \max_{u,v \in D} \|u - v\|. \quad (2.1.4)$$

In other words, knowing an overestimate of $\overline{r}(D)$, the diameter of D, we have a valid lower bound for the optimum of f in D. Obviously, the estimate $r = \|b - a\|$ can be applied in (2.1.4), since $D \subset [a, b]$. Note further that, similarly to (2.1.4), the simple upper bound $f(x) \leq z_0 + Lr$ can also be deduced.

Remark 2.1.2. If f is Lipschitz-continuous with constant L, then it is also Lipschitzian with all $L' > L$. Although, from a numerical point of view, it would be advantageous to know the smallest possible L, it is unknown in many cases. Therefore we either have to provide a 'safe' overestimate of L, or have to live with its numerical estimates (produced in the course of the optimization, or otherwise).

Remark 2.1.3. If f is continuously differentiable on an open convex set D_0, $D \subset D_0$, then it is Lipschitzian on D. To verify this assertion (stated earlier as Theorem 1.1.7), take the first order Taylor-expansion of f around an arbitrary point x_0 of D:

$$f(x) = f(x_0) + g(x_\tau)^T (x - x_0), \qquad x \in D. \quad (2.1.5)$$

Here g denotes the gradient of f, and x_τ is a suitably chosen point of the interval $[x_0, x]$. Now, applying the Cauchy-inequality to (2.1.5), we obtain

$$|f(x) - f(x_0)| = |g(x_\tau)^T (x - x_0)| \leq \|g(x_\tau)\| \, \|x - x_0\|. \quad (2.1.6)$$

Comparing (2.1.3) and (2.1.6), one can see that the maximum value of the gradient norm on D could serve as an appropriate Lipschitz-constant. Of course, finding

$$\max \|g(x)\| \quad \text{s.t.} \quad x \in D, \quad (2.1.7)$$

can be just as difficult as solving the optimization problem (2.1.1). Nevertheless, this remark in itself indicates that the class of Lipschitzian functions is broad: indeed, in many practical applications of optimization, the system description is based on continuously differentiable (smooth) and, hence, Lipschitz-continuous functions. (Recall also Theorems 1.1.6 and 1.1.8 in this context.) If, however, the Lipschitzian assumption is not valid, then—as Example 2.1.1 shows—one cannot provide a deterministic lower bound of the optimum value, on the basis of the information $\{x_i, z_i\}$, $i = 0, 1, \ldots, k$, after an arbitrary finite number of search steps.

In the following, $F_L(D)$ will denote the family of those functions f that are Lipschitz-continuous with constant L. Furthermore,

$$F(D) = \bigcup_{L>0} F_L(D) \tag{2.1.8}$$

denotes the class of all Lipschitz functions on D.

For purposes of distinction, we shall introduce the Lipschitzian global optimization problem (LGOP) as the following specification of the GOP (2.1.1):

$$\min f(x) \quad \text{s.t.} \quad x \in D, \quad \text{where} \quad f \in F(D). \tag{2.1.9}$$

Furthermore, we shall also consider the continuous global optimization problem (CGOP), which is formally identical with the GOP problem statement (2.1.1) (that is, we drop the Lipschitz assumption).

In the following chapters of Part 2, a general approach will be presented to solve the CGOP (2.1.1) or the LGOP (2.1.9). Denoting by $C(D)$ the class of continuous functions defined on D, obviously we have the inclusion

$$F(D) \subset C(D).$$

Consequently, all methods that are capable of solving the CGOP are, in principle, applicable also to the LGOP.

The subsequent exposition in Part 2 is based to a significant extent on materials from Pintér (1983b, 1986a,b,c, 1987b, 1988, 1990a, 1991b,c) and, in particular, draws on Pintér (1992a). The discussion structure, however, differs from the chronological sequence of these works, allowing for a more concise and straightforward derivation of the theoretical convergence results. In addition, some extensions to previous results, numerous additional examples, and clarifying remarks will be provided.

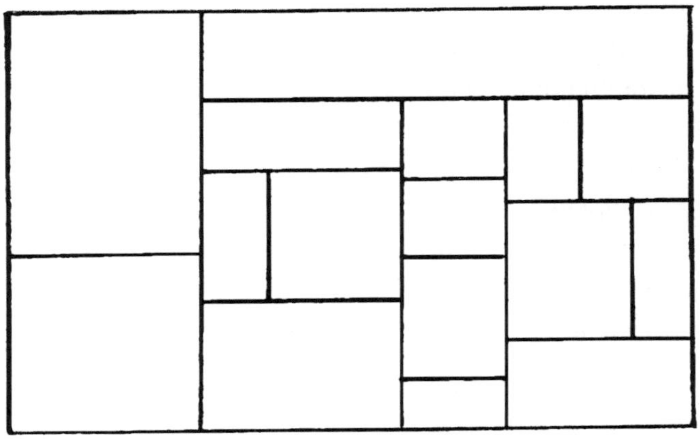

Figure 2.1.2
Robust partition of the set $D = [a, b] \subset I\!R^2$

2.1.2 Partition Algorithms: A General Scheme

Definition 2.1.2. Let D be a robust subset in $I\!R^n$, and I be a finite set of indices. A collection of sets $\{D_i : i \in I\}$ is called a robust partition of D, if D_i, $i \in I$ are all robust sets; furthermore,

$$D = \bigcup_{i \in I} D_i \quad \text{and} \quad D_i \cap D_j = \delta(D_i) \cap \delta(D_j) \qquad \text{for} \quad i, j \in I, \quad i \neq j. \tag{2.1.10}$$

In (2.1.10), $\delta(D_i)$ denotes the relative (to D) boundary of D_i.

A simple example of a robust partition is a subdivision of $D = [a, b]$ into smaller intervals D_i with disjoint interiors; see Fig 2.1.2.

Given a partition of D, one can select sample points from each partition subset D_i, in order to estimate the 'merit' of D_i for subsequent, possibly more detailed, investigation. This underlying idea leads to the following.

Definition 2.1.3. Given a robust partition of D, a partition operator P is a real-valued functional which is determined by a finite set of sample points $x_i^{(j)}$, $j = 1, \ldots, J_i$ from D_i, and the corresponding function values $z_i^{(j)} := f(x_i^{(j)})$, $j = 1, \ldots, J_i$:

$$P\{D_i, f(D_i)\} := P\{x_i^{(j)}, z_i^{(j)},\ j = 1, \ldots, J_i\} \qquad \text{for} \quad i \in I. \tag{2.1.11}$$

In the example above, sample points can be, for instance, several or all vertices of the subintervals, and/or a finite collection of their (arbitrary) interior points.

Applying the notation and definitions introduced above, we shall now formally describe a general partition algorithm scheme (PAS), which will be subsequently investigated.

Partition Algorithm Scheme (PAS)

Step 0. (Initialization) Set the iteration count $k := 0$; let $I_0 = \{0\}$ and $\{D_0\} = \{D\}$ define the initial partition.

Step 1. (Sampling) For all $i \in I_k$, select a finite number of sample points $x_i^{(j)} \in D_i$, and determine the corresponding function values $z_i^{(j)}$, $j = 1, \ldots, J_i$.

Step 2. (Aggregation) On the basis of the search information $\{x_i^{(j)} : j = 1, \ldots, J_i\}$, compute the value $P\{D_i, f(D_i)\}$ of the partition operator for all $i \in I_k$.

Step 3. (Selection) Choose an (arbitrary) index $t \in I_k$ for which the relation

$$P\{D_t, f(D_t)\} = \max_{i \in I_k} P\{D_i, f(D_i)\} \tag{2.1.12}$$

is satisfied: then the partition set D_t will be subdivided.

Step 4. (Refinement) Replace the set D_t by its finite, robust, and nontrivial partition

$$D_t = \bigcup_{l \in L_k} D_{t_l}, \qquad D_{t_l} \subset D_t, \qquad l \in L_k. \tag{2.1.13}$$

Consequently, the refined partition is given by

$$\{D_i : i \in I_{k+1}\}, \qquad \text{where } I_{k+1} = I_k \cup \{t_l : l \in L_k\} \setminus \{t\}.$$

Set $k := k + 1$; return to Step 1 (of the next iteration cycle).

Remark 2.1.4. The PAS is introduced for the purposes of theoretical convergence investigations and for the qualification of a number of (known and new) algorithms. For this reason, numerical stopping criteria have not been formally introduced. It is only noted here that such criteria can be included at several stages in the PAS, for instance in Steps 2 and 4; examples will be provided later.

Remark 2.1.5. As reflected by the selection rule (2.1.12), the value of the partition operator $P\{D_i, f(D_i)\}$ can be interpreted as expressing the estimated 'current search value' of subset D_i. For instance, if we are able to provide an estimate of the lower bound of f on all subsets D_i, then it may seem reasonable—at the actual iteration step—to devote further computational effort to exploring (that is, partitioning and sampling) that subset D_t for which the current lower bound estimate is the smallest. Naturally, an additional search effort may prove at a later stage that this choice was 'wrong' (in which case the PAS still has the capability to return and explore other subsets), or 'right' (leading to further refinement of the partition within D_t). Observe in connection with this point that, obviously, the earlier generated search points have information value at each current PAS iteration cycle. Therefore, only adaptively chosen additional sampling is necessary at each refinement step.

Before starting with a formal analysis of partition strategies, for illustrative purposes two well-known search algorithms will be reviewed: in both cases, $D = [a, b] \subset \mathbb{R}^1$, that is, we shall look at partitions of a finite one-dimensional interval, in order to find (approximate) the global minimum of a multiextremal function f.

2.1.3 Uniform Grid Search

A uniform grid search (UGS) is probably the simplest deterministic approach to solving the LGOP (2.1.9) in the special case in which $D = [a, b]$ is a finite one-dimensional interval. For purposes of illustration, and also for its role in subsequent discussions, this trivial search strategy is described below.

In UGS, select simultaneously $k + 1$ equidistant points on $[a, b]$, and evaluate f at these (see Fig. 2.1.3):

$$x_i = a + \frac{i}{k}(b - a), \quad z_i = f(x_i), \qquad i = 0, 1, \ldots, k. \tag{2.1.14}$$

For each partition subset $[x_{i-1}, x_i]$, by the Lipschitz property of f, we have the inequalities

$$\begin{aligned} &f(x) \geq z_{i-1} - L(x - x_{i-1}) \quad \text{and} \\ &f(x) \geq z_i - L(x_i - x), \quad x_{i-1} \leq x \leq x_i, \quad i = 1, \ldots, k. \end{aligned} \tag{2.1.15}$$

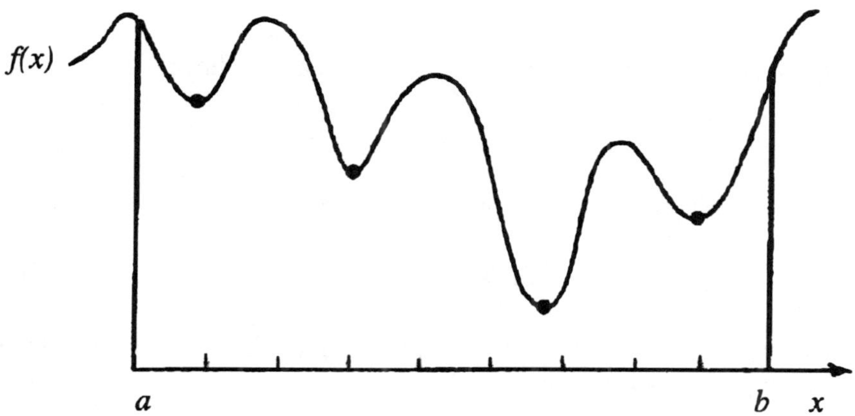

Figure 2.1.3
Uniform grid search on the one-dimensional interval
$D = [a, b]$

The intersection point of the right-hand side linear Lipschitzian minorant (underestimating) functions is

$$\underline{x}_i = \frac{x_{i-1} + x_i}{2} + \frac{z_{i-1} - z_i}{2L}, \qquad x_{i-1} \le \underline{x}_i \le x_i, \tag{2.1.16}$$

and the corresponding minorant function value (lower bound) equals

$$\begin{aligned} \underline{z}_i &= \frac{z_{i-1} + z_i}{2} - L\frac{x_i - x_{i-1}}{2} \\ &= \frac{z_{i-1} + z_i}{2} - \frac{L(b-a)}{2k}, \qquad i = 1, \ldots, k. \end{aligned} \tag{2.1.17}$$

Consequently, the UGS sample provides the following upper and lower bounds:

$$\alpha_k := \min_{i=0,1,\ldots,k} z_i \ge z^\star \tag{2.1.18}$$

and

$$\beta_k := \min_{i=1,\ldots,k} \underline{z}_i \le z^\star. \tag{2.1.19}$$

Now, the global convergence of the UGS can be easily established. For arbitrary $\varepsilon > 0$, choose $k_0 = k_0(\varepsilon)$, such that

$$\frac{L}{2k_0}(b-a) \leq \varepsilon. \tag{2.1.20}$$

Then, by (2.1.17), for each subinterval $[x_{i-1}, x_i]$ we have

$$\max\{z_{i-1}, z_i\} \geq \underline{z}_i \geq \min\{z_{i-1}, z_i\} - \varepsilon, \qquad i = 1, \ldots, k_0.$$

Therefore $k \geq k_0(\varepsilon)$ directly implies the inequalities

$$\alpha_k \leq z^\star + \varepsilon, \qquad \beta_k \geq z^\star - \varepsilon, \qquad \alpha_k - \beta_k \leq \varepsilon. \tag{2.1.21}$$

Furthermore, if $k \to \infty$, then $\varepsilon = L(b-a)/(2k) \to 0$, and the sequences $\{\alpha_k\}$ and $\{\beta_k\}$ converge to $z^\star$.

Remark 2.1.6. Inequality (2.1.20) shows the impact of the value L on the sufficient number of uniform grid points, to guarantee an ε-optimal approximate solution: namely, $k_0 \geq L(b-a)/(2\varepsilon)$ is a linear function of (the known overestimate of) L.

Although UGS is extremely simple, it is typically very inefficient: this fact will be quantitatively established later, in comparison with other partition algorithms. At this point note only that its inefficiency is due primarily to the 'passive' character of the strategy, that is, the simultaneous uniform placement of the sample points excludes the possibility of adaptively allocating the search effort. Heuristically, one would expect that a sequential search approach, in which the new sample points are chosen in accordance with a gradually refined (updated) 'picture' of the function, should generally work better. Nevertheless, as has been shown by Sukharev (1971) and Ivanov (1972), there is a case where there is no more efficient sequential search strategy than the passive UGS: this case arises when $f(x) = c$ for $x \in [a, b]$. The Sukharev–Ivanov result is based on the so-called minimax criterion, that is, UGS guarantees ε-accuracy even for the worst possible instance of the Lipschitzian function class $F_L[a, b]$. Again, one would reasonably expect that active (sequential) algorithms will perform better than UGS, except for 'pathologic' problem-instances; however, for 'flat' sections of f a passive search element can be useful—and is indeed used—as part of several adaptive global optimization algorithms (examples will be mentioned later). For

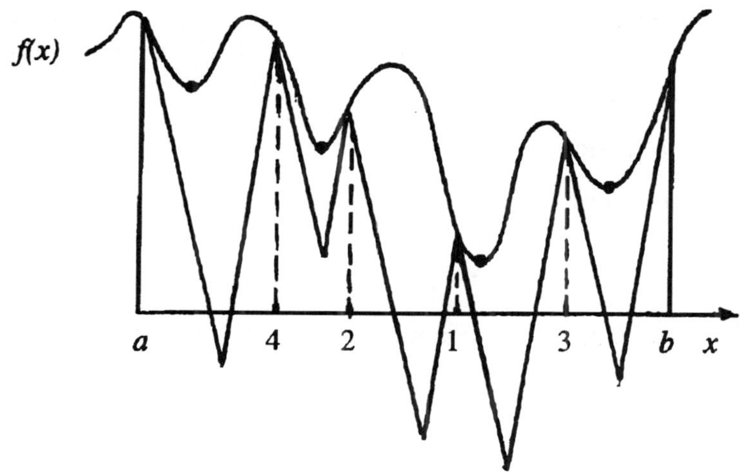

Figure 2.1.4
Piyavskii's search on the one-dimensional interval $D = [a, b]$

related additional discussions, see, for instance, also Kiefer (1953), Archetti and Betrò (1978), Schoen (1982), Basso (1985), and Sukharev (1985).

Note finally that the grid search concept can be embedded in the framework of the general PAS, in different ways. For example, one can choose a fixed integer k and, at each subsequent iteration cycle of PAS, generate a uniform k-grid on that subinterval $[x_{i-1}, x_i]$ on which the lower bound $\underline{z}_i$ is minimal (that is, it equals β_k).

2.1.4 Piyavskii's Algorithm

The univariate global optimization algorithm proposed by Piyavskii (1971) and, independently, by Shubert (1972), has been subject to extensive discussion and diverse generalizations: Hansen, Jaumard and Lu (1991a, 1992a), Hansen and Jaumard (1995) provide a comprehensive review, with numerous references. Piyavskii's method generates a sequence of search points on $[a, b]$, and gradually refines the Lipschitzian underestimating function of f; see Fig. 2.1.4 (the first four new points are shown).

Piyavskii's algorithm can be interpreted as an instance of the general PAS, according to the following description. (For purposes of clarification, the underlying motivation of the algorithm steps is also included.) Note that—in accordance with the subsequent, more general constructions—Piyavskii's original algorithm will be slightly modified: namely, instead of evaluating first the mid-point of $[a, b]$, a and b will be chosen as the first pair of sample points.

Step 0. (Initialization) Set $k := 1$; $x_0 = a$, $x_1 = b$; $z_i = f(x_i)$, $i = 0, 1$.

Initial optimum estimate:

$$\alpha_k = \min\{z_0, z_1\}.$$

Initial lower bound of optimum (see (2.1.16) and (2.1.17)):

$$\beta_k = \frac{z_0 + z_1}{2} - L\frac{x_1 - x_0}{2}.$$

Step 1. (Sampling) Assume that up to the current iteration, the—generally speaking, non-equidistant—search points and corresponding function values

$$a = x_0 \leq x_1 \leq \ldots \leq x_k = b, \quad z_i = f(x_i), \quad i = 0, 1, \ldots, k, \tag{2.1.22}$$

have been generated and ordered. On the basis of the Lipschitzian function model—lacking additional quantitative knowledge—it seems rational to select on each subinterval that point (denoted by $\underline{x}_i$) at which the anticipated Lipschitzian underestimating function attains its minimum. In other words, on each subinterval $[x_{i-1}, x_i]$, one can consider the corresponding minimal minorant function value

$$\underline{z}_i = \frac{z_{i-1} + z_i}{2} - L\frac{x_i - x_{i-1}}{2}. \tag{2.1.23}$$

Observe now that the calculation of $\underline{z}_i$ is based on the already available information (namely, $x_{i-1}, x_i, z_{i-1}, z_i$ and the value of L). Thus, there is no immediate need to take additional sample points in each subinterval $[x_{i-1}, x_i]$, unless we want to make the Lipschitzian minorant function a more refined, closer approximation of the function f along $[a, b]$. Such a refinement could be achieved, for example, taking all points $\underline{x}_i$ (see (2.1.16)) which correspond to the minimizer argument of the Lipschitzian underestimating function on $[x_{i-1}, x_i]$. This effort, however, seems 'by far too much' and, indeed, would lead to an exponential increase of the search effort.

Step 2. (Aggregation) On the basis of the above discussion, let us define

$$P_i = P\{[\boldsymbol{x}_{i-1}, \boldsymbol{x}_i], z_{i-1}, z_i\} = -\underline{z}_i.$$

In other words, the partition operator value simply equals the subinterval-specific Lipschitz minorant function value, multiplied by -1. (This transformation is purely technical, in accordance with Step 3 below.)

Step 3. (Selection) Select an index t, $1 \leq t \leq k$, for which

$$P_t = \max_{i=1,\ldots,k} P_i,$$

or, equivalently, select a subinterval on which the minorant function attains its global minimum:

$$\underline{z}_t = \min_{i=1,\ldots,k} \underline{z}_i. \tag{2.1.24}$$

Step 4. (Refinement) In accordance with the discussion preceding Step 2, at the current step it seems rational to sample (exclusively) at

$$\underline{\boldsymbol{x}}_t = \frac{\boldsymbol{x}_{t-1} + \boldsymbol{x}_t}{2} + \frac{z_{t-1} - z_t}{2L}, \tag{2.1.25}$$

where the index t is defined by (2.1.24).

Setting now $k := k + 1$, the new sample point $\underline{\boldsymbol{x}}_t$ is inserted into the (renumbered) sample sequence

$$a = \boldsymbol{x}_0 \leq \boldsymbol{x}_1 \leq \ldots \leq \boldsymbol{x}_{k+1} = b, \qquad z_i = f(\boldsymbol{x}_i), \qquad i = 0, 1, \ldots, k+1.$$

Observing that

$$\beta_{k-1} = \underline{z}_t \leq f(\underline{\boldsymbol{x}}_t) \leq \alpha_{k-1},$$

both the upper and lower bound estimates are possibly refined, as

$$\alpha_k = \min\{f(\underline{\boldsymbol{x}}_t), \alpha_{k-1}\}, \quad \text{and} \quad \beta_k \geq \underline{z}_t.$$

Step 5. (Numerical stopping criterion) Check now whether, for a fixed ε tolerance, we have

$$\alpha_k - \beta_k \leq \varepsilon. \tag{2.1.26}$$

If (2.1.26) holds, then the current optimum estimate $x_k^\star$ in

$$f(x_k^\star) = \alpha_k$$

is ε-optimal; the search is terminated. In the opposite case, the algorithm returns to Step 1 of the next cycle.

Remark 2.1.7. As can be seen, the principal advantage of Piyavskii's approach, compared to the passive UGS, lies in its adaptive character: the search always continues only on that subinterval of $[a, b]$ which seems 'most promising' at the current iteration. This strategy also implies that the search can return to any of the generated subintervals, if this is justified later, on the basis of additional function evaluations.

Remark 2.1.8. Assuming that the available value L is a guaranteed overestimate of the 'true' Lipschitz-constant, suboptimal partition subsets could be excluded from further investigations. If, at a certain iteration k, for some subinterval $[x_{i-1}, x_i]$ we have

$$\underline{z}_i > \alpha_k, \tag{2.1.27}$$

then this subinterval cannot contain globally optimal points, and hence can be deleted. We shall return later to this observation in the course of a general treatment of suboptimality deletion strategies, and also when discussing implementation aspects.

Remark 2.1.9. Recalling the Lipschitzian underestimate (2.1.23), one can observe that analogous estimation can be used, if—instead of the 'universally valid' L—a subinterval-specific, locally valid overestimate L_i is available. In this case one can write

$$\underline{z}_i = \frac{z_{i-1} + z_i}{2} - L_i \frac{x_i - x_{i-1}}{2} \leq f(x) \quad \text{for any } x \in [x_{i-1}, x_i],$$

and the Lipschitzian minorant may become significantly tighter; see Fig. 2.1.5.

We shall discuss later the important numerical consequences of this observation. It is only noted here that, as reflected by Piyavskii's partition operator

$$P_i = -\underline{z}_i = L \frac{x_i - x_{i-1}}{2} - \frac{z_{i-1} + z_i}{2},$$

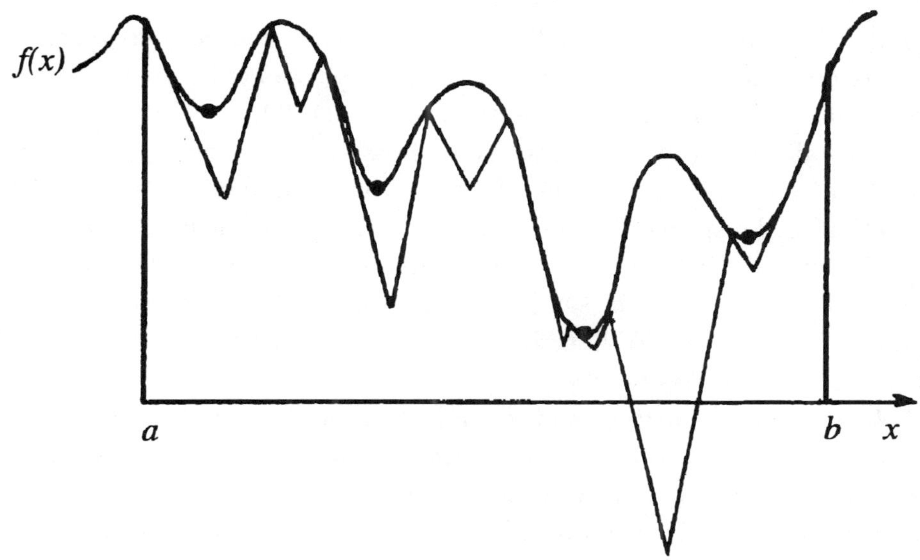

Figure 2.1.5
Subinterval-dependent Lipschitzian underestimation of function f on $D = [a, b]$

'exaggerated' estimates of the Lipschitz constant will typically lead to further exploration of the larger subintervals of the partition (whenever the expression $L(x_i - x_{i-1})/2$ dominates in calculating P_i). This comment implies that, although a safe overestimate of L is required to guarantee convergence, numerical efficiency may decrease substantially if crude overestimates are being used. This controversy is an inherent drawback of all search algorithms which rely on the Lipschitz information; note at the same time that similar reasoning would apply to all other (global) optimization approaches in which not easily available, problem-dependent information plays a significant role.

The convergence of Piyavskii's method will be established later, in the framework of a general convergence analysis which can be directly applied to a large class of partition algorithms.

2.1.5 Kushner's Algorithm

In order to indicate the broad possibilities of choosing partition strategies in global optimization, we shall now briefly discuss another one-dimensional algorithm, due to Kushner (1964). The method was put into a global optimization context again by Archetti and Betrò (1979).

This algorithm—unlike Piyavskii's—is based on an underlying statistical model of the objective function f. A very brief summary of the statistical function model approach in global optimization follows below; for additional details, consult, for example, Neimark and Strongin (1969), Mockus (1972), Mockus, Tiesis and Žilinskas (1978), Strongin (1978), Archetti and Betrò (1980), Žilinskas (1982), Boender (1984), Mockus (1989), Zhigljavsky (1991, Ch. 2.4), or Žilinskas (1992).

In stochastic function models applied in global optimization, the basic assumption is that $f(x)$ is a realization of some random function $\varphi(x,\omega)$, for a certain ω. That is, we consider an underlying probabilistic structure as defined by the Kolmogorov probability space $\{\Omega, \mathcal{A}, \mathcal{P}\}$. In this setting,

$$f(x) = \varphi(x,\omega) \quad \text{for all} \quad x \in D \quad \text{and some fixed} \quad \omega \in \Omega,$$

and

$$f \in \mathcal{F} = \{\varphi : D \otimes \Omega \to \mathbb{R}^1\}. \tag{2.1.28}$$

A frequently used model of one-dimensional, potentially multiextremal functions $f(x)$, $x \in [0,1]$ is the Wiener process. According to this model, for an arbitrary pair of points x_1, x_2 from the interval $[0,1]$, the random variable $\varphi(x_1,\omega) - \varphi(x_2,\omega)$ follows normal distribution with mean 0 and variance $\sigma^2|x_1 - x_2|$ ($\sigma > 0$ is a parameter of the function model: its actual value is usually estimated from a set of sample points). Note that the realizations of Wiener processes are continuous but nondifferentiable, with probability one (Kushner, 1964): for this reason, many random field model based methods provide corresponding algorithms to solve the (univariate) CGOP.

In most cases, the solution strategies associated with prior statistical function models are based on Bayesian principles. (For a brief exposition on the Bayesian approach, see, for example, Hillier and Lieberman, 1986; a detailed treatise is presented by Berger, 1985.) In Bayesian algorithms, the search is aimed to perform 'best' in an expected value (average) sense, considering all possible realizations of the random function $\varphi(x,\omega)$: note that this is in contrast with the 'conservative' minimax criterion based approaches mentioned earlier. The generated search information, most typically a set of sequentially generated sample points and corresponding function values $\{x_i,\ z_i\ :\ i=0,1,\ldots,k\}$, serves also to update the (posterior) function model, at each iteration.

Finding theoretically optimal Bayesian strategies is, as a rule, associated with insurmountable numerical difficulties, frequently even for one-dimensional CGOPs. Therefore—instead of 'true' k-step optimality for $k=1,2,3,\ldots$—at each search stage a one-step Bayes-optimal strategy is used. Given the search information $S_k=\{x_i,z_i:i=0,1,\ldots,k\}$, and some appropriately chosen utility function $U[\phi(x,\omega)]$, the next point is chosen in accordance with the relation

$$x_{k+1}=\arg\operatorname*{opt}_{x\in D} E\big\{U[\phi(x,\omega)|S_k]\big\}, \tag{2.1.29}$$

that is, x_{k+1} optimizes the expected utility function value at the next evaluation, given the search information S_k. One can recognize the close analogy with the philosophy of Piyavskii's search strategy which—at each step—selects the one-step optimal point corresponding to the Lipschitzian function model. An example of the general criterion form (2.1.29) can be defined as

$$x_{k+1}=\arg\max_{x\in D}\ \mathrm{P}\{\varphi(x,\omega)<f_k^\star-\varepsilon\mid S_k\}, \tag{2.1.30}$$

in which

$$f_k^\star=\min_{1\le i\le k} z_i.$$

According to (2.1.30), the new point x_{k+1} maximizes the probability of finding a new optimum estimate which is better by ε than the current optimum estimate $f_k^\star$, given S_k ($\varepsilon>0$ is a fixed parameter of the search procedure).

Without going into additional details at this point, we quote a possible partition operator derived for the Wiener process based function model. Assuming that we have the search information S_k, the partition operator corresponding to the criterion (2.1.30) can be defined (Kushner, 1964; Archetti and Betrò, 1979) as

$$P_i=\exp\left\{\frac{-2(f_k^\star-\varepsilon-z_i)(f_k^\star-\varepsilon-z_{i-1})}{\sigma^2(x_i-x_{i-1})}\right\}, \tag{2.1.31}$$

in which

$$f_k^\star = \min_{1 \le i \le k} z_i$$

is the current sample minimum, and $\varepsilon > 0$, $\sigma > 0$ are fixed model parameters.

Naturally, the partition rule (2.1.31) can also be included into a PAS, similarly to Piyavskii's rule: that, in turn, leads to the corresponding one-dimensional optimization method.

Without further analysis of algorithm examples at this point, we shall proceed now to a general qualification of adaptive partition methods for solving the CGOP, or the LGOP stated earlier. With this objective in mind, let us summarize the main conceptual steps of defining an adaptive partition strategy:

- definition of the feasible set class $D \in \mathcal{D}$ under consideration;
- definition of a prior model of the function class $\mathcal{F}$ to which the objective function f belongs;
- selection of a quantitative measure which serves to estimate the 'goodness' of an arbitrary partition subset, on the basis of the currently available search information, at each stage of the algorithm;
- determination of the PAS, specifying its sampling, aggregation, selection, and refinement steps, and—whenever possible—its lower/upper bounding procedures.

In the following chapter, we shall present a unified convergence investigation of partition algorithms in global optimization. This approach will not only lead to the straightforward establishment of convergence properties for a wide variety of specific known methods, but will also permit their nontrivial generalizations and extensions.

Chapter 2.2

Convergence Properties of Adaptive Partition Algorithms

2.2.1 Regular Partition Strategies

Let us assume that the global optimization problem CGOP (2.1.1) or LGOP (2.1.9) is to be solved by an adaptive partition strategy which, in its basic structure, follows the partition algorithm scheme (PAS) described in Section 2.1.2.

Introducing the notation $N = \{0, 1, 2, \ldots\}$, first we shall postulate a set of necessary (required) properties of the PAS.

First, let us note that by Step 4 of the PAS (recall (2.1.13)), for each nested sequence of partition subsets $\{D_{i_k}\}$, $k \in N$, either we have

$$D_{i_k} = D_{i_{k_0}} \qquad \text{for some } k_0 \text{ and all } k \geq k_0 \tag{2.2.1}$$

(if a subset is not partitioned further after finitely many iteration steps); or (in the opposite case)

$$D_{i_{k+1}} \subset D_{i_k} \qquad \text{infinitely many times.} \tag{2.2.2}$$

Definition 2.2.1. A sequence of partition subsets is called strictly nested, if (2.2.2) holds.

Definition 2.2.2. The partition algorithm is called regular, if it has the following properties P1–P4:

P1. (Perfection) Every strictly nested sequence of partition subsets converges to a single limit point $x = \bigcap_{k \in N} D_{i_k}$; the sequence $\{D_{i_k}\}$ is then called perfect.

P2. (Continuity) For every sequence of nested partition sets, independently of the sample points actually generated, we have the limit relation

$$P\{D_{i_k}, f(D_{i_k})\} \underset{k\to\infty}{\longrightarrow} P\{\lim_{k\to\infty} D_{i_k}, f(\lim_{k\to\infty} D_{i_k})\}. \tag{2.2.3}$$

Note that, if (2.2.1) holds, then (2.2.3) is often fulfilled in the form

$$P\{D_{i_k}, f(D_{i_k})\} = P\{D_{i_{k_0}}, f(D_{i_{k_0}})\} \qquad \text{for all} \quad k \geq k_0; \tag{2.2.3a}$$

while, if (2.2.2) holds and $x = \bigcap_{k \in N} D_{i_k}$, then (2.2.3) can be rewritten as

$$P\{D_{i_k}, f(D_{i_k})\} \underset{k\to\infty}{\longrightarrow} P\{x, f(x)\} := P\{\lim_{k\to\infty} D_{i_k}, f(\lim_{k\to\infty} D_{i_k})\}. \tag{2.2.3b}$$

P3. (Monotonicity) Let $\{D_{i_k}\}$ be an arbitrary nested sequence of subsets in D. Then for all points $x \in \bigcap_{k \in N} D_{i_k}$, the inequalities

$$P\{D_{i_k}, f(D_{i_k})\} > P\{x, f(x)\}, \qquad k \in N, \tag{2.2.4}$$

are valid. (Note that the value $P\{x, f(x)\}$ is defined by an arbitrary strictly nested subset-sequence which converges to the point x, see P2.)

P4. (Boundedness) The partition operator P is bounded: that is, for each partition subset $D_{i_k} \subset D$, we have

$$-\infty < P\{D_{i_k}, f(D_{i_k})\} < \infty. \tag{2.2.5}$$

Remark 2.2.1. The role of assumption P1 is evident for providing the basis for convergence investigations. Note that the PAS introduced would not automatically guarantee this property. Consider, for example, an infinite sequence of nested intervals $\{[a_k, b_k]\}$, $k \in N$ in $\mathbb{R}^n$, which contains in limit an interval $[a_\infty, b_\infty]$ of positive volume. Evidently, this sequence of intervals does not lead to a perfect partition of $[a_0, b_0]$, or in fact of any robust set in which $[a_0, b_0]$ is one of the partition subsets generated by a PAS. (Without going into further detail, one may observe that the requirement P1 is closely related to the finiteness/countability of the solution set $X^\star$, as this property clearly excludes the possibility of optimal subsets of positive volume.)

Remark 2.2.2. The continuity property P2 is also necessary for establishing convergence properties. Note that (2.2.3a) corresponds to the case in which, after iteration step k_0, the PAS does not lead to further sampling and partition of set $D_{i_{k_0}}$ (so that the information with respect to this subset will not change any more). By contrast, (2.2.3b) corresponds to the case in which an infinite sequence of nested partition subsets converges to a single point x: the corresponding search information sequence $P\{D_{i_k}, f(D_{i_k})\}$, $k \in N$, then converges to the value $P\{x, f(x)\}$ attained at the limit point x.

Remark 2.2.3. The monotonicity property (2.2.4) is in line with the informal interpretation given in Section 2.1.2, relating to the estimated 'search value' of a partition subset. Relation (2.2.4) expresses that none of the points of a subset D_{i_k} are 'more valuable' for the PAS than the entire subset itself. Note that, in certain cases, a stronger property than (2.2.4) also holds.

P3′. (Isotonicity) If D_t is a partition subset and its current partition is given by

$$D_t = \bigcup_{l \in L_k} D_{t_l}, \quad D_{t_l} \subset D_t, \quad l \in L_k,$$

then

$$P\{D_t, f(D_t)\} > P\{D_{t_l}, f(D_{t_l})\} \quad \text{for} \quad l \in L_k. \tag{2.2.4a}$$

Although it is not too restrictive and quite usual to require P3′, it suffices here to postulate the milder assumption P3: note that P3 does not exclude the case in which the estimated search value of a subset D_{i_k} is identical to that of a previously generated set which contains D_{i_k}. For distinction, a PAS having the property P3′ will be called strictly monotone. (P3′ is often postulated, for example, in interval arithmetic based global optimization strategies.)

Remark 2.2.4. The boundedness property P4 is clearly technical, to avoid manipulations with infinity in convergence arguments.

To illustrate the above properties, let us consider Piyavskii's partition algorithm for minimizing the Lipschitz function $f(x)$ on the one-dimensional interval $[a, b]$. The partition operator value corresponding to an arbitrary subinterval $[x_{i-1}, x_i]$ equals (see (2.1.23))

$$P_i = P\{[x_{i-1}, x_i], z_{i-1}, z_i\} = -\underline{z}_i = -\frac{z_{i-1} + z_i}{2} + \frac{L(x_i - x_{i-1})}{2}. \quad (2.2.6)$$

Moreover, the new point to be selected on this interval equals (see (2.1.24) and (2.1.25))

$$\underline{x}_t = \frac{x_{t-1} + x_t}{2} + \frac{z_{t-1} - z_t}{2L}, \quad (2.2.7)$$

where t is determined by the relation

$$\underline{z}_t = \min_{i=1,\ldots,k} \underline{z}_i.$$

Let us check now the fulfilment of properties P1–P4. In doing this, we shall suppose that L is a valid overestimate of the true (typically unknown) Lipschitz constant L_0. Note that, for the purposes of a purely theoretical argument, this assumption is not really restrictive (as in very few realistic problems which can be formulated as an LGOP, would the Lipschitz constant surpass, say, $10^{10^{10}}$ ...); of course, when implementing Lipschitzian global optimization algorithms, one would be pleased with more precise estimates, but we relegate the discussion of this important point to Part 3. (The earlier Remark 2.1.6 already shed some light on the effect of L on the computational effort needed.)

P1: Let us estimate the distance between the new point $\underline{x}_t$ and the corresponding subinterval endpoints:

$$\begin{aligned} \underline{x}_t - x_{t-1} &= \frac{x_{t-1} + x_t}{2} + \frac{z_{t-1} - z_t}{2L} - x_{t-1} \\ &= \frac{x_t - x_{t-1}}{2} + \frac{z_{t-1} - z_t}{2L}. \end{aligned}$$

Now, as

$$|z_{t-1} - z_t| \le L_0 \|x_{t-1} - x_t\|, \quad (2.2.8)$$

therefore, introducing the ratio $r_0 = L_0/L < 1$, (2.2.8) implies

$$\frac{|z_{t-1} - z_t|}{2L} \le \frac{L_0}{2L}\|x_{t-1} - x_t\| \le \frac{r_0}{2}(x_t - x_{t-1}).$$

Consequently, we arrive at the simple estimate

$$\underline{x}_t - x_{t-1} \le d_0(x_t - x_{t-1}) \quad \text{in which} \quad d_0 := \frac{1+r_0}{2} < 1.$$

By the obvious symmetry of (2.2.7) with respect to the role of points x_{t-1} and x_t, one can conclude now that

$$\max\{(\underline{x}_t - x_{t-1}), (x_t - \underline{x}_t)\} \le d_0(x_t - x_{t-1}). \qquad (2.2.9)$$

In other words, the new point $\underline{x}_t$ is always located in a $(1-d_0)$ contraction of interval $[x_{t-1}, x_t]$. Therefore—under appropriate parameterization $L > L_0$ —every infinite nested interval sequence generated by Piyavskii's method converges to a single point.

P2: Assume now that x is the limit point of an arbitrary strictly nested interval sequence $D_{i_j} = [x_{i-1_j}, x_{i_j}]$, $j \in N$. Since

$$\lim_{j\to\infty} \{[x_{i_j} - x_{i-1_j}]\} = 0, \quad j \in N,$$

from (2.2.6) one can directly conclude that

$$\lim_{j\to\infty} P_{i_j} = -f(x). \qquad (2.2.10)$$

(2.2.10) shows that, independently of the search point sequence generated, the limit value of Piyavskii's partition operator is $-f(x)$.

P3: Let x belong to $\bigcap_{k\in N} D_{i_k}$: then, by P1, either x is the limit point of the sequence $\{D_{i_k}\}$, or $x \in D_{i_{k_0}} = D_{i_k}$ for some k_0 and all $k \ge k_0$.

If $x = \bigcap_{k\in N} D_{i_k}$, then for an arbitrary index k,

$$P_i = P\{D_{i_k}, f(D_{i_k})\} = -\underline{z}_i > -f(x) = P\{x, f(x)\}, \qquad (2.2.11)$$

as P_i is based on a Lipschitzian underestimate of f, which has been produced using a valid overestimate L of the true Lipschitz constant L_0.

Similarly, for an arbitrary point of $D_{i_{k_0}}$, $P_{i_{k_0}}$ is based on a valid underestimate of f on this subinterval; and $P_{i_k} = P_{i_{k_0}}$ for all $k \ge k_0$, as $D_{i_{k_0}}$ was not subdivided any further throughout the subsequent iterations. Therefore (2.2.11) shows again the validity of P3 for the latter case.

P4: The boundedness property obviously follows, as the continuous function f is bounded on the finite interval $[a, b]$.

The above discussion shows that Piyavskii's algorithm can be interpreted as a regular partition strategy. Numerous further examples will be introduced later.

In the following exposition, general sufficient and necessary global convergence conditions will be established for the broad class of regular partition algorithms which follow the framework of the PAS. The previously introduced notation $X^\star$ (set of global optimizer points), X^a (set of limit points generated by a PAS algorithm-instance), and $P_i = P\{D_i, f(D_i)\}$ (partition operator values for a sequence of subsets $\{D_i\}, i \in N$) will be retained and applied below.

2.2.2 Sufficient and Necessary Convergence Conditions

For the purposes of the subsequent discussion, we shall recall the following well-known concept of mathematical analysis.

Definition 2.2.3. Let $D \subset \mathbb{R}^n$. A subset $D_1 \subset D$ is everywhere dense in D if, for an arbitrary $\varepsilon > 0$, every point x of D has in its ε-neighbourhood at least one point of D_1. (The ε-neighbourhood is defined by the intersection of D and the Euclidean sphere centered at x with radius ε.)

Note that this definition could be given in more general forms, but in the present framework it suffices to use Definition 2.2.3. Observe also that Definition 2.2.3 implies that every point of D is the accumulation point of some infinite sequence from D_1; in other words, the closure of D_1 is identical with D.

Theorem 2.2.1. *For all pairs of points $x^a \in X^a$, $x \in D \setminus X^a$, we have the inequality*

$$P\{x^a, f(x^a)\} > P\{x, f(x)\}. \tag{2.2.12}$$

Proof. Note first that, by the boundedness of D, X^a is obviously nonempty; furthermore, if $X^a = D$, then the assertion holds trivially. Take an arbitrary pair of points $x^a \in X^a$, $x \in D \setminus X^a$. Consider now for $k \in N$ the sequence of subsets $\{D_{i_k}\}$ which contain x in the course of the PAS-type iteration. By the perfection property P1, from a certain index k_0 we have

$D_{i_k} = D_{i_{k_0}}$ (since in the opposite case x would be the limit of the sample points in $\{D_{i_k}\}$). Now, exploiting the continuity (P2) and monotonicity (P3) properties, one can apply the relations

$$\begin{aligned} &P\{D_{i_k}, f(D_{i_k})\} \\ &\quad = P\{D_{i_{k_0}}, f(D_{i_{k_0}})\} > P\{x, f(x)\}, \quad \text{for } x \in D_{i_k} \text{ and } k \geq k_0. \end{aligned} \tag{2.2.13}$$

Take now the sequence of partition sets $\{D_{t_k}\}$ which converge to the limit point $x^a = \bigcap_{k \in N} D_{t_k}$. The continuity property then implies the relation

$$P\{D_{t_k}, f(D_{t_k})\} \underset{k\to\infty}{\longrightarrow} P\{x^a, f(x^a)\}. \tag{2.2.14}$$

By the selection rule of the PAS (recall (2.1.12)), for infinitely many indices k, the inequality

$$P\{D_{t_k}, f(D_{t_k})\} \geq P\{D_{i_k}, f(D_{i_k})\} \tag{2.2.15}$$

is valid (as otherwise the limit point x^a could not have been generated). Consequently, by (2.2.13)–(2.2.15), we are led to the inequality (2.2.12).

■

Remark 2.2.5. The property expressed by inequality (2.2.12) can be verbally interpreted as follows: the limit points generated by a PAS type search strategy are considered 'better'—in terms of the partition operator—than any other point which belongs to the (relative to D) complementary set of the accumulation point set X^a. As we shall see, this simply established property plays a central role in the subsequent global convergence arguments.

Corollary 2.2.1. *If the regular partition operator P has the additional property*

$$P\{x, f(x)\} = c \qquad \text{for all } x \in D \tag{2.2.16}$$

(c is a fixed constant), then $X^a = D$, that is, the corresponding PAS-type algorithm generates an everywhere dense sequence of sample points on D.

Proof. The assertion evidently follows from Theorem 2.2.1. ■

Corollary 2.2.2. *Assume that P has the property* (2.2.16). *Then the sample minimum sequence*

$$\{z_k^\star\} = \{f(x_k^\star)\}, \quad \text{where } z_k^\star := \min_{\substack{j=1,\ldots,J_i \\ i\in I_k}} z_i^{(j)}, \quad k \in N, \tag{2.2.17}$$

converges to the global optimum value $z^\star$. Equivalently, all accumulation points of the corresponding minimizer sequence $\{x_k^\star\}$ belong to $X^\star$.

Proof. Let x^a be a limit point of $\{x_k^\star\}$, and suppose indirectly that $x^a \in D \setminus X^\star$, that is, $z^a = f(x^a) > z^\star$. By Corollary 2.2.1, all points of $X^\star$ are also limit points of a corresponding sequence of sample points. But then—by continuity of f—the respective function value sequences converge to $z^\star$. This leads (from a certain index) to contradiction, concerning the definition of that subsequence of $\{x_k^\star\}$ which converges to x^a. The contradiction obtained shows that $x^a \in X^\star$, whence $z_k^\star \to z^\star$. ■

Remark 2.2.6. According to Corollaries 2.2.1 and 2.2.2, the approximation of the global solution(s) and optimum can be attained by generating (theoretically) an everywhere dense sequence of sample points in the feasible set D. Although this approach may, at first sight, seem trivial and inefficient, it needs proper consideration. First (recall Example 2.1.1), for an arbitrary continuous, but non-Lipschitz objective function f, one cannot guarantee a bounded estimate of the optimum value, after any finite number of search steps. This observation implies that for CGOPs the generation of an everywhere dense sample point set, and the application of Corollary 2.2.2, seems to be the only generally viable globally convergent deterministic solution approach. Second, an appropriately chosen partition strategy which induces everywhere dense sampling may prove to be reasonably efficient, in a numerical sense, for broad classes of multiextremal optimization problems.

Consider, for example, the algorithm of Kushner, as introduced in Section 2.1.5. This algorithm serves for globally minimizing a multiextremal function on the one-dimensional interval $[a, b]$. The partition operator value corresponding to an arbitrary subinterval $[x_{i-1}, x_i]$ is

$$P_i = \exp\left\{\frac{-2(f_k^\star - \varepsilon - z_i)(f_k^\star - \varepsilon - z_{i-1})}{\sigma^2(x_i - x_{i-1})}\right\}, \tag{2.2.18}$$

where

$$f_k^\star = \min_{1 \le i \le k} z_i$$

is the current sample minimum, and $\varepsilon > 0$, $\sigma > 0$ are fixed model parameters.

Assume now that x is the limit point of an arbitrary strictly nested interval sequence $D_{i_j} = [x_{i-1_j}, x_{i_j}]$, and that—for a properly chosen positive sequence of parameters $\{\varepsilon_k\}$, $k \in N$—we have

$$f_k^\star - \varepsilon_k - z_{i_j} < 0 \quad \text{and} \quad f_k^\star - \varepsilon_k - z_{i-1_j} < 0,$$

at least from a certain index $j \ge j_0$.

Since evidently

$$\lim_{j\to\infty} (x_{i_j} - x_{i-1_j}) = 0, \tag{2.2.19}$$

by (2.2.18) one can directly conclude that, with appropriate parameterization $\{\varepsilon_k\}$, $k \in N$,

$$\lim_{j\to\infty} P_{i_j} = 0 \tag{2.2.20}$$

is valid, independently of x. In other words, the 'search value' of every limit point generated by Kushner's algorithm equals zero, while the 'value' of all nondegenerate search intervals remains positive (for appropriate model parameterization). Recalling the subset selection rule (2.1.12) of PAS, namely,

$$P\{D_t, f(D_t)\} = \max_{i \in I_k} P\{D_i, f(D_i)\}, \qquad k \in N, \tag{2.2.21}$$

one can conclude that in the given case it will theoretically lead to an everywhere dense partition of $[a, b]$. Observe now that, by (2.2.20), this conclusion follows from Corollary 2.2.1; further examples of similar algorithms—known from the literature or new ones—will be discussed later.

Theorem 2.2.2. *For all limit points $x^a \in X^a$ of a regular partition algorithm, we have*

$$P\{x^a, f(x^a)\} = c^a \tag{2.2.22}$$

(with some fixed real value c^a). In other words, all accumulation points generated by the PAS-type algorithm have the same 'search value'.

Proof. Choose an arbitrary pair of limit points x_1^a, $x_2^a \in X^a$: if this is not possible, then the assertion is trivial. Consider now the corresponding perfect subset-sequences $\{D_{t_k}\}$ and $\{D_{i_k}\}$, such that $x_1^a = \bigcap_{k\in N} D_{t_k}$ and $x_2^a = \bigcap_{k\in N} D_{i_k}$. Then, by the selection rule (2.1.12), both inequality-sequences below are valid:

$$P\{D_{t_k}, f(D_{t_k})\} \geq P\{D_{i_k}, f(D_{i_k})\} \quad \text{for } k \in N_1 \ (N_1 \subset N \text{ infinite}),$$

and (2.2.23)

$$P\{D_{i_k}, f(D_{i_k})\} \geq P\{D_{t_k}, f(D_{t_k})\} \quad \text{for } k \in N_2 \ (N_2 \subset N \text{ infinite}).$$

Taking limits in the relation-chains (2.2.23), by the continuity property P2 of the regular PAS, we can immediately conclude (2.2.22). ∎

Definition 2.2.4. The partition operator P is strictly decreasing in limit (on D, and with respect to its function value argument), if for arbitrary points $u, v \in D$ the relation

$$f(u) > f(v) \quad \text{implies} \quad P\{u, f(u)\} < P\{v, f(v)\}. \tag{2.2.24}$$

Remark 2.2.7. According to Definition 2.2.4, if the function value at v is smaller ('better') than at u, then the limit value of the partition operator at v also has to be better than its limit at u. Recalling the subset selection rule (2.1.12), this property is quite natural to expect, unless the partition operator is designed to generate everywhere dense sample point sequences. (Recall in this context also property (2.2.16), which is clearly contradictory to (2.2.24) and, indeed, implies the generation of an everywhere dense search point set in D.)

To illustrate the fulfilment of Definition 2.2.4, let us consider again Piyavskii's algorithm. As discussed previously (see (2.2.10)), the limit of the partition operator value corresponding to an arbitrary nested subinterval sequence $[x_{i-1_j}, x_{i_j}]$, $j \in N$, which converges to point x, is given as

$$\lim_{j\to\infty} P_{i_j} = -f(x).$$

Hence, evidently the partition operator of Piyavskii's algorithm is strictly decreasing in limit. Further examples will be discussed in detail later.

Theorem 2.2.3. *Assume that P is strictly decreasing in limit. Then $X^a \subseteq X^\star$, that is, all accumulation points of the generated sample sequence solve the global optimization problem*

$$\min_{x \in D} f(x). \tag{2.2.25}$$

Proof. Let $x^a \in X^a$, $z^a = f(x^a)$, and take an arbitrary $x^\star \in X^\star \setminus \{x^a\}$ (if this is not possible, then, evidently, $X^a = X^\star = \{x^a\}$). Consider the partition set sequences for which $x^a = \bigcap_{k \in N} D_{t_k}$ and $x^\star \in \bigcap_{k \in N} D_{i_k}$. By the selection rule (2.1.12), for infinitely many indices k we have the inequality

$$P\{D_{t_k}, f(D_{t_k})\} \geq P\{D_{i_k}, f(D_{i_k})\}. \tag{2.2.26}$$

Taking limits on both sides of (2.2.26) and making use of the assumptions P1–P3, we obtain the chain of relations

$$P\{x^a, z^a) = \lim_{k \to \infty} P\{D_{t_k}, f(D_{t_k})\} \geq \lim_{k \to \infty} P\{D_{i_k}, f(D_{i_k})\} \geq P\{x^\star, z^\star\}. \tag{2.2.27}$$

From (2.2.24) and (2.2.27) one can directly conclude that $z^a \leq z^\star$, that is, $x^a \in X^\star$ necessarily holds (for each $x^a \in X^a$). ■

Let us note a simple but practically important consequence of this assertion:

Corollary 2.2.3. *Suppose that the optimization problem (2.2.25) has a unique solution: $X^\star = \{x^\star\}$. Then each realization of the PAS, governed by an operator P which is regular and strictly decreasing in limit, generates a corresponding sample point sequence that converges to $x^\star$.*

Proof. This statement immediately follows from Theorem 2.2.3. ■

Remark 2.2.8. According to Remark 2.2.7 and the last corollary, Piyavskii's algorithm—as one possible instance of a regular PAS-type search strategy—is globally convergent, in the sense of Theorem 2.2.3.

Remark 2.2.9. Theorem 2.2.3 is closely related to similar results concerning the convergence mechanism of branch-and-bound type algorithms (consult, for example, Horst, 1986, Theorem 3.1; Horst and Tuy, 1987, Theorem 2.1), and to interval arithmetic based methods (see, for instance, Ratschek and Rokne, 1988; Hansen, 1992). Note, however, that the formal introduction of a bounding operation could be omitted in the present framework, while the results mentioned are based on postulated properties of that operation and the subset selection procedure derived. Moreover, in the present framework other relevant convergence properties (namely, not only the case $X^a \subseteq X^\star$) can also be directly studied.

Definition 2.2.5. The partition operator is called translation-invariant in limit (on D, and with respect to its point argument), if for all pairs $u, v \in D$ the equality

$$f(u) = f(v) \quad \text{implies} \quad P\{u, f(u)\} = P\{v, f(v)\}. \tag{2.2.28}$$

Remark 2.2.10. Translation-invariance again fully corresponds to the informal interpretation of the subset selection rule: one can reasonably expect that points u, v with equal function value are of the same search value; this should then apply to the limit value of nested interval sequences converging to u and v, respectively.

Recall at this point that if $P\{x, f(x)\} = c$ for all x, then translation-invariance trivially holds; additionally, in this case the inclusion $X^\star \subset X^a$ is evident. The following theorem shows a much more important case with this inclusion property.

Theorem 2.2.4. *Assume that P is translation-invariant in limit and strictly decreasing in limit; that is, both* (2.2.24) *and* (2.2.28) *hold. Then $X^a = X^\star$, that is, the set of accumulation points of the generated sample sequence is identical with the solution set of the global optimization problem* (2.2.25).

Proof. As—by Theorem 2.2.3—we already know that $X^a \subseteq X^\star$, it suffices to show the reverse inclusion. Assume indirectly that an $x^\star \in X^\star$ is not a limit point. Then, by Theorems 2.2.1 and 2.2.2, we have the relations

$$P\{x^a, f(x^a)\} = c^a > P(x^\star, f(x^\star)) \quad \text{for each } x^a \in X^a.$$

But the inequality obtained is contradictory to the optimality of $x^\star$, as (2.2.24) and (2.2.28) would jointly imply that $f(x^a) < f(x^\star)$. This contradiction shows the validity of $X^\star \subseteq X^a$, that is, $X^a = X^\star$. ■

Remark 2.2.11. The sequence of Theorems 2.2.1, 2.2.3 and 2.2.4 generalizes the convergence results related to one-dimensional (Pintér, 1983b) and multivariate (Pintér, 1986a) Lipschitz optimization on interval regions D, and also extends the result of Pintér (1988). These specifications will be discussed in subsequent chapters of Part 2.

Remark 2.2.12. Comparing the assertions of Corollary 2.2.3 and Theorem 2.2.4, one can see that the latter expresses a meaningful addition (only) in the case when $X^\star$ consists of more than one point. Note, however, that the existence of multiple solutions to global optimization problems is not unusual, keeping in mind some of the typical application perspectives (finding all solutions of nonlinear equation systems or of certain approximation problems, etc.). Moreover, the requirement (2.2.28) is fairly natural and—together with the properties postulated earlier—it leads to the strongest possible convergence statement $X^a = X^\star$. (It is easy to verify that postulating only the translation-invariance property—without strict decrease with respect to the limit function value—would imply the following statement: if an arbitrary $x^\star \in X^\star$ belongs to X^a, then $X^\star \subseteq X^a$.)

Let us turn now to the analysis of necessary global convergence conditions. In this case, additional properties of regular partition algorithms will be deduced below, on the basis of their convergence type.

Theorem 2.2.5. *Assume that a PAS-type method generates an everywhere dense sequence of points in D. Then $P\{x, f(x)\} = c$, for all $x \in D$ and some fixed value c. Consequently, for regular algorithms, the statements*

$$P\{x, f(x)\} = c, \quad x \in D, \quad \text{and} \quad X^a = D$$

are equivalent.

Proof. This assertion obviously follows from Theorem 2.2.2 and Corollary 2.2.1. ∎

Theorem 2.2.6. *Assume that $X^a \subseteq X^\star$ holds for the accumulation point set X^a of a PAS-type method. Then P is strictly decreasing in limit, at least in the following (restricted) sense:*

$$P\{x^\star, z^\star\} = c^a \quad \text{for all } x^\star \in X^a \qquad (c^a \text{ is a constant}),$$

while (2.2.29)

$$P\{x, f(x)\} < c^a \quad \text{for all } x \in D \setminus X^a.$$

Proof. This statement immediately follows from Theorems 2.2.1 and 2.2.2. ■

Theorem 2.2.7. *Assume that $X^\star \subseteq X^a$ holds for the accumulation point set X^a of a PAS-type method. Then P is translation-invariant in limit, at least in the following (restricted) sense:*

$$P\{x^\star, z^\star\} = c^\star \quad \text{for all } x^\star \in X^\star \tag{2.2.30}$$

(with some fixed $c^\star$).

Proof. The assertion evidently follows from Theorem 2.2.2. ■

Corollary 2.2.4. *For regular partition algorithms, the pair of restricted translation–invariance and strict decrease properties, namely,*

$$P\{x^\star, z^\star\} = c^\star > P\{x, f(x)\} \quad \text{for all } x^\star \in X^\star \quad \text{and} \quad x \in D \setminus X^\star \tag{2.2.31}$$

is equivalent to the convergence statement $X^a = X^\star$.

Proof. This assertion follows from Theorems 2.2.6 and 2.2.7 and the observation that during the proofs of Theorems 2.2.3 and 2.2.4 only the restricted forms (2.2.29) and (2.2.30) of the translation-invariance and strict monotonic decrease properties were, in fact, exploited. ■

Remark 2.2.13. In spite of the sufficiency of (2.2.29) and (2.2.30) in the restricted form (2.2.31), it seems rational to postulate the translation-invariance and strict monotonic decrease properties in their forms given originally, as they fully correspond to the underlying informal interpretation of the operator P, see Remark 2.2.3. Besides, in general, one cannot verify translation-invariance on the (unknown) set $X^\star$.

Concluding the convergence investigations of Chapter 2.2, one can summarize its main results by the following three statements.

i. If the PAS is regular, then

$$P\{x^a, f(x^a)\} = c^a > P\{x, f(x)\} \text{ for all } x^a \in X^a \text{ and } x \in D \setminus X^a$$
$$(c^a \text{ is fixed}).$$

ii. For a regular PAS, $X^a = D$ is equivalent to the relation $P\{x, f(x)\} = c$ (required for all $x \in D$, and some fixed c).

iii. For a regular PAS, $X^a = X^\star$ is equivalent to the chain of relations $P\{x^\star, z^\star\} = c^\star > P\{x, f(x)\}$ (required for all pairs $x^\star \in X^\star$, $x \in D \setminus X^\star$, and some fixed value $c^\star$).

In other words, regular partition operators in global optimization may generate two types of sample point sequences. In the first case, the sample point sequence is everywhere dense in D and, hence, the sample minimum and minimizer sequences provide an approximation of the globally optimal solution. As discussed previously, such strategies can be applied to solve CGOPs without the Lipschitz-continuity assumption: for (merely) continuous problems one cannot expect a stronger type of general deterministic convergence statement.

The second case is related to PAS-type algorithms which generate a sample point sequence with the strong convergence property expressed by the relation $X^a = X^\star$. Recall that, in terms of the partition operator, this necessitates a clear distinction between optimal and suboptimal limit points. As indicated by the example of Piyavskii's algorithm, such methods will typically rely on some additional quantitative information (such as, for instance, the Lipschitz property and a valid overestimate of the Lipschitz constant), besides the postulated continuity of the objective function f. We shall return to this point later, when analysing other concrete algorithm forms.

In the following chapters of Part 2 we shall demonstrate that the classification presented leads to a direct conclusion of the convergence properties of a number of known univariate partition algorithms. Moreover, one can propose straightforward generalizations and modifications of these algorithms which have the same convergence properties in the multivariate case, under various specifications of the feasible set D.

Chapter 2.3

Partition Algorithms on Intervals

2.3.1 Convergence of Univariate Global Optimization Algorithms

In the simplest and most frequently studied special case of the general GOP, D is a one-dimensional finite interval. Let $D = [a, b]$, $-\infty < a < b < \infty$, and f a (possibly) multiextremal continuous or Lipschitz function defined on $[a, b]$. Applying the notation introduced in Chapter 2.1, the corresponding problem statements are

$$\min_{a \leq x \leq b} f(x), \quad \text{where } f \in C([a, b]); \tag{2.3.1}$$

and

$$\min_{a \leq x \leq b} f(x), \quad \text{where } f \in F([a, b]). \tag{2.3.2}$$

In the past three decades, a large number of global optimization algorithms were suggested to solve the CGOP (2.3.1) or LGOP (2.3.2). As discussed earlier, all these methods are based on the assumption that f belongs to $C([a, b])$ or $F([a, b])$ (or a proper subset of these broad function families, possibly with additional postulated properties). As the algorithm proceeds and search information is generated, the general, a priori 'picture' of the objective function f is combined with the sample to derive sequentially updated posterior models (approximations) of f: recall the methods of Piyavskii and Kushner, as examples of this approach.

Since the underlying prior function models and search strategies may reflect markedly different assumptions, one can expect a wide variety of corresponding algorithms. Nevertheless, many of the algorithms referred to can be described as PAS-type search strategies: this fact allows the straightforward verification of their convergence properties, and will also make possible their diverse generalizations and extensions. It is important to emphasize in this context that the underlying function models and search principles of the originally proposed algorithms are, typically, tractable only in the univariate case. Therefore, their proper multidimensional extensions were not available until recently: in fact, several such extensions will be presented for the first time in Chapter 2.4. It should be noted that the new multivariate algorithms are not based on the direct generalization of the one-dimensional function models. Rather, they are derived on the basis of the convergence investigations of Chapter 2.2, which were clearly of topological character and, as such, independent of the dimensionality of the optimization problem.

To illustrate this approach, first a number of known univariate algorithms will be collected. All of these methods can be interpreted as being of PAS type. Similarly to Piyavskii's or Kushner's algorithm, they lead to a sequential partition of $[a, b]$ into subintervals $D_i = [x_{i-1}, x_i]$, $i = i(k)$, $k \in N$. We will apply again the notation $z_i = f(x_i)$, and write simply

$$P_i(D_i, f(D_i)) = P_i\{x_{i-1}, z_{i-1}, x_i, z_i\}. \tag{2.3.3}$$

The sequential order of the subintervals D_i depends, of course, on the iteration index k: $D_i = D_{i_k}$, but this fact will not always be explicitly reflected by the notation.

Remark 2.3.1. The value of P_i in (2.3.3) may also depend on other—often problem-specific—parameters, such as, for example, the (estimated) Lipschitz-constant in Piyavskii's method, or parameters ε and σ in Kushner's algorithm. This fact, however, is not reflected by the generic notation of (2.3.3).

Remark 2.3.2. Before presenting the partition rules of a number of algorithms, let us make a simplifying assumption. This is based on the observation that the PAS can be simply realized in the univariate case, if a single new point (viz., the interval midpoint) is chosen in each subinterval which is being partitioned at the refinement step. This bisection does not always correspond to the originally proposed search strategies, but the simplification will not influence the conclusions on convergence properties. In order to improve the numerical efficiency of the search, one can always generate alternative (or additional) search points: this aspect will be discussed later in detail. Consequently, the corresponding univariate specification of the PAS has the following form:

Partition Algorithm Scheme for One-Dimensional Global Optimization

Step 0. (Initialization) Set $k := 0$ (iteration count); $I_0 := \{1\}$ (initial index set of active partition sets); $\{D_0\} := \{[a,b]\} = \{[x_0, x_1]\}$ (initial list of partition sets). Determine $z_0 = f(x_0)$ and $z_1 = f(x_1)$.

Step 1. (Sampling) For all $i \in I_k$, sample points are the subinterval endpoints x_{i-1} and x_i, with corresponding function values z_{i-1} and z_i. Additionally, one can choose the subinterval midpoint

$$xm_i = \frac{x_{i-1} + x_i}{2}. \tag{2.3.4}$$

(In fact, it is not necessary to actually sample f at xm_i, except on the subinterval chosen for partition refinement: this will be done at Step 4.)

Step 2. (Aggregation) For each $i \in I_k$, calculate the value of the partition operator

$$P_i\{x_{i-1}, z_{i-1}, x_i, z_i\}.$$

Step 3. (Selection) Choose an arbitrary index $t \in I_k$, for which the relation

$$P_t\{x_{t-1}, z_{t-1}, x_t, z_t\} = \max_{i \in I_k} P_i\{x_{i-1}, z_{i-1}, x_i, z_i\} \tag{2.3.5}$$

is satisfied; then $[x_{t-1}, x_t]$ will be subdivided in the current iteration cycle.

Step 4. (Refinement) Taking the midpoint xm_t defined by (2.3.4), $D_t = [x_{t-1}, x_t]$ is replaced by the two half-intervals

$$D_{t+1} = [x_{t-1}, xm_t] \quad \text{and} \quad D_{t+2} = [xm_t, x_t], \quad t = t(k). \tag{2.3.6}$$

The refined partition, hence, is defined by the collection of intervals

$$\{D_i : i \in I_{k+1}\} = \{[x_{i-1}, x_i] : i \in I_{k+1}\},$$

in which

$$I_{k+1} = I_k \cup \{t+1, t+2\} \setminus \{t\}. \tag{2.3.7}$$

Set $k := k + 1$; return to Step 1 (of the next iteration cycle).

Remark 2.3.3. Because of the midpoint selection rule, every strictly nested subinterval-sequence $\{[x_{i-1}, x_i], i = i(k)\}$ generated by the PAS, will obviously converge to a single limit point

$$x = \bigcap_{k \in N} [x_{i-1}, x_i], \qquad i = i(k).$$

In other words, the PAS satisfies the perfection property P1. In order to guarantee the applicability of the convergence results presented in Chapter 2.2, we shall also postulate the properties P2–P4. This leads to the following specific assumptions (recall the detailed statement of these conditions as given in Section 2.2.1):

P2. (Continuity) If $x = \bigcap_{k \in N} [x_{i-1}, x_i]$, $i = i(k)$, then

$$\lim_{k \to \infty} P_i\{x_{i-1}, z_{i-1}, x_i, z_i\} = P\{x, z\}, \quad z = f(x), \quad i = i(k). \tag{2.3.8}$$

P3. (Monotonicity) If $x = \bigcap_{k \in N} [x_{i-1}, x_i]$, then, for all $i = i(k)$, $k \in N$,

$$P_i\{x_{i-1}, z_{i-1}, x_i, z_i\} > P\{x, z\}. \tag{2.3.9}$$

P4. (Boundedness)

$$-\infty < P_i\{x_{i-1}, z_{i-1}, x_{i_1}, z_i\} < \infty. \tag{2.3.10}$$

Partition algorithms which satisfy assumptions P1 to P4, will be called (univariate) regular.

To illustrate the use (and verification) of these assumptions, we shall present now a number of one-dimensional partition algorithms, known from the literature. For the sake of completeness, the methods of Kushner and Piyavskii will also be listed here. Details with regard to the derivation of these partition strategies can be found in the original works cited; see also Strongin (1978), Archetti and Betrò (1979, 1980), Žilinskas (1982, 1986), Boender (1984), Mockus (1989), or Zhigljavsky (1991) for additional discussions of function models in global optimization.

Examples of One-Dimensional PAS Type Algorithms: Partition Operators (Interval Selection Rules)

Kushner algorithm (1964)

$$P_i = \exp\Big(-2\,\frac{(f_k^\star - \varepsilon - z_i)(f_k^\star - \varepsilon - z_{i-1})}{\sigma^2(x_i - x_{i-1})}\Big). \tag{2.3.11}$$

In (2.3.11), $\varepsilon > 0$, $\sigma > 0$ are fixed function model parameters, and $f_k^\star = \min_{0\le i\le k} z_i$ is the current sample minimum.

Piyavskii algorithm (1967)

$$P_i = L\frac{x_i - x_{i-1}}{2} - \frac{z_{i-1} + z_i}{2}. \tag{2.3.12}$$

Here $L > 0$ is a known overestimate of the exact Lipschitz-constant.

Strongin algorithm (1973)

$$P_i = m(x_i - x_{i-1}) + \frac{(z_i - z_{i-1})^2}{m(x_i - x_{i-1})} - 2(z_{i-1} + z_i). \tag{2.3.13}$$

Here $m > 2L$ is a function model parameter, to be estimated adaptively from the generated sample information $\{x_i, z_i\}$, $i = i(k)$, $k \in N$.

Mockus–Tiesis–Žilinskas algorithm (1978)

$$P_i = \max_{x_{i-1}\le x\le x_i} \sigma_i(x)\,\mathcal{N}\left(\frac{f_{k_0} - \mu_i(x)}{\sigma_i(x)}\right). \tag{2.3.14}$$

In (2.3.14), f_{k_0} is the minimum estimate based on an initial equidistant sample of $k_0 + 1 \le k$ elements on $[a, b]$, and $\sigma > 0$ is a function model parameter. The expressions below define the functions $\mu_i(x)$ and $\sigma_i(x)$, for arbitrary $x_{i-1} \le x \le x_i$:

$$\mu_i(x) = \frac{z_i(x - x_{i-1}) + z_{i-1}(x_i - x)}{x_i - x_{i-1}}$$

$$\sigma_i(x) = \frac{\sigma^2(x - x_{i-1})(x_i - x)}{x_i - x_{i-1}}.$$

Finally, $\mathcal{N}(\cdot)$ denotes the standard normal probability distribution function (p.d.f.).

Žilinskas algorithm (1981)

$$P_i = \mathcal{N}\Big(\frac{\underline{z}^\star_{k_0}(x_i - x_{i-1}) - z_i(a_i - x_{i-1}) - z_{i-1}(x_i - a_i)}{\sigma[(a_i - x_{i-1})(x_i - a_i)(x_i - x_{i-1})]^{1/2}}\Big). \tag{2.3.15}$$

In (2.3.15), $f_{k_0} = \min_{0 \le i \le k_0} z_i$ is the sample minimum based on an initial sample of $k_0 + 1$ points, $\underline{z}^\star_{k_o} = f_{k_0} - \gamma$, and $\gamma > 0$, $\sigma > 0$ are model parameters; furthermore,

$$a_i = \frac{x_{i-1} + x_i}{2} - \frac{(x_i - x_{i-1})(z_i - z_{i-1})}{2[(z_i + z_{i-1} - 2\,\underline{z}^\star_{k_0})]},$$

and $\mathcal{N}(\cdot)$ is the standard normal p.d.f.

Boender algorithm (1984)

$$\begin{aligned} P_i = &- \min\{z_{i-1}, z_i\} \\ &+ (x_i - x_{i-1})\sqrt{\frac{\pi}{2}}\;\sigma\exp\Big(\frac{(z_i - z_{i-1})^2}{2\sigma^2(x_i - x_{i-1})}\Big)\mathcal{N}\Big(\frac{-|z_{i-1} - z_i|}{\sigma^2(x_i - x_{i-1})}\Big). \end{aligned} \tag{2.3.16}$$

Again, $\sigma > 0$ is a function model parameter, and $\mathcal{N}(\cdot)$ is defined as earlier.

Remark 2.3.4. Quite a few further examples could be added to this list; a number of these can be considered as a modification and/or extension of Piyavskii's method, and will be reviewed briefly below.

Evtushenko (1971) proposed an ordered sequential search method, making use of the current best function value and the Lipschitz information, to determine at each iteration the largest valid step size which cannot lead to 'stepping over' the optimum point. This idea leads to a non-uniform, sequentially derived sample on interval $[a, b]$.

A somewhat similar idea leads to the algorithm of Bromberg and Chang (1992), in which linear (not necessarily Lipschitzian) lower bounds are constructed sequentially to generate an approximation of the optimum value (within a given tolerance $\varepsilon > 0$) in a finite number of steps.

Strongin (1973)—instead of using the exact value of L—applied its adaptively updated estimate, which is a multiple of the largest slope of the line segment connecting successive search points. In this way, the Lipschitzian minorant function may change as the iteration proceeds (and so does the partition operator, see (2.3.13)).

The algorithm of Timonov (1977) is close in spirit to Piyavskii's method. He suggested choosing all function evaluation points among those of a fixed (passive) uniform grid search strategy (discussed in Section 2.1.3). Therefore, instead of choosing the next sample point as the minimizer of the current Lipschitzian minorant function, one can take the current 'best' point of the uniform grid according to the minorant. Note that, to approximate this rule, it suffices to compare the two grid points which fall nearest to the left and to the right of the Piyavskii minimizer point of the minorant function.

A similar idea is proposed by Schoen (1982), who selects evaluation points among those of a dynamically applied passive search strategy: at each step, the passive (grid) search is applied only on that subinterval which contains the Piyavskii minimizer point.

Basso (1982) also suggested a modification of Piyavskii's method, based on the simultaneous bipartition of all subintervals which may still contain better points than the current optimum estimate: these subintervals can be found by comparing the current optimum estimate and the subinterval-specific Lipschitzian underestimate of the minimum.

Galperin's (1985) search method in the univariate case leads to the following strategy: at each iteration step, subdivide all subintervals of the current list into $p > 1$ equal ones, and then select a sample point (for example, the midpoint) in each new subinterval. Obviously, this 'cautious' strategy can be adapted to derive a Basso type algorithm; see also Galperin (1988, 1993) for modifications.

Shen and Zhu (1987) also considered a modification of Piyavskii's algorithm, providing a (computationally less expensive, but also less precise) minorant function: this minorant is based on considering (only) the midpoint of each subinterval.

Jones, Perttunen and Stuckman (1993) recently proposed an interesting modification of the Piyavskii–Shubert algorithm—without assuming knowledge of an appropriate Lipschitz constant. Their method associates, to each partition subinterval, a function value (measured at the interval centre), and the half-diagonal length. The subintervals are represented by these two values as points on the plane. Now—at each iteration—such subintervals are partitioned further, whose corresponding characteristic point lies on the 'lower right' part of the convex hull of all points representing subintervals.

Breiman and Cutler (1993) and Baritompa and Cutler (1994) propose adaptive partition strategies in which both first and second order information is utilized to customize the bounding (aggregation and subsequent selection) operations in their PAS-type algorithms.

Without going into detail, note that not all of the methods listed can be conceived as regular partition strategies. For instance, Evtushenko's algorithm always generates a sequence of subintervals in a well-defined order (the next subset is adjacent from the right to the previous subinterval). As another example, the Jones–Perttunen–Stuckman algorithm is not based on a scalar valued partition operator: hence, it does not fit directly into the PAS framework.

Concluding this discussion, note finally that Hansen, Jaumard and Lu (1991a, 1992a,b), and Hansen and Jaumard (1995) analyse in detail most of the approaches listed above, together with several other methods. They also suggest a two-phase algorithm which combines Piyavskii's strategy with locally applied Fibonacci search (Kiefer, 1953); moreover, they accomplish a test problem set based numerical comparison of a number of one-dimensional optimization algorithms. Additional examples of univariate optimization methods are discussed, for example, in Zhigljavsky (1991, Ch. 2).

Similarly to the convergence analysis of Kushner's and Piyavskii's search methods, we shall examine now formulas (2.3.13) to (2.3.16). Although these, at first sight, may look quite different—exploiting the results in Chapter 2.2—their essential common and distinctive features, including their convergence type, can be directly established.

First, (2.3.13)–(2.3.16) obviously define partition operators which are continuous (P2) and bounded from below (P4). Below we shall verify also their monotonicity (P3). The monotonicity of the operators (2.3.14) and (2.3.15) can be simply established for an arbitrary continuous objective function, as these functions are positive for an arbitrary nondegenerate subinterval of $[a, b]$, while their limit value equals zero for all $x \in [a, b]$ (see below).

Now, let us look at the partition operators (2.3.13) and (2.3.16). First, recalling that in (2.3.13) $m > 2L$, and L is a valid (overestimate of the 'true') Lipschitz-constant, one can compare (2.3.12) and (2.3.13). This leads, for any $x \in [x_{i-1}, x_i]$ and $z = f(x)$, to the following chain of relations:

$$\begin{aligned} P\{x, z\} &= -4z \leq 2L(x_i - x_{i-1}) - 2(z_{i-1} + z_i) \\ &< m(x_i - x_{i-1}) - 2(z_{i-1} + z_i) + \frac{(z_{i-1} - z_i)^2}{m(x_i - x_{i-1})} = P_i, \end{aligned} \tag{2.3.17}$$

see (2.3.13). The chain of relations (2.3.17) shows the monotonicity of Strongin's partition operator.

Consider now (2.3.16), assuming that f is Lipschitzian with constant L. Observe first that for an arbitrary x in $[x_{i-1}, x_i]$, $P\{x, z\} = -z$. Moreover, let us suppose that $(x_i - x_{i-1}) < 1$ (simply valid, for example, if $[a, b]$ is transformed into $[0, 1]$), and that L satisfies the inequality

$$L < 2\sigma(\pi/2)^{\frac{1}{2}} \exp\Big(\frac{(z_i - z_{i-1})^2}{2\sigma^2(x_i - x_{i-1})}\Big)\mathcal{N}\Big(\frac{-|z_{i-1} - z_i|}{\sigma^2(x_i - x_{i-1})}\Big). \tag{2.3.18}$$

This is an implicit condition on $\sigma = \sigma(L)$, and will be certainly met if $\sigma \to \infty$. Then, by (2.3.18), one can directly conclude that

$$P\{x, z\} = -z \leq \frac{1}{2}[L(x_i - x_{i-1}) - (z_{i-1} + z_i)] < P_i,$$

see (2.3.16). In other words, if f is Lipschitzian, then—with appropriate parameterization—Boender's partition operator is also monotonic (P3). Note in this context that (2.3.12), (2.3.13) and (2.3.16) can be made even isotonic (P3$'$).

Concluding the above discussion, one can see that all partition operators listed in (2.3.11)–(2.3.16) are regular (the regularity of (2..3.11) and (2.3.12) was shown earlier).

Having applied Corollary 2.2.1, we saw earlier that Kushner's algorithm (theoretically) leads to an everywhere dense sample point set in $[a, b]$. Now, for both (2.3.14) and (2.3.15), it is not difficult to verify that

$$\lim_{i \to \infty} P_i = P\{x, z\} = 0, \quad \text{where } x \in \bigcap_{k \in N} [x_{i-1}, x_i], \quad i = i(k). \tag{2.3.19}$$

For (2.3.14), the limit relation (2.3.19) is obviously valid, since $\sigma_i(x)$ tends to zero, if $[x_i - x_{i-1}] \to 0$, and $\mathcal{N}\{\cdot\} \le 1$ for all arguments.

In the case of (2.3.15), after straightforward algebraic transformations one arrives at the following equivalent expression:

$$P_i = \mathcal{N}\left\{\frac{-8(z_{i-1} - \underline{z}^\star_{k_0})(z_i - \underline{z}^\star_{k_0}) - (z_i - z_{i-1})^2}{\sigma(x_i - x_{i-1})^{\frac{1}{2}}[16(z_{i-1} - \underline{z}^\star_{k_0})(z_i - \underline{z}^\star_{k_0}) + 3(z_i - z_{i-1})^2]^{\frac{1}{2}}}\right\}. \tag{2.3.20}$$

Now, if $[x_i - x_{i-1}] \to 0$ and $z = f(x)$, then—assuming that $\underline{z}^\star_{k_0}$ is a valid underestimate of the minimum—(2.3.20) converges to the same limit as $\mathcal{N}\{-2(z - \underline{z}^\star_{k_0})/\sigma(x_i - x_{i-1})^{\frac{1}{2}}\}$: thus the relation (2.3.19) holds also for the algorithm of Žilinskas.

Consequently, by Corollaries 2.2.1 and 2.2.2, one can conclude that the corresponding PAS-type algorithms of Mockus, Tiesis and Žilinskas (1978) and Žilinskas (1981) have the same type of convergence property as Kushner's method. That is, they all generate an everywhere dense search point set, and the sample minimum and minimizer sequences provide estimates for the global optimum and (an) optimizer. Note that—similarly to Kushner's method—the two other algorithms mentioned can be applied to solve both the one-dimensional CGOP and LGOP.

Proceeding now to the other three partition operators (2.3.12), (2.3.13), and (2.3.16), let us recall first of all that Piyavskii's algorithm satisfies the requirements of both Theorem 2.2.3 (strict monotonic decrease in limit), and Theorem 2.2.4 (translation-invariance in limit). Therefore it is convergent in the sense of the relation $X^a = X^\star$.

Now let us look at the operators in the algorithms of Strongin and Boender, respectively. As for (2.3.13) we have

$$\lim_{k\to\infty} P_i = P\{x, z\} = -4z, \quad \text{where } x \in \bigcap_{k\in N} [x_{i-1}, x_i], \quad i = i(k), \tag{2.3.21}$$

and for (2.3.16)

$$\lim_{k\to\infty} P_i = P\{x, z\} = -z, \quad \text{where } x \in \bigcap_{k\in N} [x_{i-1}, x_i], \quad i = i(k), \tag{2.3.22}$$

therefore both partition operators are evidently translation-invariant and strictly decreasing in limit. Consequently, the convergence property $X^a =$

X^* can be directly concluded for both Strongin's and Boender's methods. These methods—similarly to Piyavskii's—can be applied to solve the LGOP if their parameterization is chosen in accordance with (a proper overestimate of) the Lipschitz constant of the objective function f.

It is to be emphasized at this point that the methodology applied here to verify the convergence properties of the listed (or other) partition strategies is apparently independent of the construction of the underlying function models. Moreover, the direct ('shortcut') approach introduced will allow us to suggest straightforward modifications and extensions of convergent partition algorithms, while preserving their established convergence, and without the actual construction of a corresponding function model. There is an obvious advantage to this approach, as the function models referred to are typically derived with considerable effort and, in most cases, they cannot be directly generalized to the multidimensional case. A number of algorithm modifications and generalization options will be discussed in detail in subsequent chapters of Parts 2 and 3.

2.3.2 An Efficiency Estimate

The objective of this section is to quantify in a sense the general qualitative convergence results of the previous chapter. Namely, we shall show that, in solving the univariate global optimization problem

$$\min f(x) = f(x^*) = z^*, \quad \text{s.t.} \quad x \in [a, b], \tag{2.3.23}$$

the total sampling effort is bounded on all subintervals of $[a, b]$ which do not contain globally optimal points. It is obvious that this type of statement can be applicable only to those PAS-type algorithms which are convergent in the sense of $X^a = X^*$, and not to those strategies which lead to everywhere dense sample point sequences. (Recall in this context the properties of these partition strategies, as discussed in Chapter 2.2 and Section 2.3.1.) For purposes of the subsequent analysis, we shall suppose that the translation-invariance and strict monotonic decrease properties of the partition operator P now hold not only in limit, but also for nondegenerate subintervals of $[a, b]$. In other words, if

$$[x_{i-1}, x_i] \subset [a, b], \quad [x_{i-1} + v, x_i + v] \subset [a, b] \quad (0 < v \leq b - x_i),$$

and

$$z_{i-1} = f(x_{i-1}), \quad z_i = f(x_i), \quad \delta_1 \geq 0, \quad \delta_2 \geq 0, \quad \delta_1 + \delta_2 > 0,$$

then we postulate the relations

$$P\{x_{i-1}, z_{i-1}, x_i, z_i\} = P\{x_{i-1} + v, z_{i-1}, x_i + v, z_i\} \tag{2.3.24}$$

(translation-invariance), and

$$P\{x_{i-1}, z_{i-1}, x_i, z_i\} < P\{x_{i-1}, z_{i-1} - \delta_1, x_i, z_i - \delta_2\} \tag{2.3.25}$$

(monotonic decrease).

Remark 2.3.5. The relation (2.3.24) implies that two subintervals in $[a, b]$, at the current iteration of the PAS-type method, have the same 'search value' if they are of equal length, and the function values at the interval endpoints are, respectively, identical. Relation (2.3.25) postulates that if we have two search subintervals of the same length, but for the first of them at least one endpoint function value is strictly better, than for the second (while the other function value is not worse), then the first interval has more current 'search value' than the second. To illustrate these properties, recall Piyavskii's partition operator

$$P_i = L\frac{x_i - x_{i-1}}{2} - \frac{z_{i-1} + z_i}{2}.$$

Evidently, this operator is both translation-invariant and strictly monotonically decreasing, not only in limit but also in the sense of (2.3.24) and (2.3.25).

Definition 2.3.1. Let $[\alpha, \beta]$ denote an arbitrary subinterval of $[a, b]$, and let $\{x_k\}$ be the search point sequence generated by a PAS-type algorithm. If $\nu[\alpha, \beta]$ denotes the overall number of those search points which are selected from $[\alpha, \beta]$, then the density of the sequence $\{x_k\}$ on $[\alpha, \beta]$ is defined by the ratio

$$r[\alpha, \beta] = \frac{\nu[\alpha, \beta]}{\beta - \alpha}. \tag{2.3.26}$$

Remark 2.3.6. If the PAS algorithm would theoretically generate an everywhere dense sample sequence, then obviously $r[\alpha, \beta] = \infty$, for each nondegenerate subinterval $[\alpha, \beta]$.

To make the following discussion somewhat simpler, it will be assumed again that the PAS-type algorithm in question generates a singular new point on the selected subinterval $[x_{t-1}, x_t]$ following the midpoint selection rule

$$x^{k+1} = \frac{x_{t-1} + x_t}{2}, \quad t = t(k), \quad k \in N. \tag{2.3.27}$$

Theorem 2.3.1. *Let the objective function f in (2.3.23) be Lipschitzian. Assume that this LGOP is solved by a PAS-type algorithm, and for a certain subinterval $[\alpha, \beta] \subset [a, b]$ we have*

$$f(x) \geq z^\star + \delta \text{ for all } x \in [\alpha, \beta], \text{ where } \delta = \delta[\alpha, \beta] > 0. \tag{2.3.28}$$

Now, if there exists a minimal interval width $w = w(\delta, z^\star)$, $0 < w \leq \beta - \alpha$, such that for all intervals $[x, x + w] \subset [\alpha, \beta]$ we have

$$P\{x, z^\star + \delta, x + w, z^\star + \delta\} \geq P\{x^\star, z^\star\}, \tag{2.3.29}$$

then

$$\nu[\alpha, \beta] \leq \lceil 2(\beta - \alpha)/w \rceil + 1. \tag{2.3.30}$$

In (2.3.30), $\lceil u \rceil$ denotes the smallest integer which is not less than the positive real number u.

Proof. Let $[x_{i-1}, x_i]$ be a subinterval of $[\alpha, \beta]$ with length $v < w$, and assume that this subinterval has been generated by the PAS-type algorithm applied to solve (2.3.23). Since we have

$$\min\{z_{i-1}, z_i\} \geq z^\star + \delta,$$

by the minimality of w (see (2.3.29)), and the postulated properties (2.3.24)–(2.3.25) of the partition operator P, we can write

$$\begin{aligned} P\{x_{i-1}, z_{i-1}, x_i, z_i\} &\leq P\{x_{i-1}, z^\star + \delta, x_i, z^\star + \delta\} \\ &= P\{x_{i-1}, z^\star + \delta, x_{i-1} + v, z^\star + \delta\} < P\{x^\star, z^\star\} \\ &< P\{x_{j-1}, x_j, z_{j-1}, z_j\}. \end{aligned} \tag{2.3.31}$$

In (2.3.31), $[x_{j-1}, x_j]$, $j = j(k)$, $k \in N$, denotes that partition subinterval which contains an arbitrary globally optimal $x^\star \in X^\star$. The chain of relations (2.3.31) implies that—by the selection rule (2.1.12) of the PAS—the algorithm will not select $[x_{i-1}, x_i]$ for further partition (bisection) throughout the remaining iterations. In other words, if the length of an arbitrary subinterval in $[a, b]$ is less than w, and the endpoint function values are not better than $z^\star + \delta$, then this interval can be discarded from further search. Moreover, by virtue of the midpoint selection rule (2.3.27), at most a singular point can be generated on subintervals of $[\alpha, \beta]$ which are of length w. This observation leads to the estimate (2.3.30). ∎

Corollary 2.3.1. *Under the conditions of Theorem 2.3.1, the density of the search point sequence on $[\alpha, \beta]$ is bounded by the inequality*

$$r[\alpha,\beta] \le \frac{\left\lceil \frac{2(\beta-\alpha)}{w} \right\rceil + 1}{\beta - \alpha}. \tag{2.3.32}$$

Proof. This statement follows immediately from (2.3.30). ∎

Remark 2.3.7. Corollary 2.3.1 generalizes Strongin's (1978) efficiency estimate, valid in relation to his specific algorithm (recall the partition operator defined by (2.3.13)).

To specify the result of Corollary 2.3.1 to the Lipschitzian case, suppose now that $f \in F_L[a,b]$. If for a certain subinterval $[x_{i-1}, x_i] \subset [a,b]$ we have

$$\min\{z_{i-1}, z_i\} \ge z^\star + \delta \quad (\delta > 0), \qquad \text{and} \quad x_i - x_{i-1} < \frac{2\delta}{L},$$

then—by the definition of Lipschitz-continuity—this interval cannot contain any globally optimal point. Therefore, in accordance with the definition of w in Theorem 2.3.1 (see (2.3.29)), one can see that $w \ge 2\delta/L$ must hold. Using this estimate, (2.3.32) immediately leads to

Corollary 2.3.2. *Assume that $f \in F_L[a,b]$. Then, under the conditions of Theorem 2.3.1, one can derive the Lipschitz-constant dependent density estimate*

$$r[\alpha,\beta] \le \frac{\left\lceil \frac{L(\beta-\alpha)}{\delta} \right\rceil + 1}{\beta - \alpha}. \quad ∎ \tag{2.3.33}$$

Remark 2.3.8. Remark 2.3.6, Theorem 2.3.1 and Corollaries 2.3.1–2.3.2 provide a clear quantitative distinction between two PAS-type methods, if one of them is convergent in the sense of the relation $X^a = X^\star$, and the other has the property $X^a = D$. In addition, from Corollary 2.3.2 one can see that—for the class of globally convergent methods considered—the density estimate is essentially a linear function of the Lipschitz-constant L, and it is inversely proportional to δ. Note that δ can be interpreted here as the parameter of the precision required in solving the LGOP.

Remark 2.3.9. Zabinsky and Smith (1992) provide a theoretical analysis of pure adaptive random search for minimizing a Lipschitz function on a convex, compact set $D \subset \mathbb{R}^n$. (For earlier work, to establish such theoretical strategies, consult Chichinadze, 1967, 1969; Galperin and Zheng, 1987; and Patel, Smith and Zabinsky, 1988.) For that conceptual algorithm-class (which requires generating random points uniformly distributed over the nested improving level sets of f, at each iteration), they show that the expected number of iterations, to achieve a given solution accuracy, can be bounded by a logarithmic function of the Lipschitz-constant. This result is not in contrast with those of the present section, however; rather, it shows the essential dependence of theoretical efficiency estimates on the postulated properties of the problem and algorithm-classes analysed.

Chapter 2.4

Partition Algorithms on Multidimensional Intervals

2.4.1 Introduction

The present chapter is devoted to the analysis of adaptive partition strategies to solve the multivariate global optimization problem

$$\min_{x \in D} f(x), \tag{2.4.1}$$

where

$$D = [a, b] \subset I\!R^n, \qquad a = (a_1, \ldots, a_n), \qquad b = (b_1, \ldots, b_n),$$
$$-\infty < a_j < b_j < \infty, \qquad j = 1 \ldots, n.$$

Obviously, (2.4.1) is a special case of the general GOP stated in Section 2.1.1, if we suppose the continuity or Lipschitz-continuity of f. As earlier, $X^\star$ denotes the set of globally optimal solutions to (2.4.1), and $z^\star = f(x^\star)$ for $x^\star \in X^\star$.

The problem statement (2.4.1) is frequently met in practical applications of global optimization. As a number of such applications will be discussed in detail later, we shall mention only one example here. In calibrating scientific or engineering system models, the modeller often has a fairly sound idea about the possible range of the critical model parameters, but no further information may be directly available. If the system model in question is nonlinear, and the suitability of its parameterizations is assessed by actually running the (complex) numerical system model, then the model calibration problem frequently leads to a CGOP or LGOP, in the form (2.4.1).

It is expected that with the increase of dimensionality n in (2.4.1), the computational demand will increase substantially. To illustrate this point, recall the simple grid search approach (UGS, as discussed in Section 2.1.3). The multivariate version of UGS is based on selecting simultaneously $k+1$ equidistant points for each coordinate

$$x_{ji} = a_j + \frac{i}{k}(b_j - a_j), \qquad j = 1, \ldots, n, \qquad i = 0, 1, \ldots, k, \tag{2.4.2}$$

and evaluating f at all possible combinations $x = x(i_1, \ldots, i_n) = (x_{1i_1}, \ldots, x_{ni_n})$ of these. The number of grid points will be of order k^n, and thus it can be huge, even for modest problem dimensionality: for instance, to attain 10^{-3} relative precision by UGS, with respect to each coordinate on an interval $[a, b] \subset \mathbb{R}^4$, would need $O(10^{12})$ evaluations of the objective function.

Because of the obvious computational challenge posed by (2.4.1) and other (possibly much more complicated) multidimensional GOPs, a number of attempts have been proposed to reduce (2.4.1) to the one-dimensional case: below we shall briefly review two such approaches.

Piyavskii (1967, 1972) suggests a nested optimization scheme. Following this idea, (2.4.1) is formally replaced by the equivalent problem

$$\min_{x_1 \in [a_1, b_1]} \Big\{ \min_{x_2 \in [a_2, b_2](x_1)} \cdots \{ \min_{x_n \in [a_n, b_n](x_1, \ldots, x_{n-1})} f(x)\} \cdots \Big\}. \tag{2.4.3}$$

In (2.4.3), $[a_j, b_j](x_1, \ldots, x_{j-1})$ denotes the one-dimensional interval $[a_j, b_j]$, given the coordinates $x_1, \ldots, x_{j-1}$; the corresponding nested univariate minimization problem is to be solved (theoretically) for all possible given values of $(x_1, \ldots, x_{j-1})$. This way, the initial multivariate GOP is, at least in principle, reduced to solving an (infinite) sequence of nested one-dimensional optimization problems. Without a detailed discussion of this approach at this point, observe that even the approximate numerical realization of this approach needs a grid search type construction, and hence it is probably not readily implementable, except for very low dimensional ($n = 2, 3$) problems.

The second dimensionality reduction approach (see, e.g., Butz, 1968; Strongin, 1978, 1984, 1992) is based on the (approximate) mapping of the multidimensional interval $[0, 1]^n \subset \mathbb{R}^n$ into $[0, 1] \subset \mathbb{R}^1$. (Having such a

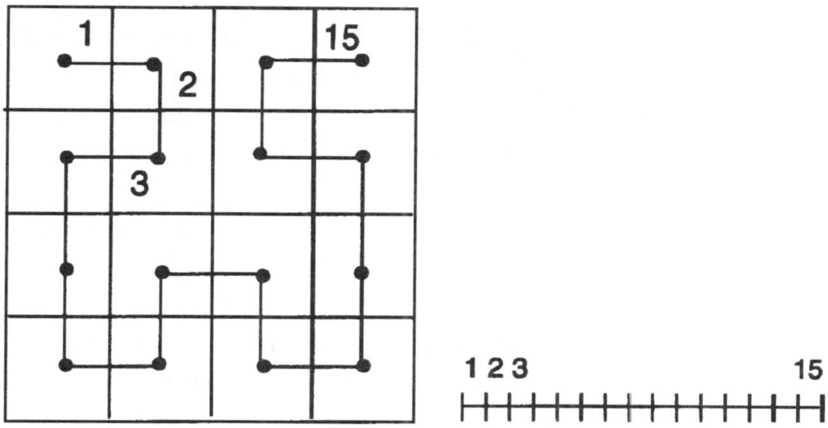

Figure 2.4.1
A Peano curve and the resulting mapping of $[0,1]^n$ into $[0,1]$
$(n = 2, m = 2)$

method, simple linear transformations make possible the mapping of an arbitrary finite multivariate interval into an arbitrary finite real interval.) The mapping proposed by Butz and Strongin is based on the Peano curve which is obtained as follows. First, partition $[0,1]^n$ into 2^{mn} hypercubes of identical size: the length of an edge of each small cube equals $1/2^m$, the integer $m > 1$ will determine the accuracy of the approximation. Now connect the centres of the small hypercubes by a continuous curve which consists of straight line segments (many such constructions are possible; see one of them in Fig. 2.4.1).

Now subdivide the real interval $[0,1]$ into $2^{mn} - 1$ parts, and then map consecutive segments of the connecting curve in $[0,1]^n$ into consecutive segments of $[0,1]$. This way, global optimization on $[0,1]^n$ is approximated by solving the same GOP on the one-dimensional Peano curve. Obviously—in order to make the Peano curve approximation valid—this approach necessitates a quantitative assumption (such as, for example, Lipschitz-continuity) with regard to the objective function. Second, as Fig. 2.4.1 illustrates, 'close' points of the multidimensional region $[0,1]^n$ can become 'distant' in the univariate image set $[0,1]$. Undoubtedly, this circumstance has a negative effect on all implementations: for example, the number of nearly optimal points may drastically increase as a result of the Peano transformation. Strongin (1978), Sergeyev and Grishagin (1994b) report the application of Peano curve transformations for solving a number of two-dimensional trigonometric test problems. In order to improve the distance-preservation during Peano transformations, Strongin (1992) also proposes using a set of joint space-filling curves.

The two approaches reviewed above indicate that the appropriate multi-dimensional extension of univariate approaches is far from being a trivial issue.

The objective of the following discussion is, therefore, to present a general class of adaptive partition strategies which can be applied to solve (2.4.1): for instance, the one-dimensional global optimization strategies treated in Chapter 2.3 can be directly extended to the multivariate case. This generalization is based on the dimensionality-free convergence investigations of Chapter 2.2: for this reason, the approach is essentially independent of the original one-dimensional function class models which served as the basis of the corresponding univariate global optimization algorithms. It should be noted at this point that diverse, more specific generalizations and diverse extensions of Piyavskii's univariate algorithm are discussed, for example, by Mayne and Polak (1984), Galperin (1985, 1987, 1988), Mladineo (1986, 1992a,b), Strigul (1987), Nefedov (1987), Shepilov (1987), Meewella and Mayne (1988, 1989), Korotkich (1989), Wood (1991, 1992) and Zhang, Wood and Baritompa (1993). Additionally, semi-heuristic multivariate extensions of one-dimensional statistical function models are reviewed and discussed, for example, by Mockus (1989) and Žilinskas (1982, 1992). The subsequent treatise draws on Pintér (1986a,b, 1988, 1992a).

We shall first specify the adaptive partition algorithm scheme (PAS) of Chapter 2.1 for the case of generating rectangular subdivisions of $[a,b] \subset \mathbb{R}^n$. In the following exposition

$$P_i\{D_i, f(D_i)\} \tag{2.4.4}$$

will denote the value of the partition operator for an arbitrary subinterval $D_i \subset [a,b]$. Again, we shall assume that the value of P_i is determined by a finite sample of search points selected from the set D_i and the corresponding values of f at these points. Recall that some other model-dependent parameters (such as, for example, the Lipschitz-constant) can play a role in determining the value of P_i. The sequential order of the subintervals D_i generated by a PAS-type algorithm depends, of course, on the iteration index $k : D_i = D_{i_k}$, $k \in N$, but this fact will not always be explicitly reflected by the notation.

Analogously to Remark 2.3.2, we shall make a simplifying assumption with regard to the partition refinement steps of the PAS. Namely, it will be supposed that these are realized as follows. For $D_t = [xl_t, xu_t]$, its central point

$$xc_t = \frac{xl_t + xu_t}{2} \tag{2.4.5}$$

is chosen to define a bisection on each subinterval D_t which is partitioned at the refinement step. The bisection can be realized, halving D_t by an arbitrary hyperplane which passes through xc_t and is parallel to one of the facets of the subinterval D_t. Note that this does not have to be the case in the numerical implementation of the actual search strategies, but the simplification above makes it slightly easier to verify convergence properties. (For illustration of a general rectangular partition of $[a, b]$, recall Fig. 2.1.2.)

The corresponding PAS has the following basic structure.

Rectangular Partition Algorithm Scheme for Multidimensional Global Optimization (RPAS)

Step 0. (Initialization) Set $k := 0$ (iteration count); $I_0 = \{0\}$, $\{D_0\} = \{[a, b]\}$. Introducing the notation $xl_0 = a$ and $xu_0 = b$, determine $zl_0 = f(xl_0)$ and $zu_0 = f(xu_0)$, and (possibly) a finite number of further function values taken at some other vertices or interior points of $[xl_0, xu_0]$.

Step 1. (Sampling) At each iteration k, we have a collection of partition subintervals $D_i = [xl_i, xu_i]$, $i \in I_k$, $k \in N$. In each D_i, sample points are taken: these include the 'lower left' and 'upper right' vertices xl_i and xu_i; the corresponding function values are zl_i and zu_i. Additionally, finitely many further sample points can be selected from all current subintervals D_i.

Step 2. (Aggregation) For each $i \in I_k$, $k \in N$, determine the value of the partition operator $P_i\{D_i, f(D_i)\}$.

Step 3. (Selection) Choose an arbitrary index t for which the relation

$$P_t\{D_t, f(D_t)\} = \max_{i \in I_k} P_i\{D_i, f(D_i)\} \tag{2.4.6}$$

is satisfied; then subinterval $D_t = [xl_t, xu_t]$ will be subdivided at this iteration cycle.

Step 4. (Refinement) Taking the central point xc_t defined by (2.4.5), the interval $[xl_t, xu_t]$ is replaced by two half-intervals D_{t+1} and D_{t+2}, $t = t(k)$, of equal volume.

The refined partition, therefore, is given by the collection of sets

$$\{D_i : i \in I_{k+1}\}, \qquad I_{k+1} = I_k \cup \{t+1, t+2\} \setminus \{t\}. \tag{2.4.7}$$

Set $k := k + 1$; return to Step 1 (of the next iteration cycle).

Remark 2.4.1. Because of the halving bisection at Step 4, every strictly nested subinterval-sequence generated by the RPAS will obviously converge to a single limit point

$$x = \bigcap_{k \in N} D_i, \qquad i = i(k). \tag{2.4.8}$$

In other words, the RPAS satisfies the perfection property P1. In order to guarantee the applicability of the convergence results presented in Chapter 2.2, we shall also postulate the properties P2–P4. This leads to the following specific assumptions (again, recall the detailed statement of these conditions as discussed in Section 2.2.1):

P2. (Continuity) If (2.4.8) holds, then

$$\lim_{k \to \infty} P_i\{D_i, f(D_i)\} = P\{x, z\}, \quad z = f(x), \quad i = i(k), \quad k \in N. \tag{2.4.9}$$

P3. (Monotonicity) If (2.4.8) holds, then for each embedding subinterval D_i,

$$P_i\{D_i, f(D_i)\} > P\{x, z\}. \tag{2.4.10}$$

P4. (Boundedness)

$$-\infty < P_i\{D_i, f(D_i)\} < \infty. \tag{2.4.11}$$

Algorithms which satisfy assumptions P1 to P4, are called regular (multivariate) rectangular partition strategies. Notice that the one-dimensional specification of P1–P4 fully corresponds to the conditions postulated in Chapter 2.3.

To illustrate the conceptual RPAS, two multivariate partition algorithms will be presented below. As mentioned earlier, these methods are based on direct generalizations of known univariate global optimization strategies, applying the results of Chapter 2.2. Given the RPAS, only the corresponding new partition operators need to be provided. Let us emphasize that the operators will serve only for illustration, as numerical efficiency aspects will be discussed in detail later: at that point, we shall 'streamline' these operators, and shall present others as well.

First, let us introduce a generalization of Kushner's method to the multidimensional case. We shall take a finite sample of points from D_i, including the lower left and upper right vertices $\boldsymbol{xl}_i$ and $\boldsymbol{xu}_i$. If these sample points and the corresponding function values are $\{\boldsymbol{x}_i(m), z_i(m) : m = 1, \ldots, M_i\}$, then one can define the partition operator P_i as

$$P_i = \exp\left\{ \frac{-\prod_{m=1}^{M_i} (z_i(m) + \gamma - f_k^\star)}{\sigma^2 \|\boldsymbol{xu}_i - \boldsymbol{xl}_i\|} \right\}. \tag{2.4.12}$$

Similarly to Kushner's original method, in (2.4.12), γ and σ are parameters of the algorithm (typically preassigned, or estimated from the sample), and $f_k^\star$ is a fixed global optimum estimate.

As a second example, we shall provide an extended multidimensional Piyavskii-type partition operator. Consider now the set of all vertices in D_i and the function values taken at them. This results in the sample information $\{\boldsymbol{x}_i(m), z_i(m) : m = 1, \ldots, 2^n\}$. Define now the generalized Piyavskii operator by

$$P_i = L\|\boldsymbol{xu}_i - \boldsymbol{xl}_i\| - \frac{1}{2^n} \sum_{m=1}^{2^n} z_i(m). \tag{2.4.13}$$

In the next section we will establish the convergence properties of the multidimensional RPAS. Specifically, we will show the convergence of both partition strategies which are based on the operators (2.4.12) and (2.4.13), respectively.

2.4.2 Convergence of Multivariate Rectangular Partition Methods

Assuming the regularity of a multivariate rectangular partition strategy, and making use of the results in Chapter 2.2, one can readily establish convergence of the RPAS. The sufficient convergence conditions can be summarized by the following assertions.

Theorem 2.4.1. *Given the regular multivariate rectangular partition operator P, for the corresponding RPAS two types of convergence (limit properties) can be established:*

a. *If $P\{x,z\} = c$ for $x \in [a,b]$, $z = f(x)$ (c is fixed), then $X^a = [a,b]$.*

b. *If $P\{x^\star, z^\star\} = c^\star > P\{x,z\}$ for $x^\star \in X^\star$ and $x \in D \setminus X^\star$ ($c^\star$ is fixed), then $X^a = X^\star$.* ■

To illustrate the application of this general result, let us first analyse formula (2.4.12). Evidently, this partition operator is bounded (P4). Assuming now that

$$x = \bigcap_i [xl_i, xu_i], \qquad i = i(k), \qquad k \in N,$$

obviously $||xu_i - xl_i|| \to 0$. Consequently—for suitably large $\gamma > 0$—the limit value of (2.4.12) is zero for all $x \in D$, while it is positive for each nondegenerate interval $[xl_i, xu_i]$. This shows both the continuity (P2) and monotonicity (P3) of the generalized Kushner partition operator. Consequently, this operator is regular.

Since the relation $P\{x,z\} = 0$ is valid for all $x \in [a,b]$, it allows the direct application of Theorem 2.4.1 (recalling Corollaries 2.2.1 and 2.2.2). According to these results, the generalized Kushner algorithm governed by the operator (2.4.12) theoretically leads to an everywhere dense set of sample points in $[a,b]$; furthermore, the sample minimum and minimizer sequences provide estimates of the global optimum and (one) optimal solution to the GOP (2.4.1). Note again that in order to apply the generalized Kushner algorithm, the objective function in (2.4.1) can be merely continuous, without the additional Lipschitz assumption.

Suppose now that (2.4.1) is an LGOP. We shall verify the multivariate Piyavskii algorithm, as determined by the partition operator (2.4.13) in which L is a valid overestimate of the Lipschitz-constant of f. First, note that for an arbitrary limit point x, that is,

$$x = \bigcap_i [xl_i, xu_i], \;\; i = i(k), \;\; k \in N, \;\; \text{and} \;\; z = f(x),$$

we have

$$P\{x, z\} = -z. \tag{2.4.14}$$

This shows that P_i is continuous (P2). Moreover, as f is a Lipschitz function, for any $x \in [x_{i-1}, x_i]$ we have

$$|z - z_i(m)| \le L\|x - x_i(m)\|, \qquad m = 1, \ldots, 2^n. \tag{2.4.15}$$

Denote now the global minimizer and minimum of $f(x)$ on D_i by $x_i^\star = x_i^\star(D_i)$ and $z_i^\star = z_i^\star(D_i)$, respectively. Substituting these into (2.4.15), we arrive at the set of inequalities

$$z_i(m) - z_i^\star \le L\|x_i^\star - x_i(m)\|, \qquad m = 1, \ldots, 2^n. \tag{2.4.16}$$

Adding up these relations, for each $x \in D_i$ we can write the following chain of inequalities:

$$\begin{aligned} P\{x, z\} &= -z \le -z_i^\star \\ &\le \frac{1}{2^n} \sum_{m=1}^{2^n} L\|x_i^\star - x_i(m)\| - \frac{1}{2^n} \sum_{m=1}^{2^n} z_i(m) \\ &< L\|xu_i - xl_i\| - \frac{1}{2^n} \sum_{m=1}^{2^n} z_i(m) = P_i. \end{aligned} \tag{2.4.17}$$

This shows the monotonicity (P3) of the multidimensional Piyavskii operator (2.4.13). Consequently, Piyavskii's generalized rectangular partition operator is regular.

In addition to this, the relation $P\{x, z\} = -z$ shows that the multivariate Piyavskii operator satisfies the requirements of both Theorem 2.2.3 (strict monotonic decrease in limit), and Theorem 2.2.4 (translation-invariance in limit). Consequently, the corresponding RPAS-type algorithm is convergent in the sense of the relation $X^a = X^\star$.

As mentioned earlier, it is not difficult to prove the convergence of other similar extensions of one-dimensional partition algorithms. For instance, the univariate methods of Strongin (1973), Mockus, Tiesis and Žilinskas (1978), Žilinskas (1981) and Boender (1984), presented in Chapter 2.3 (recall formulas (2.3.13)–(2.3.16)) can also be generalized to the multidimensional case. Moreover, on the basis of the previous discussion, it is not difficult to see that these generalizations can be defined in different ways. Computationally efficient generalizations will be considered later in detail; at that point, we shall discuss multivariate extensions of all operators (2.3.11)–(2.3.16).

2.4.3 An Efficiency Estimate

The purpose of the following discussion is to derive a straightforward multivariate extension of the results presented in Section 2.3.2. Namely, it will be shown that in solving the global optimization problem

$$\min f(x) = f(x^\star) = z^\star \quad \text{s.t.} \quad x \in [a,b] \subset I\!R^n, \tag{2.4.18}$$

the overall sampling effort of a broad class of RPAS-type methods is bounded on such subintervals of $[a,b]$ that do not contain points of $X^\star$. Similarly to the univariate case, this type of statement is related only to such RPAS algorithms which are convergent in the sense of $X^a = X^\star$.

We shall suppose again that both the translation invariance and strict monotonic decrease properties of the partition operator P hold not only in limit, but also for nondegenerate subintervals of $[a,b]$. In other words, if

$$[xl_i, xu_i] \subset [a,b], \quad [xl_i + v, xu_i + v] \subset [a,b], \quad v = (v_1, \ldots, v_n), \quad v_j \geq 0,$$

and

$$z_i(m) = f(x_i(m)), \quad m = 1, \ldots, 2^n,$$

$$\delta = \{\delta_i(m) : m = 1, \ldots, 2^n\}, \qquad \delta_i(m) \geq 0, \qquad \sum_{m=1}^{2^n} \delta_i(m) > 0,$$

then we postulate the relations

$$P\{x_i(m), z_i(m)\} = P\{x_i(m) + v, z_i(m)\} \quad \text{(translation-invariance)} \tag{2.4.19}$$

and

$$P\{x_i(m), z_i(m)\} < P\{x_i(m), z_i(m) - \delta\} \quad \text{(monotonic decrease).} \tag{2.4.20}$$

The relation (2.4.19) implies that two subintervals in $[a,b]$, at the current iteration cycle, have the same 'search value' if they can be translated into each other, and the function values at the corresponding interval vertices are identical. Additionally, (2.4.20) postulates that if two search subintervals are of the same size and shape, but one of them has at least one vertex at which the function value is strictly better than its counterpart—and all other function values are not worse than their counterparts—then this interval has more (actual) 'search value' than the second one.

To illustrate these concepts, recall the generalized Piyavskii partition operator (2.4.13)

$$P_i = L||xu_i - xl_i|| - \frac{1}{2^n}\sum_{m=1}^{2^n} z_i(m).$$

It is immediately seen that this operator is both translation-invariant and strictly monotonically decreasing.

Definition 2.4.1. Let $[\alpha, \beta] \subset I\!R^n$ denote an arbitrary subinterval of $[a, b]$; and let $\{x_k\}$ denote the search point sequence generated by an RPAS-type algorithm. Now, if $\nu[\alpha, \beta]$ denotes the number of those search points which are selected from $[\alpha, \beta]$ during RPAS iterations, then the density of the sequence $\{x_k\}$ on $[\alpha, \beta]$ is defined by the ratio

$$r[\alpha, \beta] = \frac{\nu[\alpha, \beta]}{V[\alpha, \beta]}. \tag{2.4.21}$$

In (2.4.21), $V[\alpha, \beta] = \prod_{j=1}^{n}(\beta_j - \alpha_j)$ denotes the volume of subinterval $[\alpha, \beta]$.

Remark 2.4.2. Note that if the RPAS-type algorithm would theoretically generate an everywhere dense sample sequence, then obviously $r[\alpha, \beta] = \infty$, for each nondegenerate subinterval $[\alpha, \beta]$.

For the sake of simplicity, it will be assumed again that the sampling steps of the studied RPAS-type algorithm are based on generating a singular new point on the selected subinterval $[xl_t, xu_t]$, following the midpoint selection rule

$$xc_t = \frac{xl_t + xu_t}{2}, \qquad t = t(k), \qquad k \in N. \tag{2.4.22}$$

Then the partition refinement is done by halving $[xl_t, xu_t]$ via a hyperplane which passes through xc_t and is parallel to one of the facets of $[xl_t, xu_t]$.

For an arbitrary real n-vector v with positive components $v_1, \ldots, v_n$, consider the partition operator value $P\{[\alpha, \alpha + v], (z^\star + \delta)\}$: by definition, this corresponds to the interval $[\alpha, \alpha + v]$, assuming that all vertex function values equal $z^\star + \delta$.

Theorem 2.4.2. *Let the objective function f in (2.4.18) be Lipschitzian. Suppose that this LGOP is solved by an RPAS-type algorithm, and that for a certain subinterval $[\alpha, \beta] \subset [a, b] \subset \mathbb{R}^n$ we have the inequality*

$$f(x) \geq z^\star + \delta \text{ for all } x \in [\alpha, \beta], \quad \text{where } \delta = \delta[\alpha, \beta] > 0. \tag{2.4.23}$$

Now, assume that there exists a componentwise minimal interval width vector $w = w(\delta, z^\star)$, $0 < w_j \leq \beta_j - \alpha_j$, $j = 1, \ldots, n$, such that for all intervals $[\alpha, \alpha + w] \subset [\alpha, \beta]$ we have

$$P\{[\alpha, \alpha + w], (z^\star + \delta)\} \geq P\{x^\star, z^\star\}. \tag{2.4.24}$$

Then the following inequality holds:

$$\nu[\alpha, \beta] \leq \prod_{j=1}^{n} \Big\{ \Big\lceil \frac{2(\beta_j - \alpha_j)}{w_j} \Big\rceil + 1 \Big\}. \tag{2.4.25}$$

Again, $\lceil u \rceil$ denotes the smallest integer which is not less than the positive real number u.

Remark 2.4.3. The translation-invariance property (2.4.19) guarantees that in (2.4.24) α could be replaced by an arbitrary n-vector $x \in [a, b]$.

Proof. Let $D_i = [xl_i, xu_i]$ be a subinterval of $[\alpha, \beta]$ with interval width vector $v < w$ (componentwise), and assume that this subinterval has been generated by the RPAS-type algorithm applied to solve (2.4.18). By supposition, we have the vertex function values

$$z_i(m) \geq z^\star + \delta, \qquad m = 1, \ldots, 2^n. \tag{2.4.26}$$

Now, by the minimality of w (see (2.4.24)), and the postulated properties of the partition operator P, we can write

$$\begin{aligned} P\{x_i(m), z_i(m)\} &\leq P\{x_i(m), (z^\star + \delta)\} \qquad\qquad (2.4.27)\\ &= P\{[xl_i, xl_i + v], (z^\star + \delta)\} < P\{x^\star, z^\star\} < P\{x_j(m), z_j(m)\}. \end{aligned}$$

In (2.4.27), $[xl_j, xu_j] := \{x_j(m)\}$, $j = j(k)$, $k \in N$, denotes that partition subinterval which contains an arbitrary global solution $x^\star \in X^\star$. The chain of relations (2.4.27) implies that—by the selection rule (2.1.12) of the

PAS—the algorithm will not select $[xl_i, xu_i]$ for further partition (bisection) throughout the remaining iterations. In other words, if the edge length vector of an arbitrary subinterval in $[a, b]$ is less than w (componentwise), and the vertex function values are not better than $z^\star + \delta$, then this interval can be discarded from all further investigations. Moreover, in virtue of the midpoint selection rule (2.4.22), componentwise at most a single point can be generated on subintervals of $[\alpha_j, \beta_j]$ which are of size w_j: the number of such points is at most $\lceil 2(\beta_j - \alpha_j)/w_j \rceil + 1$, $j = 1, \ldots, n$. This observation leads directly to the estimate (2.4.25). ■

Corollary 2.4.1. *Under the conditions of Theorem 2.4.2, the density of the search point sequence on $[\alpha, \beta]$ is bounded by the inequality*

$$r[\alpha, \beta] \leq \prod_{j=1}^{n} \left\{ \frac{\lceil \frac{2(\beta_j - \alpha_j)}{w_j} \rceil + 1}{\beta_j - \alpha_j} \right\}. \tag{2.4.28}$$

Proof. This statement follows immediately from (2.4.25). ■

Remark 2.4.4. As $w_j < \beta_j - \alpha_j$, we can write

$$\frac{\lceil \frac{2(\beta_j - \alpha_j)}{w_j} \rceil + 1}{\beta_j - \alpha_j} < \frac{3}{w_j}, \quad j = 1, \ldots, n. \tag{2.4.29}$$

Therefore, the right-hand side in (2.4.28) can be approximated by the (simpler) estimate $\prod_{j=1}^{n}(3/w_j)$. This form clearly shows the impact of the components of w on the efficiency measure $r[\alpha, \beta]$. The estimate obtained also indicates that the (guaranteed) search effort is still an exponential function of the problem dimensionality n.

Let us specify now the result of Corollary 2.3.1 to the Lipschitzian case $f \in F_L[a, b]$. Consider an n-dimensional hypercube $C^n \subset [a, b]$ which has edges of the length $v_j = v_1$, $j = 2, \ldots, n$. Assume that the function values at the vertices of C^n all equal $z^\star + \delta$; then, obviously, there is no point in C^n at which $f(x) < z^\star + \delta - Lv_1\sqrt{n}/2$. Consequently, if $v_1 < 2\delta/(L\sqrt{n})$, then C^n cannot contain any globally optimal point.

Therefore, in accordance with the definition of w (see (2.4.24)), one can see that, for $w = (w_1, \ldots, w_n)$, $w_j = w_1$, $j = 2, \ldots, n$, $w_1 \geq 2\delta/(L\sqrt{n})$ must hold. Applying this estimate, (2.4.28) leads to

Corollary 2.4.2. *Assume that $f \in F_L[a,b]$. Then, under the conditions of Theorem 2.4.1, one can derive the Lipschitz-constant dependent density estimate*

$$r[\alpha,\beta] \leq \prod_{j=1}^{n}\left\{\frac{\left\lceil\frac{L\sqrt{n}(\beta_j-\alpha_j)}{\delta}\right\rceil+1}{\beta_j-\alpha_j}\right\}. \qquad \blacksquare \quad (2.4.30)$$

Remark 2.4.5. In complete analogy with the one-dimensional case, Theorem 2.4.2 and Corollaries 2.4.1–2.4.2 provide a clear quantitative distinction between the two classes of RPAS-type methods discussed earlier. Specifically, Corollary 2.4.2 shows that, for algorithms with the strong convergence property $X^\star = X^a$, in Lipschitzian GOPs the total number of points generated on suboptimal (as expressed by (2.4.26)) intervals is finite. At the same time, the guaranteed density estimate is essentially an exponential function in n of both the Lipschitz-constant L and $1/\delta$. This obviously follows also from the estimate (2.4.29), which now leads to

$$r[\alpha,\beta] \leq (3L\sqrt{n}/2\delta)^n. \qquad (2.4.31)$$

These estimates indicate, again, the inherent complexity of the LGOP in terms of the strong dependence on the available estimate of the Lipschitz constant L, the precision required (δ), and problem dimensionality n. Note additionally that one can compare this result with that of Zabinsky and Smith (1992) which—for a theoretical random search algorithm-class—indicates a more favourable efficiency estimate. Specifically, the expected number of (theoretically realizable) random search steps logarithmically depends on L, and linearly on n. Again (see Remark 2.3.9), this is not in contradiction to the results presented here, and shows only the essential dependence on the postulated model of the optimization problem, and on the solution strategy.

2.4.4 Decomposable Partition Operators

In contrast with the underlying objective function models suggested previously in the context of one-dimensional global optimization, the present approach is essentially independent of the problem dimensionality. This fact makes possible the 'direct' definition of globally convergent multivariate partition algorithms without explicitly constructing a corresponding n-dimensional function model. To illustrate this approach, two general families of multivariate rectangular partition algorithms will be introduced below.

We shall consider the partition operators

$$P_i = P_i\{D_i, f(D_i)\}, \quad i = i(k), \quad k \in N, \tag{2.4.32}$$

defined in the context of a global search strategy which follows the RPAS.

Definition 2.4.2. P_i in (2.4.32) is a decomposable multiplicative partition operator, if it can be written in the form

$$P_i\{D_i, f(D_i)\} = P_i^1\{\{x_i(m)\}\} \times P_i^2\{\{z_i(m)\}\}, \tag{2.4.33}$$
$$m = 1, \ldots, M_i, \quad i = i(k), \quad k \in N.$$

That is, P_i can be expressed as the product of two factors, which depend only on the sample points chosen and the corresponding objective function values, respectively.

Definition 2.4.3. P_i in (2.4.32) is a decomposable additive partition operator, if it can be written in the form

$$P_i\{D_i, f(D_i)\} = P_i^1\{\{x_i(m)\}\} + P_i^2\{\{z_i(m)\}\}, \tag{2.4.34}$$
$$m = 1, \ldots, M_i, \quad i = i(k), \quad k \in N.$$

That is, P_i can be expressed as the sum of two terms: the first term depends only on the sample points chosen from the partition subset, while the second depends only on the corresponding set of objective function values.

Remark 2.4.6. In accordance with the earlier discussion of partition operators, P_i^1 and P_i^2 may depend also on several function model and procedural parameters, without introducing additional notation to reflect this fact. Moreover, throughout the following discussion we shall also assume the regularity of the decomposable partition operators (see the related discussions in Sections 2.2.1 and 2.4.1). In particular, we shall suppose that regularity holds, as postulated by the requirements in Remark 2.4.1 (see (2.4.8)–(2.4.11)).

Remark 2.4.7. Recalling the one-dimensional partition operators (2.3.11)–(2.3.16), one can immediately see that Piyavskii's algorithm is based on a decomposable additive partition operator; moreover, Kushner's operator is of the decomposable multiplicative type. Observe at the same time that none of the other univariate algorithms cited there directly meet the requirements of Definitions 2.4.2 and 2.4.3.

The general convergence qualification provided in Chapter 2.2, and summarized by Theorem 2.4.1, leads to two general classes of globally convergent (regular multivariate rectangular) partition algorithms, which have the following convergence properties:

a. If $P\{x, z\} = c$ for $x \in [a, b]$, $z = f(x)$ (c is fixed), then $X^a = [a, b]$.

b. If $P\{x^\star, z^\star\} = c^\star > P\{x, z\}$ for $x^\star \in X^\star$ and $x \in D \setminus X^\star$ ($c^\star$ is fixed), then $X^a = X^\star$.

Applying these results to the general classes of partition operators introduced above, one can directly prove the following assertions.

Theorem 2.4.3. *Suppose that the regular decomposable multiplicative partition operator*

$$P_i\{D_i, f(D_i)\} = P_i^1\{\{x_i(m)\}\} \times P_i^2\Big\{\{z_i(m)\}\Big\}$$

has the following property:

$$x = \bigcap_k [xl_i, xu_i], \quad i = i(k), \quad k \in N$$

implies

$$P_i^1\{\{x_i(m)\}\} \underset{k}{\to} 0, \quad \text{or} \quad P_i^2\{\{z_i(m)\}\} \underset{k}{\to} 0. \tag{2.4.35}$$

Then, applying the corresponding RPAS-type algorithm to solve the CGOP or LGOP (2.4.1), it generates an everywhere dense set of points in $[a, b]$; moreover, all accumulation points of the minimizer sequence are in $X^\star$.

Proof. It is evident that (2.4.35) implies the condition stated in assertion (a) of Theorem 2.4.1; therefore the respective convergence statement holds. ∎

Theorem 2.4.4. *Suppose now that the regular decomposable additive partition operator*

$$P_i\{D_i, f(D_i)\} = P_i^1\{\{x_i(m)\}\} + P_i^2\{\{z_i(m)\}\}$$

has the following properties:

$$x = \bigcap_k [xl_i, xu_i], \quad i = i(k), \quad k \in N$$

implies

$$P_i^1\{\{x_i(m)\}\} \underset{k}{\rightarrow} 0 \quad \text{and} \quad P_i^2\{\{z_i(m)\}\} \underset{k}{\rightarrow} P_i^2(z), \qquad (2.4.36)$$

in which $z = f(x)$, *and* $P_i^2(z)$ *is strictly monotonically decreasing in* z.

Then, applying the corresponding RPAS-type algorithm to solve the LGOP (2.4.1), *it generates a set of limit points* X^a *in* $[a, b]$, *identical to* $X^\star$.

Proof. Again, it is evidently seen that the properties postulated in (2.4.36) imply the condition in assertion (b) of Theorem 2.4.1; therefore, the respective convergence statement holds. ∎

As an example, we shall apply these results to show a simple general method of constructing globally convergent decomposable multivariate partition operators. Further—possibly much more 'economical'—implementations will be discussed later.

Define

$$P_i^1\{x_i(m)\} = P_i^1\{||xu_i - xl_i||_p\}, \qquad 1 \le p \le \infty, \qquad (2.4.37)$$

$$P_i^2\{z_i(m)\} = P_i^2\Big\{\frac{1}{2^n}\sum_{m=1}^{2^n} z_i(m)\Big\}. \qquad (2.4.38)$$

That is, P_i^1 is a function of the l_p-norm of subinterval $[xl_i, xu_i]$ for some given p; and P_i^2 is a function of the sum of the function values at the vertices of $[xl_i, xu_i]$.

We shall assume now that P_i^1 is continuous, strictly monotonically increasing, and its limit value $P_i^1\{x\} = P_i^1\{||0||\} = 0$ for $x \in [a, b]$; moreover, that P_i^2 is continuous and strictly monotonically decreasing in limit. Then, P_i is continuous in all arguments $\{x_i(m), z_i(m)\}$; its limit value at $x = \bigcap_k [xl_i, xu_i]$ is $P_i^2\{z\}$, $z = f(x)$, and hence, it is translation-invariant in x and monotonically decreasing in z.

Now, if we consider decomposable multiplicative partition operators constructed using these component function-types, then obviously Theorem

2.4.3 can be applied in a straightforward manner. In other words, the corresponding algorithm will generate an everywhere dense set of search points in $[a, b]$. For instance, simple partition operator instances can be of the form

$$P_i = \|xu_i - xl_i\|_p \cdot \Big\{-\frac{1}{2^n}\sum_{m=1}^{2^n} z_i(m)\Big\} \tag{2.4.39}$$

or

$$P_i = \|xu_i - xl_i\|_p \cdot \exp\Big\{-\frac{1}{2^n}\sum_{m=1}^{2^n} z_i(m)\Big\}. \tag{2.4.40}$$

Let us consider now decomposable additive partition operators constructed using these component function-types. To establish regularity, we need to establish the monotonicity property, as specified in Remark 2.4.1.

If $x = \bigcap_k [xl_i, xu_i]$, $i = i(k)$, $k \in N$ holds, then for each embedding subinterval D_i we shall require that

$$P_i\{D_i, f(D_i)\} > P\{x, z\}. \tag{2.4.41}$$

To verify the fulfilment of this condition, additional assumptions are necessary. First of all, we shall suppose that the objective function f is Lipschitz-continuous on $[a, b]$, and L is a proper (over)estimate of the 'true' Lipschitz constant. Moreover, let P_i^2 be also Lipschitzian with constant L_2 (for all i), and have the form

$$P_i^2\Big\{\frac{1}{2^n}\sum_{m=1}^{2^n} z_i(m)\Big\}. \tag{2.4.42}$$

Recalling the notation $z_i^\star = f(x_i^\star) = \min\{f(x) : x \in [xl_i, xu_i]\}$, $i = i(k)$, one can derive the following chain of relations, valid for each $x \in [xl_i, xu_i]$, $z = f(x)$:

$$P_i^2\{z_i^\star\} - P_i^2\Big\{\frac{1}{2^n}\sum_{m=1}^{2^n} z_i(m)\Big\} \le L_2\Big(\frac{1}{2^n}\sum_{m=1}^{2^n} z_i(m) - z_i^\star\Big) \tag{2.4.43}$$

$$= \frac{1}{2^n}L_2\Big\{\sum_{m=1}^{2^n}(z_i(m) - z_i^\star)\Big\} \le \frac{1}{2^n}LL_2\sum_{m=1}^{2^n}\|x_i(m) - x_i^\star\|$$

$$< \frac{1}{2^n}LL_2\sum_{m=1}^{2^n}\|xu_i - xl_i\| = LL_2\|xu_i - xl_i\|.$$

Consequently, if P_i^1 is defined simply by the l_2-norm of $D_i = [\boldsymbol{xl}_i, \boldsymbol{xu}_i]$, as

$$P_i^1\{D_i\} = K\|\boldsymbol{xu}_i - \boldsymbol{xl}_i\|, \quad \text{and} \quad K > LL_2, \qquad (2.4.44)$$

then, since

$$P\{\boldsymbol{x}, z\} = P_i^2\{z\} \leq P_i^2\{z_i^\star\} \quad \text{for } \boldsymbol{x} \in [\boldsymbol{xl}_i, \boldsymbol{xu}_i],$$

(2.4.43) and (2.4.44) imply the inequality (2.4.41). That is to say, the appropriately parameterized decomposable additive partition operator

$$P_i = K\|\boldsymbol{xu}_i - \boldsymbol{xl}_i\| + P_i^2\left\{\frac{1}{2^n}\sum_{m=1}^{2^n} z_i(m)\right\}, \qquad K > LL_2, \qquad (2.4.45)$$

is indeed regular. Note that, for example, the multivariate Piyavskii operator (2.4.13)

$$P_i = L\|\boldsymbol{xu}_i - \boldsymbol{xl}_i\| - \frac{1}{2^n}\sum_{m=1}^{2^n} z_i(m)$$

is evidently a special case of (2.4.45).

Applying now Theorem 2.4.4, one can immediately establish the global convergence of all RPAS-type strategies based on the operator form (2.4.45): note that convergence holds in the strong sense of $X^a = X^\star$.

The generality of these examples shows that one has a significant degree of freedom in defining decomposable partition strategies. Further examples, considering also numerical implementation aspects, will be discussed later in detail.

Chapter 2.5

Simplex Partition Strategies

2.5.1 Introduction

The case of simplex feasible regions is common in a number of important practical applications of global optimization: examples will be discussed later, in Part 4. Typically, the simplex plays a similar role to the interval feasible set in the previous chapters of Part 2; that is, it defines the explicit ('vague') bounds related to the globally optimal solution which is sought. According to this interpretation, one would expect again that the optimal solution is, typically, an interior point of the simplex feasible set.

The subject of this chapter is to propose a class of adaptive subdivision and search strategies to solve multiextremal optimization problems on simplicial feasible regions. This will be accomplished by specifying the general PAS to this problem. (The exposition draws primarily on Pintér (1991c), and on Chapter 2.2.)

Definition 2.5.1. Let $v_0, v_1, \ldots, v_n \in I\!R^n$ be given vectors. These vectors determine the n-dimensional simplex

$$\begin{aligned} S &= S(v_0, v_1, \ldots, v_n) \qquad (2.5.1) \\ &= \Big\{ x \in I\!R^n : x = \sum_{j=0}^{n} \alpha_j v_j,\ \alpha_j \geq 0 \quad j = 0, 1, \ldots, n,\ \sum_{j=0}^{n} \alpha_j = 1 \Big\}. \end{aligned}$$

We shall assume throughout that the simplex S is nondegenerate, that is, it has a positive volume in $I\!R^n$. This condition is equivalent to requiring that all n-collections of the vectors

$$C_i = \{v_0 - v_i, \ldots, v_{i-1} - v_i, v_{i+1} - v_i, \ldots, v_n - v_i\}, \qquad i = 0, 1, \ldots, n,$$

form a corresponding basis in $\mathbb{R}^n$.

The vertex-set $\{v_0, v_1, \ldots, v_n\}$ will also be denoted by $V(S)$. Specifically, if $v_0 = 0 \in \mathbb{R}^n$ is the origin, and $v_1 = e_1, \ldots, v_n = e_n$ are the coordinatewise unit vectors, then

$$\begin{aligned} S^0 &= S^0(0, e_1, \ldots, e_n) \\ &= \left\{x \in \mathbb{R}^n : x = (\alpha_1, \ldots, \alpha_n),\ \alpha_j \geq 0\ j = 1, \ldots, n,\ \sum_{j=1}^{n} \alpha_j \leq 1\right\} \end{aligned} \tag{2.5.2}$$

is the unit simplex in $\mathbb{R}^n$.

Consider now the global optimization problem

$$\min f(x) \qquad \text{s.t.} \quad x \in S, \tag{2.5.3}$$

assuming that f is Lipschitzian on the n-dimensional simplex S. As earlier, we shall assume that a valid overestimate of the (minimal) Lipschitz-constant is available.

To solve (2.5.3), the general partition algorithm scheme introduced in Chapter 2.2 will be specified as follows. (For illustration, see Fig. 2.5.1.)

Simplex Partition Algorithm Scheme for Multidimensional Global Optimization (SPAS)

Step 0. (Initialization) Set the iteration count $k := 0$; define the initial (trivial) simplex partition by $S_0 := S$, and the current set of partition subsets as $U_0 = \{S_0\}$. Evaluate f at selected (or all) vertices of S_0. Determine the incumbent solution $x_0^\star \in V(S_0)$, $z_0^\star = f(x_0^\star)$, by choosing the vertex at which the objective function value is minimal. Determine also a lower bound estimate $\beta_0 = l(S_0)$ for the minimum of f. (Note that lower bounds can be calculated on the basis of the available function evaluations at several vertices and the Lipschitz property of f; we shall return to the discussion of this point somewhat later.)

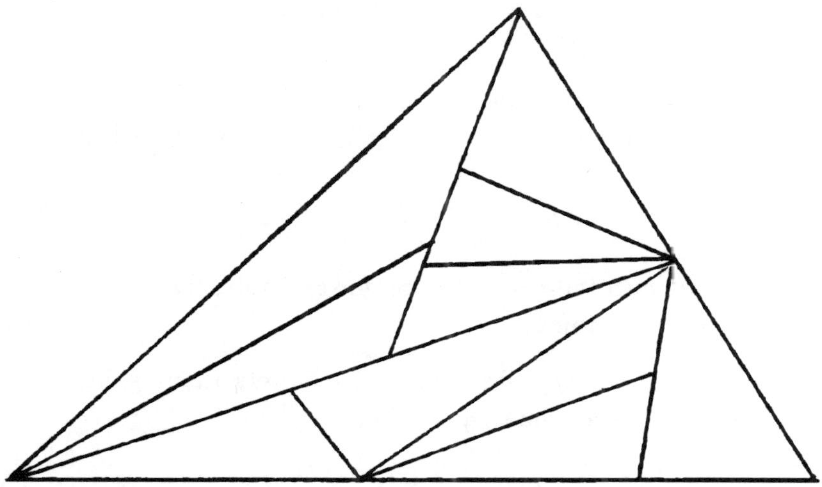

Figure 2.5.1
A simplex partition in $\mathbb{R}^2$

At each subsequent iteration cycle $k = 1, 2, 3, \ldots$, there exists a current list of adaptively generated, disjoint (with respect to their interior) robust subsimplices

$$U_k = \{S_1, \ldots, S_W\}, \qquad W = W(k), \tag{2.5.4}$$

that jointly cover the actual search region of interest. A current estimate of the global optimizer $x_k^\star \in V(S_1) \cup V(S_2) \ldots \cup V(S_W)$ and optimum $z_k^\star = f(x_k^\star)$ is also available. Moreover, on the basis of the Lipschitz property of f and the available function evaluations, lower bounds

$$l(S_r) \leq f(x) \text{ for all } x \in S_r, \qquad r = 1, \ldots, W, \quad W = W(k) \tag{2.5.5}$$

are determined for each subset.

Consider now the following steps at iteration cycle k, $k \in N$.

Step 1. (Suboptimality elimination) Delete all partition subsets which cannot contain a better point than the incumbent solution: if

$$l(S_r) > z_k^\star, \tag{2.5.6}$$

then obviously S_r can be excluded from further investigation. Denote by R_k the collection of the remaining subsets.

Step 2. (Selection for next partition) For all $S_r \in R_k$, consider the corresponding lower bound estimate $l(S_r)$. Select index t for which this bound is minimal:

$$l(S_t) = \min_{S_r \in R_k} l(S_r). \tag{2.5.7}$$

(If there exist several such indices t, choose, for example, the smallest one.) Then S_t will be partitioned next.

Step 3. (Partition refinement) Replace S_t by finitely many subsimplices that determine its nontrivial, robust partition

$$S_t = \bigcup_{j \in J_k} S_{t_j}. \tag{2.5.8}$$

Step 4. (Evaluation and bound refinement) The new partition is

$$U_{k+1} = \left(R_k \setminus S_t\right) \cup \left(\bigcup_{j \in J_k} S_{t_j}\right). \tag{2.5.9}$$

Evaluate now f at selected (or all) new vertices of the new simplices generated in Step 3. Determine the incumbent solution $x_k^\star$ and the optimum estimate $z_k^\star$. Derive (update) valid lower bounds on each subsimplex of the current partition

$$l(S_r) \leq f(x), \quad x \in S_r, \quad S_r \in U_{k+1}. \tag{2.5.10}$$

Update also the estimated lower bound of the optimum

$$\beta_k = \min_{S_r \in U_{k+1}} l(S_r). \tag{2.5.11}$$

Return to Step 1 (start iteration cycle $k+1$).

Remark 2.5.1. Note that the algorithm-scheme introduced here is more specific than the general PAS of Section 2.1.2. Since Lipschitzian bounding and deletion operations are directly introduced, exploiting the Lipschitz property of f, the partition operator value now simply equals the subset-specific lower bound multiplied by -1. Observe also that at Step 3 a numerical stopping criterion may be included. For instance, if, for a given $\varepsilon > 0$ and k, we have

$$z_k^\star \leq \beta_k + \varepsilon, \qquad (2.5.12)$$

then the incumbent solution provides an ε-optimal estimate of the global optimum.

In the remainder of this chapter we shall further specify the simplicial PAS introduced, and then establish its global convergence.

2.5.2 Convergence of Simplicial Algorithms

Let us analyse now the conditions which are sufficient to imply convergence of the SPAS to the set of global solutions $X^\star$. To this end, we shall recall the general regularity conditions (from Section 2.2.1), and—as an example—show a corresponding specification of the SPAS.

P1. (Perfection) Every strictly nested, infinite sequence of partition subsets $\{S_j,\ j = j(k),\ k \in N\}$ converges to a single limit point

$$x = \bigcap_j \{S_j\}, \qquad j = j(k), \qquad k \in N. \qquad (2.5.13)$$

Evidently, this property is simply satisfied, if, for example, at Step 3 only two simplices are generated, by halving bisection of the set S_t: $S_t = S_{t_1} \cup S_{t_2}$. As discussed earlier, this bisection can be defined, for example, by the midpoint xm_t of (one of) the longest edge(s) of S_t. Observe that, according to this scheme, xm_t will be the single new vertex of the current subsimplex set U_{k+1} in (2.5.9): this means that (minimally) only one new function evaluation is needed at each iteration.

Let us state also the three other regularity properties (see the PAS as defined in Section 2.2.1).

P2. (Continuity) For each strictly nested subsimplex-sequence, the corresponding sequence of partition operator values converges to the limit value at the accumulation point x. (In case of increasingly sharper lower bounds $l(S_r)$, the limit value equals $f(x)$.)

P3. (Monotonicity) For each $x \in \cap_j S_j$, the partition operator P satisfies the inequality

$$P\{S_j, f(S_j)\} > P\{x, f(x)\}, \qquad j = j(k), \qquad k \in N. \qquad (2.5.14)$$

P4. (Boundedness) The partition operator P is bounded from below and from above.

These properties will be established as an immediate consequence of a simplicial lower bounding procedure, specified below.

Let us consider first an arbitrary point x of the simplex S. Denoting by z_j the function values $f(v_j)$, $j = 0, 1, \ldots, n$, the Lipschitz property of f implies the inequalities

$$|f(x) - z_j| \leq L\|x - v_j\|, \qquad j = 0, 1, \ldots, n. \tag{2.5.15}$$

Taking now a globally optimal point $x^\star$ from S, (2.5.15) implies the lower bounds

$$z^\star = f(x^\star) \geq z_j - L\|x^\star - v_j\|, \qquad j = 0, 1, \ldots, n. \tag{2.5.16}$$

Summing up these inequalities, one arrives at the relation

$$z^\star \geq \frac{1}{n+1}\Big\{\sum_{j=0}^{n} z_j - L\sum_{j=0}^{n}\|x^\star - v_j\|\Big\}. \tag{2.5.17}$$

Naturally, as the location of $x^\star$ is typically unknown, one has to provide a further estimate for the second sum in (2.5.17). Observing now that S is a convex set, and

$$C(x) = \sum_{j=0}^{n}\|x - v_j\| \tag{2.5.18}$$

is a convex function, we can directly conclude that its maximum is attained at one of the vertices of S (see, for example, Horst and Tuy, 1990, Ch. 1.2, Theorem 1.1). Consequently, one can apply the bound

$$\sum_{j=0}^{n}\|x^\star - v_j\| \leq \max_{0\leq i\leq n}\sum_{j=0}^{n}\|v_i - v_j\|. \tag{2.5.19}$$

Note that (2.5.19) is the sharpest possible bound, as $x^\star$ theoretically may coincide with one of the vertices of S.

Let us take now an arbitrary subsimplex S_r in S, determined by its vertices $V(S_r) = \{v_{rj},\ j = 0, 1, \ldots, n\}$. Introducing $z_{rj} = f(x_{rj})$, on the basis of the above discussion one can conclude that, for example, the expression

$$l(S_r) = \frac{1}{n+1}\Big\{\sum_{j=0}^{n} z_{rj} - L \max_{v_{ri}\in V(S_r)} \sum_{j=0}^{n} \|v_{ri} - v_{rj}\|\Big\} \tag{2.5.20}$$

is a valid minorant of f on S_r. Consequently,

$$P(S_r) = -l(S_r) \tag{2.5.21}$$

can serve as the partition operator associated with the SPAS.

Indeed, continuity (P2) follows, as for each strictly nested subsimplex-sequence, the corresponding sequence of partition operator values (2.5.20) converges to $z = f(x)$, x being the accumulation point. Monotonicity (P3) directly follows by the argument which leads to (2.5.20). Finally, boundedness (P4) simply follows by the finiteness of f on $S = S_0$:

$$\infty > P\{S_r, f(S_r)\} > P\{x, f(x)\} = -f(x) > -\max_{y\in S} f(y). \tag{2.5.22}$$

Concluding the discussion, note finally that

$$P\{S_r, f(S_r)\} = -\frac{1}{n+1}\Big\{\sum_{j=0}^{n} z_{rj} - L \max_{v_{ri}\in V(S_r)} \sum_{j=0}^{n} \|v_{ri} - v_{rj}\|\Big\} \tag{2.5.23}$$

is obviously translation-invariant and strictly monotonically decreasing in limit. Therefore the SPAS specified with this partition operator is globally convergent in the sense of Theorem 2.2.4: that is, the set of accumulation points X^a generated by the simplicial partition strategy is identical to the set $X^\star$ of global optimizers.

Moreover, it can also be immediately seen that (2.5.23) is a decomposable additive partition operator, in complete analogy to the definition presented in Section 2.4.4. In order to allow for further generalizations, we shall extend Definitions 2.4.2 and 2.4.3, for the case of arbitrary partition subsets.

Definition 2.5.2. Consider a partition algorithm scheme governed by the operator P. P is a (general) decomposable additive partition operator, if—for an arbitrary partition subset D_i—it has the form

$$P_i\{D_i, f(D_i)\} = P_i^1\{D_i\} + P_i^2\{f(D_i)\}, \quad i = i(k), \quad k \in N. \tag{2.5.24}$$

That is, P_i can be written as the sum of two terms: the first one depends only on the sample points chosen from the partition subset, and the second depends only on the corresponding set of objective function values. Both P_i^1 and P_i^2 may depend on additional algorithm parameters.

Definition 2.5.3. P is a (general) decomposable multiplicative partition operator, if it can be written as

$$P_i\{D_i, f(D_i)\} = P_i^1\{D_i\} \times P_i^2\{f(D_i)\}, \quad i = i(k), \quad k \in N. \tag{2.5.25}$$

That is, P_i can be expressed as the product of two factors, which depend only on the sample points chosen and the corresponding objective function values, respectively (while both P_i^1 and P_i^2 may depend on additional algorithm parameters).

To illustrate the latter definition, let us consider the decomposable multiplicative simplex partition operator defined, with respect to the subsimplex S_r, as

$$P\{S_r, f(S_r)\} = \exp\Big\{-\frac{1}{n+1}\sum_{j=0}^{n}[z_{rj} - \underline{z}]\Big\} \times \Big\{\sum_{j=0}^{n} \|v_{r,j+1} - v_{rj}\|\Big\}. \tag{2.5.26}$$

In (2.5.26), $\underline{z}$ is a strict lower estimate of the optimum value (a procedure parameter), and $v_{r,n+1} := v_{r0}$.

It is elementary to verify that (2.5.26) satisfies the regularity requirements. Specifically, for all $x \in S$, the limit of (2.5.26) equals

$$P\{x, f(x)\} = 0. \tag{2.5.27}$$

Therefore, by Corollary 2.2.1, the SPAS based on this operator will theoretically generate an everywhere dense sequence of sample points on S; furthermore, by Corollary 2.2.2, the sample minimum and the corresponding minimizer sequence provide an estimate of the globally optimal solution. Evidently, the Lipschitzian assumption can be dropped in this case, as the operator (2.5.26) does not rely on L: in other words, (2.5.26) can serve to solve problem (2.5.1), when f is (merely) continuous.

Some possible applications and numerical examples related to simplex partition strategies will be presented in Part 4.

Chapter 2.6

Partition Methods on General Convex and Star Sets

2.6.1 Introduction

In this chapter we shall investigate the global optimization problem, specialized as

$$\min_{x \in D} f(x), \quad \text{where} \quad f \in F(D) = \bigcup_{L>0} F_L(D). \tag{2.6.1}$$

In (2.6.1), $D \subset \mathbb{R}^n$ is a robust, bounded and convex subset of the real n-space, and the Lipschitzian objective function $f : D \to \mathbb{R}^1$ is (possibly) multiextremal on the set D. Observe that this problem is significantly more general than the case of interval or simplex feasible regions studied in previous chapters of Part 2: consequently, it is not obvious how a PAS type strategy could be directly realized for solving (2.6.1).

Note that a frequently investigated special case of (2.6.1) is the concave minimization problem (CMP), in which f is a (quasi)concave function (recall Chapter 1.1). Following the seminal work of Tuy (1964), this problem has been investigated extensively, see, for example, Martos (1975), Hoffman (1981), Forgó (1988), or the survey of Horst (1984b). As is known, the CMP has a very special structure, namely, all optimal solutions (the elements of X^*) lie on the boundary $\delta(D)$ of the set D. This property allows the efficient use of general algorithmic concepts, such as, for example, outer approximation (discussed in Chapter 1.2), tailored to the problem at hand.

Below we shall investigate the LGOP (2.6.1) in its full generality. The solution approach is based on embedding D into an 'easily tractable' feasible

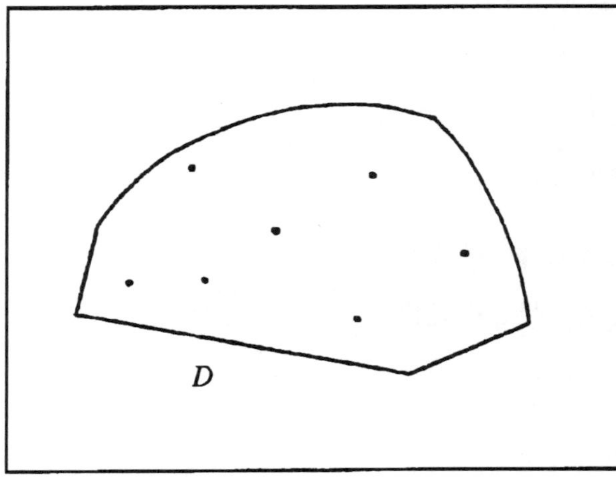

Figure 2.6.1
Global optimization problem on a general convex feasible set in $\mathbb{R}^2$: The feasible set D, an embedding interval $[a, b]$, and several local minima of the objective function (dots) are shown

set such as a rectangle $[a, b]$ (or a simplex S), and then extending (redefining) f on $[a, b]$, as suggested in Pintér (1986c). Note that the present approach is essentially based on the Lipschitz assumption, and it cannot be directly transferred to handle general CGOPs. This point will be treated in detail later; in Part 3 other stochastic approaches to solve CGOPs will also be discussed.

2.6.2 Lipschitzian Extension of the Objective Function

Let us take an arbitrary finite interval which includes D (see Fig. 2.6.1):

$$D \subseteq [a, b], \qquad [a, b] \subset \mathbb{R}^n. \tag{2.6.2}$$

In connection with (2.6.2), note that—besides the constraints $x \in D$—explicit componentwise bounds $a_i \leq x_i \leq b_i$, $i = 1, \ldots, n$, are frequently known. In the opposite case, one can determine, for example, the tightest embedding interval by solving $2n$ separate convex programming problems which will determine

$$a_i = \min_{x \in D} x_i \quad \text{and} \quad b_i = \max_{x \in D} x_i, \quad \text{for } i = 1, \ldots, n. \tag{2.6.3}$$

(For appropriate solution approaches to (2.6.3), see, for example, Stoer and Witzgall, 1970.) Observe that in cases when several difficult (nonlinear, or even just algorithmically given convex) constraints are also present in the definition of D, they can be simply dropped when solving the problems (2.6.3): this leads to 'suboptimal' but still appropriate embeddings.

In the following, we shall define a suitable Lipschitzian extension of the function $f \in F(D)$ on $[a, b]$: then the results of Chapter 2.4 can be applied in a straightforward manner. In other words, denoting the extension of f by $f_e = f_{e,[a,b]}$, problem (2.6.1) will be replaced by

$$\min_{a \le x \le b} f_e(x). \tag{2.6.4}$$

A proper extension f_e should exactly preserve the set of globally optimal solutions X^*.

We shall define f_e as follows: first choose an arbitrary fixed interior point x^0 of the (robust) set D. Note that, frequently, such point(s) x^0 are known a priori. In the opposite case, for finding interior points of robust convex sets given as $D = \{x \in I\!R^n : g_i(x) \le 0,\ i = 1, \ldots, I\}$, one can apply, for example, the following auxiliary problem statement: find the pair $\{x^0, \varepsilon^0\}$, $\varepsilon^0 > 0$, which solves

$$\max \varepsilon \quad \text{s.t.} \quad g_i(x) + \varepsilon \le 0, \qquad i = 1, \ldots, I.$$

For details on solving this problem, consult again Stoer and Witzgall (1970).

Define now

$$f_e(x) = f(x_h) + L\|x - x_h\| \quad \text{for } x \in [a, b], \tag{2.6.5}$$

where $L = L(D)$ is a valid (over)estimate of the Lipschitz-constant of f on D, and

$$\begin{aligned} x_h &= x_h(x) = \lambda_h x + (1 - \lambda_h) x^0, \\ \lambda_h &= \lambda_h(x) = \max\{0 \le \lambda \le 1 : \lambda x + (1 - \lambda) x^0 \in D\}. \end{aligned} \tag{2.6.6}$$

In other words, $x_h = x$, if x belongs to the feasible set D; in the opposite case, x_h is defined by the intersection of the interval $x^0 + \lambda(x - x^0)$, $0 \leq \lambda \leq 1$, with $\delta(D)$, the boundary of D.

Remark 2.6.1. By the convexity of D, x_h is unique; therefore (2.6.5) is unambiguously defined. (Note that λ_h can be found, for example, by a simple bisection procedure on the interval $x^0 + \lambda(x - x^0)$, $0 \leq \lambda \leq 1$.) Obviously, x_h—therefore, also, the extended function $f_e(x)$—depends on the prefixed point x^0, although, for simplicity of notation, this dependence will not be explicitly referred to below.

Remark 2.6.2. By virtue of (2.6.5)–(2.6.6), for all feasible points we have

$$f_e(x) = f(x), \quad x \in D; \tag{2.6.7}$$

moreover, for an arbitrary infeasible point x, we have

$$f_e(x) > f_e(x_h), \quad \text{where } x_h = x_h(x), \quad \text{and } x \in [a, b] \setminus D. \tag{2.6.8}$$

Therefore, the global optimization problems (2.6.1) and (2.6.4) are equivalent in terms of their optimal solution set X^* and optimum z^*. Note also that if x^0 is fixed and $D \subseteq D_1 \subseteq D_2$, D_1 and D_2 being n-dimensional intervals, then the extension f_e on D_1 is independent of D_2. For the sake of brevity, we shall apply the notation $D_0 = [a, b]$.

Lemma 2.6.1. *There exists a (minimal) constant $C > 0$ such that the relation*

$$\|x_h(x^1) - x_h(x^2)\| \leq C\|x^1 - x^2\| \tag{2.6.9}$$

is valid for all pairs $x^1, x^2 \in D_0$.

Proof. First, we shall verify that the projection operator $x_h(x)$ is continuous on D_0, that is, for arbitrary sequences of points $\{x^{1,k}\} \subset D$ and $\{x^{2,k}\} \subset D$, $k \in N$, $\lim_k \|x^{1,k} - x^{2,k}\| = 0$ implies

$$\lim_k \|x_h(x^{1,k}) - x_h(x^{2,k})\| = 0. \tag{2.6.10}$$

Define

$$x = \lim_k x^{1,k} = \lim_k x^{2,k} \qquad \text{and} \qquad \overline{x}_h = x_h(x), \quad x \in D_0. \tag{2.6.11}$$

If x belongs to the interior of D, then, from a certain iteration index k_0, the following relations are valid:

$$x_h^{1,k} := x_h(x^{1,k}) = x^{1,k} \quad \text{and} \quad x_h^{2,k} := x_h(x^{2,k}) = x^{2,k};$$

hence, obviously,

$$x_h^1 := \lim_k x_h^{1,k} = x \quad \text{and} \quad x_h^2 := \lim_k x_h^{2,k} = x.$$

Consider now the case where $x \in D_0 \setminus \operatorname{int} D$. Assuming indirectly the discontinuity of the function x_h, let x_h^1 and x_h^2 be two different accumulation points of the sequences $\{x_h^{1,k}\}$ and $\{x_h^{2,k}\}$, respectively. Since $\{x_h^{1,k}\}$ and $\{x_h^{2,k}\}$ are sequences of boundary points, therefore both x_h^1 and x_h^2 are also boundary points of D. Taking now x and $\overline{x}_h$ by (2.6.11), we have (at least) either $\overline{x}_h \neq x_h^1$ or $\overline{x}_h \neq x_h^2$. In other words, there exist (at least) two boundary points which—by the projection x_h—would correspond to x. But this clearly contradicts the postulated convexity of D, as all points between $x_0 \in \operatorname{int}(D)$ and any boundary point of D belong to $\operatorname{int}(D)$, that is, x_h^1 and x_h^2 cannot be simultaneously respective boundary points.

The contradiction obtained shows that the operator $x_h = x_h(x)$ is indeed continuous on D_0. Now, on the basis of this continuity and the compactness of D_0—applying standard analytical argument—one can conclude that $x_h(x)$ is also uniformly continuous on D_0.

Define now the function

$$g(x^1, x^2) = \frac{\|x_h^1 - x_h^2\|}{\|x^1 - x^2\|} \quad \text{for } x^1, x^2 \in D_0, \quad x^1 \neq x^2,$$

where $x_h^j = x_h(x^j)$, $j = 1, 2$. As x_h is uniformly continuous on D_0 and the norm function is also uniformly continuous, thus g is also continuous on the set

$$\{(x^1, x^2) \in D_0 \times D_0,\ x^1 \neq x^2\}.$$

Moreover, the function g can be continuously extended to $D_0 \times D_0$, defining, for all $x \in D_0$,

$$g(x, x) := \limsup_{\substack{x_k^{1,k} \to x \\ x^{2,k} \to x}} g(x^{1,k}, x^{2,k}), \quad x^{1,k}, x^{2,k} \in D_0 \times D_0, \quad x^{1,k} \neq x^{2,k}.$$

Now, by the compactness of $D_0 \times D_0$, applying the existence theorem of Weierstrass, g attains its finite maximum on $D_0 \times D_0$. Consequently, there exists a (minimal) constant $C > 0$, for which the inequality

$$g(\boldsymbol{x}_1, \boldsymbol{x}_2) \leq C$$

holds for all $\boldsymbol{x}_1, \boldsymbol{x}_2 \in D_0$. The lemma is proved. ■

Remark 2.6.3. The minimal value of C for which (2.6.9) is valid obviously depends on the choice of $\boldsymbol{x}^0$ and the geometry (curvature) of $\delta(D)$. This fact, however, will not be indicated by using additional notation.

Lemma 2.6.2. *Assume that $f \in F_L(D)$, with a properly (over)estimated Lipschitz-constant L. Let the extension f_e of f on D_0 be defined by (2.6.5)–(2.6.6).*

Then

$$f_e \in F_{L_1}(D_0), \quad \text{where } L_1 = L(2C+1) \tag{2.6.12}$$

(C is defined by (2.6.9)). That is, the extended function f_e is also Lipschitz-continuous on the n-dimensional interval D_0.

Proof. Choose two arbitrary points $\boldsymbol{x}^1$ and $\boldsymbol{x}^2$ from D_0. By (2.6.5), one can write

$$f_e(\boldsymbol{x}^1) - f_e(\boldsymbol{x}^2) = f(\boldsymbol{x}_h^1) + L\|\boldsymbol{x}^1 - \boldsymbol{x}_h^1\| - f(\boldsymbol{x}_h^2) - L\|\boldsymbol{x}^2 - \boldsymbol{x}_h^2\|,$$

whence

$$|f_e(\boldsymbol{x}^1) - f_e(\boldsymbol{x}^2)| \leq |f(\boldsymbol{x}_h^1) - f(\boldsymbol{x}_h^2)| + L|\|\boldsymbol{x}^1 - \boldsymbol{x}_h^1\| - \|\boldsymbol{x}^2 - \boldsymbol{x}_h^2\||.$$

Now, since $f \in F_L(D_0)$, applying also the result of Lemma 2.6.1, we have

$$|f(\boldsymbol{x}_h^1) - f(\boldsymbol{x}_h^2)| \leq L\|\boldsymbol{x}_h^1 - \boldsymbol{x}_h^2\| \leq LC\|\boldsymbol{x}^1 - \boldsymbol{x}^2\|$$

and

$$\begin{aligned} |\|\boldsymbol{x}^1 - \boldsymbol{x}_h^1\| - \|\boldsymbol{x}^2 - \boldsymbol{x}_h^2\|| &\leq \|\boldsymbol{x}^1 - \boldsymbol{x}_h^1 - \boldsymbol{x}^2 + \boldsymbol{x}_h^2\| \\ &\leq \|\boldsymbol{x}^1 - \boldsymbol{x}^2\| + \|\boldsymbol{x}_h^1 - \boldsymbol{x}_h^2\| \leq (1+C)\|\boldsymbol{x}^1 - \boldsymbol{x}^2\|. \end{aligned}$$

Consequently,

$$|f_e(\boldsymbol{x}^1) - f_e(\boldsymbol{x}^2)| \leq L(C + C + 1)\|\boldsymbol{x}^1 - \boldsymbol{x}^2\|.$$

The assertion (2.6.12) is proved. ■

The general convergence results presented in Chapter 2.2, and made more specific in Chapter 2.4 (recall statement (b) in Theorem 2.4.1), now lead directly to

Theorem 2.6.1. *Assume that the global optimization problem* (2.6.1) *is transformed into the form* (2.6.4), *in which the Lipschitzian extension of the objective function f on D_0 is defined by* (2.6.5)–(2.6.6). *Now, if* (2.6.4) *is solved by an arbitrary PAS-type regular partition algorithm on the interval* $[a, b]$, *and the partition operator P satisfies the requirements*

$$P\{x^\star, f(x^\star)\} = c^\star > P\{x, f(x)\} \quad \textit{for } x^\star \in X^\star,\ x \in D \setminus X^\star$$

($c^\star$ is fixed), then $X^a = X^\star$. In other words, the set of accumulation points of the search point sequence generated is identical to the set of globally optimal solutions of problem (2.6.1). ∎

Remark 2.6.4. For the case of a (merely) continuous f in (2.6.1), statement (a) of Theorem 2.4.1 could be applied. This would imply the approximation of the globally optimal solution by generating (theoretically) an everywhere dense sequence of points on $[a, b]$; retaining the feasible ($x \in D$) points from this sequence leads to a method of approximating the globally optimal solution. As mentioned previously, we shall return to the analysis of CGOPs in Part 3, and will discuss other solution approaches to this case.

Remark 2.6.5. In the remaining three sections of this chapter numerical aspects of realizing the present approach will be analysed. During these investigations it will be tacitly assumed that

- an interior point of D is available;
- an embedding interval $D_0 = [a, b]$ is known; and
- a rectangular partition method can be efficiently applied on D_0.

(Some pertinent numerical issues are postponed to Part 3.) Hence, the principal problem here lies in the extension of f to D_0, because—throughout the iteration steps—the algorithm has to generate the values of f_e.

The case of linearly constrained sets is considered first, this being followed by the case of general (possibly nonlinear) convex constraints and that of star-shaped (special nonconvex) regions.

2.6.3 Linearly Constrained Feasible Sets

Suppose now that problem (2.6.1) has the following specific form:

$$\min_{x \in D} f(x), \quad \text{in which} \tag{2.6.13}$$

$$D \subset D_0 = [a, b] \subset I\!R^n,$$
$$D = \{x : c_i x \le d_i,\ i = 1, \ldots, I\},$$
$$(a, b, c_i \in R^n,\ \ d_i \in I\!R^1\ \ i = 1, \ldots, I \text{ are given, and } c_i x := c_i^T x);$$

that is, D is a (nonempty) compact polyhedral set. We shall also postulate the robustness of D.

Select x^0, a fixed interior point of D. Taking now an arbitrary point $x \in D_0$, we shall define the index sets

$$I_1 = I_1(x) = \{i : c_i x > d_i\}, \quad I_2 = I_2(x) = \{i : c_i x \le d_i\}.$$

Obviously, the relation $I_1(x) \neq \emptyset$ is equivalent to $x \in D_0 \setminus D$. For such x the intersection point $x_{h,i}$, $i \in I_1$, of the hyperplane $c_i y = d_i$ $(y \in I\!R^n)$ and the interval $x^0 + \lambda(x - x^0)$, $0 \le \lambda \le 1$, is given by

$$x_{h,i} = x^0 + \lambda_i(x - x^0), \text{ where } \lambda_i = \frac{d_i - c_i x^0}{c_i(x - x^0)}, \qquad i \in I_1.$$

Thus, the point $x_h = x_h(x)$ is determined by

$$x_h = x^0 + \lambda_i^*(x)(x - x^0), \text{ where } \lambda_i^*(x) = \min_{i \in I_1} \frac{d_i - c_i x^0}{c_i(x - x^0)}. \tag{2.6.14}$$

Summing up the aforesaid, the extension of f on the set D_0 can be defined by the following formula:

$$f_e(x) = \begin{cases} f(x_h) + L[1 - \lambda_i^*(x)]\|x - x^0\|, & \text{if } I_1(x) \neq \emptyset; \\ f(x), & \text{if } I_1(x) = \emptyset. \end{cases} \tag{2.6.15}$$

Formulas (2.6.14) and (2.6.15) indicate that the determination of the values $f_e(x)$ for $x \in D_0$ does not need much computational effort, unless the number of linear constraints is large. Therefore all regular PAS-type globally convergent algorithms can be immediately applied to solve (2.6.13).

Remark 2.6.6. Obviously, the size ('tightness') of D_0 will play a role in the numerical performance of the algorithmic realizations of the present approach, as 'tighter' embeddings are expected to lead to fewer function evaluations. Recall in this context Remark 2.6.1, which makes possible the construction of the 'tightest' embedding, by solving $2n$ auxiliary convex programming problems. This note applies also to the general convex case, discussed below.

2.6.4 Nonlinearly Constrained Convex Sets

Suppose now that in

$$\begin{aligned} &\min_{x \in D} f(x) \\ &D \subset D_0 = [a, b] \subset \mathbb{R}^n, \\ &D = \{x \in DL : f_j(x) \le 0 \;\; j = 1, \ldots, J\}, \\ &(a, b \in \mathbb{R}^n \text{ are given}). \end{aligned} \tag{2.6.16}$$

In (2.6.16), DL denotes the (possibly absent additional) polyhedral constraint set given by a finite number of inequalities (recall (2.6.13)). We shall assume that the functions f_j are nonlinear, convex, and continuously differentiable; moreover, that an interior point x^0 of D is known.

Remark 2.6.7. There exist well-known optimization algorithms in which linear approximations of nonlinear constraints are applied. As an example, the classical supporting hyperplane method of Veinott (1967) can be mentioned: this method generates an adaptively refined sequence of polyhedral outer approximations to the set D. Hence, solving (approximately) the resulting linearly constrained problems (using, for example, the approach of the previous section), a convergent hierarchical procedure can be constructed. Obviously, the basic numerical difficulty lies in the solution of the approximating problems to properly increasing accuracy, as the selection of suitable stopping rules remains a nontrivial issue. A simplifying 'shortcut' is, of course, to accept a fixed linear approximation of D; in this case, however, one has to accept an approximate solution of uncertain quality. Note that, applying simultaneous inner and outer approximations of D, one can obtain corresponding upper and lower bounds of the optimum. We will not study these issues here in any further detail, as they lead out of the main subject of this chapter—and, in fact, of the chosen scope of the present book. (One can recall the related brief discussion of relaxation strategies in Chapter 1.2.)

We shall consider now direct applications of the Lipschitzian extension method, without applying linearization. A natural way of defining $f_e(x)$,

$x \in D_0$, can be given analogously to (2.6.14)–(2.6.15). Let us consider first the nonlinear constraints and define the function

$$G(\lambda, x) = \max_{1 \leq j \leq J} f_j(x^0 + \lambda(x - x^0)).$$

Determine now $\lambda_{h,1} = \lambda_{h,1}(x)$ as the unique solution of the (univariate) equation $G(\lambda, x) = 0$. In other words, this defines the maximal value of λ such that the point $x^0 + \lambda_{h,1}(x - x^0)$ satisfies all nonlinear constraints f_j, (at least) one of them with equality:

$$\lambda_{h,1} = \min_j \lambda_j \text{ in which } \lambda_j \text{ solves } f_j(x^0 + \lambda_j(x - x^0)) = 0, \quad j = 1, \ldots, J.$$

Evidently, for convex functions f_j, simple bisection can be applied again to find λ_j, $j = 1, \ldots, J$.

Considering now also the linear constraints, one can determine

$$\lambda_{h,2} = \lambda_{h,2}(x) = \min_{i \in I_1} \frac{d_i - c_i x^0}{c_i(x - x^0)},$$

that is, $x^0 + \lambda_{h,2}(x - x^0)$ meets all linear constraints, and at least one of them is met exactly. Then one can choose $\lambda_h = \min(\lambda_{h,1}, \lambda_{h,2})$ and write

$$f_e(x) = f(x_h) + L\|x - x_h\| \qquad \text{for } x \in [a, b], \tag{2.6.17}$$

where

$$x_h = x_h(x) = \lambda_h x + (1 - \lambda_h)x^0.$$

Note in passing that in this way again $x_h = x$, if x belongs to the feasible set D; in the opposite case, x_h is defined by the (unique) intersection of the interval $x^0 + \lambda(x - x^0)$, $0 \leq \lambda \leq 1$, with the boundary $\delta(D)$.

Remark 2.6.8. Without going into detail, note that this approach is, in principle, applicable also in the case where the functions f_j are not necessarily convex (consequently, D may also be nonconvex). In this case, $f_e(x)$ can be defined as the lower envelope of $f(x)$ composed with the linear norm terms $\|x - x_h\|$, corresponding to those line segments of $[x^0, x]$ which belong to $D_0 \setminus D$. The numerical tractability of such procedures will, however, depend significantly on the structural complexity of the feasible set D.

Remark 2.6.9. Let us mention briefly still another approach, which is based on the use of penalty functions. In this framework, (2.6.16) can be substituted by the following linearly constrained problem:

$$\min f_p(x, q),$$

where

$$f_p(x, q) := f(x) + \sum_{j=1}^{J} q_j \max[0, f_j(x)], \tag{2.6.18}$$

and

$$x \in DL \subseteq D_0 = [a, b].$$

In (2.6.18) all nonlinear constraints are taken into account via the modified objective function $f_p(x, q)$; $q = (q_1, \ldots, q_J)$ denotes the vector of nonnegative penalty coefficients. For details of the penalty approach, one can refer, for example, to classical works of Zangwill (1967), Fiacco and McCormick (1968), or to the survey of Fletcher (1983). The penalty function method will be discussed in more detail later (in Chapter 3.3), as it can be applied under more general conditions.

In the next section an extension of the present approach to certain nonconvex regions is summarized. Chapter 2.7 will offer a much more general methodology for global optimization on arbitrary nonconvex Lipschitzian regions.

2.6.5 General Star-Shaped Sets

Let us take an n-dimensional sphere $S^n(x^0, r)$ with centre x^0 and radius r; now consider all radii r, emanating from x^0 to the boundary $\delta(S^n)$.

Definition 2.6.1. A star transformation of S^n is defined by a continuous function $x_h = x_h(r)$, which maps all points of the radius

$$\{x^0 + \alpha r : 0 \le \alpha \le 1\}$$

into a corresponding line segment

$$\{x^0 + \alpha x_h(r) : 0 \le \alpha \le 1\} \quad \text{of the ray } \{x^0 + \alpha r : 0 \le \alpha\}.$$

In this way, x_h maps $\delta(S^n)$ into the boundary $\delta(D)$ of the image set D.

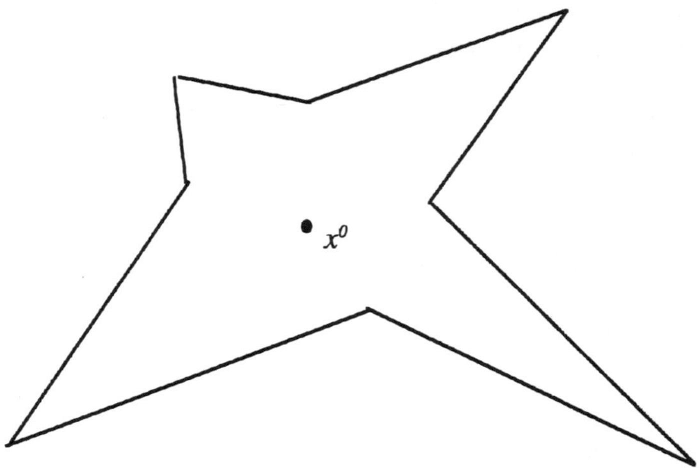

Figure 2.6.2
A nonconvex star set in $\mathbb{R}^2$

Definition 2.6.2. A star set D is defined by a corresponding star transformation of S^n. The point x^0 is the star point of D.

Remark 2.6.10. Note that, inter alia, all convex sets satisfy the above definition, that is, they are star sets. Additionally, a general family of nonconvex sets can be defined as star regions, see Fig. 2.6.2.

Remark 2.6.11. Evidently, by their construction, all star regions D have the following property: if x^0 is the star point, then all rays emanating from x^0 have a single intersection point with the set boundary $\delta(D)$. Therefore, considering now the global optimization problem (2.6.1) on the star set D, the extension of the objective function from such a set to an arbitrary embedding interval $D_0 = [a, b]$ can be accomplished in full analogy to the approach in Section 2.6.1. Namely, Lemmas 2.6.1 and 2.6.2 and Theorem 2.6.1 can be applied to this case as well.

Chapter 2.7

Partition Strategies in General Lipschitz Optimization

2.7.1 Introduction

In the preceding chapter an approach was proposed to solve multiextremal optimization problems on convex (robust and compact) sets, with a direct extension applicable to star domains. To realize this method, a simply implementable extension of the objective function to an interval embedding of the feasible set was suggested. Evidently, the numerical tractability of this approach depends crucially on the uniqueness of the boundary point $x_h = x_h(x)$ on the segment $x^0 + \lambda(x - x^0)$, ($x^0 \in \text{int}(D)$ given, $x \in D_0 \setminus D$ arbitrary point, $0 \leq \lambda \leq 1$).

In the case of more general mathematical programming problems, including several (possibly) nonconvex constraints, the straightforward application of the approach of Chapter 2.6 may encounter substantial difficulties. Note that these complications are only partly due to the fact that nonconvex constraints may lead to an a priori unknown number of boundary points $x_h(x)$ on $x^0 + \lambda(x - x^0)$, for any given x of $D_0 \setminus D$. For this reason, a different algorithmic approach will be presented below, which is aimed at solving global optimization problems on 'complicated'—nonconvex, possibly even disconnected—feasible sets. This approach is essentially based on the general ideas presented in Chapter 2.2. The generic partition algorithm scheme (PAS) is, however, extended in order to make possible the elimination of certain partition subsets, due to their suboptimality and/or infeasibility. (Recall in this context—as a fairly simple special case—the simplex partition strategy discussed in Chapter 2.5.) The elimination operations will be based on the branch-and-bound (B&B) concept, frequently applied to handle difficult (for example, integer or more specific global) optimization problems. General principles of B&B strategies, in the context of integer programming, are discussed, for example, by Lawler and Wood (1969), Mitten (1970), Glover (1978), Daellenbach, George and Nickle (1983, Ch. 18); Hillier and Lieberman (1986, Ch. 13); or Winston (1994, Ch. 9); recall also the brief description of B&B in Chapter 1.2. Horst (1986, 1988), Horst and Tuy (1987, 1990, Ch. IV) provide a detailed discussion of convergent B&B algorithms in global optimization, with numerous references. One should also mention here the obvious connection between B&B and interval arithmetic based global optimization procedures: on the latter subject, see, for example, Hansen (1979, 1980, 1992), Moore (1979, 1988), Ratschek (1985), Fujii, Ichida and Ozasa (1986), Ratschek and Rokne (1987, 1988), Moore and Ratschek (1988), Moore, Hansen and Leclerc (1992), Csendes and Pintér (1993a). The material of the present chapter originates from Pintér (1988); the exposition, however, draws also on Chapters 2.2 and 2.4, extending the partition algorithm frameworks discussed there.

2.7.2 A Branch-and-Bound Algorithm Scheme for Lipschitzian Optimization

Consider the global optimization problem in the general form

$$\min_{x \in D} f(x), \quad \text{where } D = \{f_j(x) \leq 0,\ j = 1, \ldots, J\} \subset \mathbb{R}^n. \tag{2.7.1}$$

With respect to the structure of (2.7.1), we shall make the following assumptions:

- the feasible region D is embedded into a finite (known) interval region $D_0 = [a, b]$;
- the functions f_j, $j = 0, 1, \ldots, J$, $f_0 := f$, are all defined on (or are appropriately extended to) the set D_0;
- f_j, $j = 0, 1, \ldots, J$, are Lipschitz-continuous, and belong to the corresponding function-classes $f_j \in F_{L_j}(D_0)$, with known (over)estimates of their (minimal) Lipschitz-constant L_j;

- if $D \neq \emptyset$, then it is the union of a finite number of nonempty robust sets $D^{(i)}$, $i = 1, \ldots, I$;
- the set $X^\star$ of globally optimal solutions is, at most, countable.

Remark 2.7.1. Since D is bounded and closed, problem (2.7.1) has a (not necessarily unique) globally optimal solution—unless the feasible set D is empty. Because of its practical relevance, we shall not exclude this case a priori (one may think of, for example, the problem of finding the not necessarily existing solution of a system of nonlinear equations). We shall require, however, that if $D \neq \emptyset$, then $D = \bigcup_{i=1}^{I} D^{(i)}$, and all sets $D^{(i)}$ are closures of nonempty open sets. Furthermore, the countability of $X^\star$ includes the practically most important cases, in which $X^\star$ is finite, or it consists of just one point $x^\star$. Evidently, it is also assumed that the mentioned extension of the objective and constraints to D_0 must not lead to additional globally optimal points.

Remark 2.7.2. As discussed earlier (in Chapter 2.1), in the absence of the postulated Lipschitzian structure, deterministic estimates of the optimum value $z^\star = f(x^\star)$ cannot—as a rule—be provided on the basis of generating any finite sequence of search points on D. One can immediately add to this observation that, in case of infeasible search points, the quantification of infeasibilities also necessitates the use of Lipschitzian (or stronger analytical) information.

Remark 2.7.3. The finiteness of the embedding interval region $[a, b]$ is not essential in order to apply B&B methodology. In many cases, however, lower and upper bounds on the decision variables are readily available, and the inclusion of this assumption makes possible the direct extension of the approach presented in the preceding chapters, see in particular Chapter 2.4.

In spite of the structural restrictions listed, (2.7.1) describes a fairly general class of global optimization problems: for illustration, see Fig. 2.7.1. The structural requirements listed above exclude, however, strict equality constrained problems (their exclusion is motivated primarily by numerical reasons).

The branch-and-bound approach is (theoretically) applicable to many specific subclasses of the problem (2.7.1). Obviously—in the presence of additional specific structural information—more 'customized' algorithms may exist than the 'general purpose' B&B procedure described below.

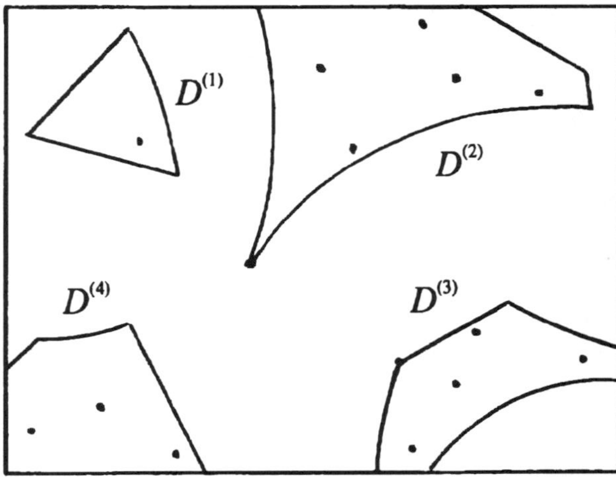

Figure 2.7.1
Global optimization under general Lipschitz constraints in $\mathbb{R}^2$: The feasible region $D = D^{(1)} \cup D^{(2)} \cup D^{(3)} \cup D^{(4)} \subset [a,b] \subset \mathbb{R}^2$, and several local minima of the objective function (dots) are shown

Partition Algorithm Scheme for Solving Lipschitzian Optimization Problems

Step 0. (Initialization) Set $k := 0$ (iteration count), and define the following initial settings:

- $FD_0 := D$ (the active, that is, non-fathomed subset of the feasible set);
- $D_0 := [a,b] = [xl_0, xu_0]$, $FD_0 \subset D_0$ (embedding of the feasible set);
- $I_0 := \{0\}$ (index set of active partition subsets which jointly cover FD_0);
- $\underline{D}_k := \{D_0\}$ (collection of active partition subsets covering FD_0).

On the basis of a finite set of sample points taken from D_0, determine also

- S_0 (the feasible subset of search points);
- the current estimate of the optimal solution and optimum, $x_0^\star$ and α_0:

$$\alpha_0 = f(x_0^\star) = \min_{x \in S_0} f(x) \quad (\alpha_0 := +\infty, \text{ and } x_0^\star \text{ is undefined, if } S_0 = \emptyset); \tag{2.7.2}$$

- the current Lipschitzian lower bound β_0 of f on D: for instance, one can define

$$\beta_0 = \beta_0^0 := \max_{x \in S_0} f(x) - L_0\|xu_0 - xl_0\|; \quad (\beta_0 := -\infty, \ \text{if } S_0 = \emptyset); \tag{2.7.3}$$

- lower bounds of the constraint functions on D_0

$$\beta_0^j \leq f_j(x) \ \text{for all } x \in D_0, \qquad j = 1, \ldots, J; \tag{2.7.4}$$

  these can be determined analogously to (2.7.3) on the basis of the sample S_0.

In general, at the beginning of each iteration cycle k, $k = 1, 2, 3, \ldots$, the following information is available:

- FD_{k-1} (the active subset of the feasible set D);
- I_{k-1} (index set of active partition subsets which cover FD_{k-1});
- $\underline{D}_{k-1} = \{D_{k-1,i}, \ i \in I_{k-1}\}$ (a current finite rectangular partition which embeds FD_{k-1}):

$$\begin{aligned} &D_{k-1,i} = [xl_i, xu_i], \\ &\text{int}(D_{k-1,i_1}) \cap \text{int}(D_{k-1,i_2}) = \emptyset; \ \text{for all } i_1, i_2 \in I_{k-1}, \\ &FD_{k-1} \subseteq \bigcup_{i \in I_{k-1}} D_{k-1,i} \subseteq D_0. \end{aligned} \tag{2.7.5}$$

  Note that in (2.7.5) index i will typically depend on k, that is, $i = i(k-1)$, but explicit reference to the iteration index k will be omitted when this does not lead to ambiguity;
- $S_{k-1} \subset D$ (the—possibly empty—set of feasible sample points generated so far);
- a current estimate of the optimal solution $x_{k-1}^\star \in S_{k-1}$, and optimum $\alpha_{k-1} = f(x_{k-1}^\star)$ (again, $\alpha_{k-1} := +\infty$, and $x_{k-1}^\star$ is undefined, if $S_{k-1} = \emptyset$);
- for each active partition subset $i \in I_{k-1}$, corresponding subinterval-specific lower bounds of the objective and constraint functions

$$\begin{aligned} &\beta_{k-1,i}^j \leq f_j(x) \ \text{for all } x \in D_{k-1,i}, \qquad j = 0, 1, \ldots, J, \\ &\beta_{k-1,i} := \beta_{k-1,i}^0, \ \text{and} \end{aligned}$$

- a lower bound of the optimum value $z^\star$,

$$\beta_{k-1} := \min_{i \in I_{k-1}} \beta_{k-1,i}.$$

Again, these lower bounds can be determined, for instance, analogously to (2.7.3).

In each iteration cycle k, the steps below are executed:

Step 1. (Infeasibility elimination) Delete all currently active subsets $D_{k-1,i} \in \underline{D}_{k-1}$, which cannot contain feasible points. This deletion is accomplished on the basis of the Lipschitzian lower bounding operation: if, for some $j = 1, \ldots, J$, $\beta^j_{k-1,i} > 0$, then obviously $D_{k-1,i}$ can be excluded from further search.

Step 2. (Feasibility check) If all sets of the actual partition $\underline{D}_{k-1}$ are eliminated, then problem (2.7.1) has no feasible solution ($D = \emptyset$): the partition algorithm terminates.

Step 3. (Suboptimality elimination) Delete all subsets $D_{k-1,i} \in \underline{D}_{k-1}$ which cannot contain globally optimal point(s): if

$$\beta_{k-1,i} > \alpha_{k-1} = f(x^\star_{k-1}),$$

then $D_{k-1,i}$ can be fathomed. Update the index set I_{k-1}.

Step 4. (Selection for partition) Let $\underline{R}_{k-1} \subseteq \underline{D}_{k-1}$ denote the set of remaining active partition subintervals. Choose a subset $D_{k-1,t} \in \underline{R}_{k-1}$ for further partition. In accordance with the general PAS of Chapter 2.2—but specifying that to the present case—one can select $D_{k-1,t} \in \underline{R}_{k-1}$, if the current lower bound estimate is minimal on that subset:

$$\beta_{k-1,t} = \min_{D_{k-1,i} \in \underline{R}_{k-1}} \beta_{k-1,i}, \qquad i \in I_{k-1}. \tag{2.7.6}$$

If there exist more than a singular index which satisfies the relation (2.7.6), then t can be arbitrarily selected among them.

Step 5. (Partition refinement) The selected subinterval $D_{k-1,t}$ is partitioned into a finite number of interval subsets:

$$D_{k-1,t} = \bigcup_{l \in E_k} D_{k-1,t_l} \qquad (E_k \text{ is the corresponding finite set of indices}). \tag{2.7.7}$$

Hence, introducing

$$I_k = (I_{k-1} \cup E_k) \setminus \{t\} \quad \text{and} \quad \underline{D}_{k-1,t} = \{D_{k-1,t_l},\ l \in E_k\},$$

the new partition becomes

$$\underline{D}_k = \{D_{k,i},\ i \in I_k\} := \{(\underline{R}_{k-1} \cup \underline{D}_{k-1,t}) \setminus D_{k-1,t}\}. \tag{2.7.8}$$

Step 6. (Bound refinement) Take additional sample points from the new partition subsets. If the set of feasible points is increased by the above partition, then determine (update) S_k. Refine, if possible, the current estimate of the optimum. In addition, the lower bound $\beta_{k-1,t}$ can be replaced by subset specific bounds β_{k-1,t_l}, $l \in E_k$, and some other lower bounds may also be refined on the basis of the added sample points. Consequently, all Lipschitzian bounds are either left unchanged, or become 'tighter', as a result of the partition refinement:

$$\begin{aligned}
&\alpha_k = f(x_k^\star) \le \alpha_{k-1} \\
&\qquad\qquad (x_k^\star := x_{k-1}^\star, \text{ if no better incumbent solution was found});\\
&\beta_{k,i}^j \ge \beta_{k-1,i}^j \ \text{ for all } i \in I_k, \qquad j = 0, 1, \ldots, J;\\
&\beta_k := \min_{i \in I_k} \beta_{k,i} \ge \beta_{k-1}, \qquad\qquad (\beta_{k,i} = \beta_{k,i}^0).
\end{aligned}$$

Step 7. (ε-optimality check) By the construction described, $\beta_k \le \alpha_k$, for all $k \in N$. Now, if $\alpha_k - \beta_k < \varepsilon$ ($\varepsilon > 0$ is a given, sufficiently small tolerance parameter), then the algorithm can be terminated: an ε-approximate solution is provided by

$$x_k^\star = \arg\min_{x \in S_k} f(x).$$

In the opposite case—setting $k := k + 1$—the search procedure continues at Step 1 of the next iteration cycle.

Remark 2.7.4. In general, estimates of the form

$$g(x) \ge \underline{g} := \max_{x \in S} g(x) - L\|xu - xl\|$$

in which S is the set of feasible sample points, and $g \in F_L(D)$ is an arbitrary Lipschitz function, do not provide a very sophisticated Lipschitzian bound of $g(x)$ on $[xl, xu]$. Note, however, that the discrepancy $g(x) - \underline{g}$ tends to zero, if $\|xu - xl\| \to 0$. Moreover, (2.7.3) is easy to compute and it already guarantees theoretical convergence. For simplifying the convergence analysis, we shall assume that the subinterval-specific Lipschitz bounds are based on the defining vertices xl_i and xu_i of $D_{k-1,i} \in \underline{D}_{k-1}$. Since, for each $x \in [xl, xu]$, $g(x) \geq g(xl) - L\|x - xl\|$ and $g(x) \geq g(xu) - L\|xu - x\|$, the lower bounds in (2.7.3)–(2.7.4) will be simply derived as

$$\beta^j_{k,i} := \max\{f_j(xl_i), f_j(xu_i)\} - L_j\|xu_i - xl_i\| \leq f_j(x), \\ \text{for all } x \in D_{k-1,i}, \; j = 0, 1, \ldots, J, \quad k \in N. \tag{2.7.9}$$

The subject of providing tighter bounds will be discussed in Chapter 3.2.

Remark 2.7.5. In the B&B scheme presented above, the feasibility of search points in all generated partition sets is not required. The main reasons for this are that the feasible set D itself may or may not be empty, and it may be difficult to find feasible points in a partition subset, even if $D_{k-1,i} \cap D \neq \emptyset$, as this set may have a fairly 'complicated' topology. (Recall Fig. 2.7.1, which shows a nonconvex set D, and imagine the rectangular partition of the embedding interval D_0.) Therefore the B&B scheme described will not seek directly for feasible points: this search is assisted indirectly, by subset elimination (in Steps 1 and 3) and the sequential refinement of the 'most promising' subintervals.

Remark 2.7.6. The presented 'prototype' B&B procedure can be implemented in many variations, by providing a more detailed description of (and/or modifying) its elimination, partition, and bounding operations. The above description is based on Pintér (1988), but other implementations are, obviously, possible (see, for example, Horst and Tuy (1990, Chs. IV and XI.2, and related references therein). In particular, the subset selection in Step 4 could be based on an arbitrary regular partition operator which exploits the Lipschitz information. In this case, (2.7.6) is replaced by the general—partition operator based—subset selection rule: choose $D_{k-1,t}$ such that

$$P(D_{k-1,t}) = \max_{D_{k-1,i} \in \underline{R}_{k-1}} P(D_{k-1,i}). \tag{2.7.10}$$

Furthermore, the partition refinement (Step 5) can be done, for example, by simple halving bisection or more sophisticated methods; for details, recall the corresponding discussions from Chapters 2.2 and 2.4. It is worth emphasizing, however, that the above detailed specifications of the B&B scheme (already) guarantee the global convergence properties required.

Remark 2.7.7. In Step 7, ε-optimality serves as a numerical criterion for terminating the search. Evidently, other usual stopping criteria could also be included. For instance, in the context of one-dimensional search, Strongin (1978) applies the following criterion: if the selected subinterval is sufficiently small, then the search is terminated. In the present case, this can be expressed, for example, as follows: terminate search if $\|xu_t - xl_t\| < \delta$ ($\delta > 0$ fixed search accuracy parameter). Further stopping criteria will be discussed in Part 3.

Note finally that, in the following proof of theoretical convergence, numerical termination criteria will be omitted (setting $\delta = \varepsilon = 0$).

2.7.3 Convergence

Theorem 2.7.1. *Suppose that the following conditions hold:*

i. The global optimization problem (2.7.1) has the properties listed in Section 2.7.1; namely,

- *the feasible region $D = \emptyset$, or D is the union of finitely many robust sets, $D \subset D_0 = [a, b]$;*
- *the functions f_j, $j = 0, 1, \ldots, J$, ($f_0 := f$) are Lipschitz-continuous: $f_j \in F_{L_j}(D_0)$; and all Lipschitz-constants L_j are known, or properly (over)estimated.*

ii. The B&B scheme is realized according to the description in Section 2.7.2 (so that it exploits the Lipschitzian problem structure).

Then

1. *If the feasible set D is empty, then this fact is concluded after a finite number of iteration steps.*
2. *If D is nonempty, and the algorithm terminates at a certain step k, then $x_k^\star \in X^\star$, that is, the optimum estimate found belongs to the set of globally optimal solutions.*

3. *If D is nonempty, and the algorithm generates an infinite number of search points, then the (nonempty) set X^a of the accumulation points of $\{S_k\}$, $k = 1, 2, 3, \ldots$, is identical to set $X^\star$. Specifically, if $X^\star = \{x^\star\}$, then $\{x_k^\star\} \to x^\star$.*

Proof

1. Assertion (1) will be proven indirectly, by showing the validity of statement (3). (We shall prove that if the B&B algorithm scheme generates an infinite sequence of points, then the feasible set D cannot be empty.)

2. If the feasible set D is nonempty, then $X^\star \neq \emptyset$ also holds. Hence, the infeasibility elimination Step 1 cannot exclude all subintervals generated in $D_0 = [a, b]$. The remaining stopping criterion of the B&B scheme in Step 7 becomes activated (with $\varepsilon = 0$) only in the case where, for a certain iteration index k, we have $\alpha_k = \beta_k$. In other words, a feasible point $x_k^\star$ has been found which satisfies the chain of relations

$$\alpha_k = f(x_k^\star) = \beta_k = \min_{D_{k,i} \in \underline{D}_k} \beta_{k,i}.$$

Considering the method of constructing upper and lower bounds (recall (2.7.2), (2.7.3) and (2.7.9)), this situation can only occur if at iteration k there exists a vertex of a generated subinterval which happens to be a globally optimal solution of problem (2.7.1).

3a. We shall prove now the inclusion $X^a \subset D$. If the B&B algorithm generates an infinite sequence of points, then the set of accumulation points, X^a, is certainly nonempty. We shall prove first that all elements of X^a are feasible.

Assuming the contrary, take an $x^a \in X^a$ such that there exists at least an index $j = j(x^a)$ from $1 \leq j \leq J$, for which we have

$$f_j(x^a) = \varphi = \varphi(x^a) > 0. \tag{2.7.11}$$

Consider now the sequence of subintervals $[xl_t, xu_t]$, $t = t(k)$, $k \in N$, which contain x^a in cycle k. By the partition refinement rule, we know that the sequence $\{[xl_t, xu_t]\}$ was selected infinitely many times for subpartition, and

$$xl_t \underset{k}{\rightarrow} x^a, \qquad ||xu_t - xl_t|| \underset{k}{\rightarrow} 0. \tag{2.7.12}$$

Take now the lower bounds (2.7.9):

$$\beta^j_{k,i} := \max\{f_j(xl_i), f_j(xu_i)\} - L_j||xu_i - xl_i|| \leq f_j(x)$$
$$\text{for all } x \in D_{k-1,i}, \quad j = 0, 1, \ldots, J, \quad k \in N.$$

By (2.7.11) and (2.7.12), one can conclude now that from some index $t_0 = t_0(\varphi, x^a)$, $\beta^j_{k,i} \geq \varphi/2$, $j = j(x^a)$, $t \geq t_0$. But then, by the infeasibility elimination Step 1, the interval $D_{k-1,t_0} = [xl_{t_0}, xu_{t_0}]$ was to be excluded from further search; consequently, $x^a \in D_{k-1,t_0}$ cannot belong to X^a. The contradiction obtained shows that, indeed, $X^a \subset D$.

Note that this argument also leads to the conclusion $D \neq \emptyset$, proving indirectly assertion 1) of the theorem.

3b. Finally, we prove the identity $X^a = X^\star$. We shall show that the accumulation points are exactly the globally optimal solutions of (2.7.1). (As $D \neq \emptyset$, the set $X^\star$ is also nonempty.) Again, by the subinterval selection rule and the contraction property of the refinement operation, we can make use of the relations (2.7.12). Considering now the specific form (2.7.9) of the subset selector function in (2.7.6), for an arbitrarily chosen $x^a \in X^a$, we have the limit relation

$$\lim_{k\to\infty} \beta_{k-1,t} = z^a := f(x^a), \qquad t = t(k), \qquad k \in N. \tag{2.7.13}$$

Let us take now those subintervals $[xl_i, xu_i]$, $i = i(k)$, $k \in N$, which contain an arbitrary fixed global optimizer point $x^\star$. As f is Lipschitzian with constant L_0, applying (2.7.9) we have the estimate

$$z^\star \geq \max\{f(xl_i), f(xu_i)\} - L_0||xu_i - xl_i|| = \beta_{k-1,i}. \tag{2.7.14}$$

Considering now (2.7.6), (2.7.13) and (2.7.14), one can write

$$z^a = \lim_{k\to\infty} \beta_{k-1,t} \leq \lim_{i\to\infty} \beta_{k-1,i} \leq z^\star. \tag{2.7.15}$$

Since obviously $z^a \geq z^\star$, by (2.7.15) one can directly conclude that $z^a = z^\star$, that is, x^a is an optimal solution. This shows that $X^a \subset X^\star$. Moreover, $\lim_{k\to\infty} \beta_{k-1,t} < z^\star$ will hold, whenever $||xu_i - xl_i||$ does not converge to zero. This leads again, by (2.7.15), to contradiction. The contradiction obtained implies that all elements $x^\star$ of the set $X^\star$ are also limit points (that is, $X^\star \subset X^a$). Consequently, in case (3) of Theorem 2.7.1, we have indeed the relation $X^a = X^\star$. The theorem is proved. ∎

Remark 2.7.8. Checking through the steps in the proof of Theorem 2.7.1, it is not difficult to verify that the special partition selection rule (2.7.6) and the halving bisection mentioned can naturally be replaced by corresponding operations of an arbitrary regular partition strategy which satisfies the general assumptions of Theorem 2.4.1 or Theorem 2.4.4. This observation immediately leads to the following general convergence results. (For details, recall the discussion in Chapter 2.4.)

Corollary 2.7.1. *Suppose that, besides the structural requirements (i) of Theorem 2.7.1, the B&B scheme is governed by a regular partition operator P which satisfies the conditions (b) of Theorem 2.4.1:*

$$P\{x^\star, z^\star\} = c^\star > P\{x, z\} \text{ for all } x^\star \in X^\star \text{ and } x \in D \setminus X^\star$$

($c^\star$ is a fixed value). Then the assertions of Theorem 2.7.1 will be valid for the corresponding realization (instance) of the B&B algorithm scheme. ∎

Corollary 2.7.2. *Suppose that, besides the structural requirements (i) of Theorem 2.7.1, the B&B scheme is governed by a regular decomposable additive partition operator P which satisfies the conditions of Theorem 2.4.4:*

$$P\{D_i, f(D_i)\} = P_i^1\{x_i(m)\} + P_i^2\{z_i(m)\}, \qquad m = 1, \ldots, M_i;$$

$$x = \bigcap_k [xl_i, xu_i], \; i = i(k), \; k \in N \text{ implies } P_i^1\{x_i(m)\} \underset{k}{\to} 0;$$

$$x = \bigcap_k [xl_i, xu_i], \; i = i(k), \; k \in N \text{ implies } P_i^2\{z_i(m)\} \underset{k}{\to} P_i^2(z), \; z = f(x);$$

$P_i^2(z)$ is strictly monotonically decreasing in z.

Then the assertions of Theorem 2.7.1 will be valid for the corresponding realization (instance) of the B&B algorithm scheme. ∎

Remark 2.7.9. As referred to earlier, it is very simple to extend the other type of convergence results of Chapter 2.4, and derive conditions which assure the generation of everywhere dense sets of search points on D_0 (such that $X^a = D_0$ holds). Although this approach could be applied also to solve constrained optimization problems under (mere) continuity assumptions, it will not be discussed. Other—computationally more promising—strategies for solving the general class of continuous global optimization problems will be analysed in Chapters 3.3 and 3.6–3.7.

Remark 2.7.10. Due to their generic structural features, the B&B methods presented can be naturally adapted to solve discrete global optimization problems. (As mentioned in Chapter 1.2, they were originally developed to solve exactly such problems.) A prototype rectangular partition algorithm (operating on the embedded regular lattice) is discussed by Benson, Erenguc and Horst (1990). They illustrate the concept by proposing an algorithm instance for minimizing a concave function on the set of integers contained by a given polytope. Without going into detail, note that following the general framework of that paper, and drawing on the ideas of the present chapter, the global optimization of continuous or Lipschitz functions on discrete sets can also be established.

Remark 2.7.11. Closing the discussion of general Lipschitz optimization problems, we refer to the approach of Thach and Tuy (1987, 1990). They establish a transformation between Lipschitz and differential convex problems, then propose branch-and-bound and relaxation strategies to solve the d.c. problem derived. The approach is theoretically elegant, and may lead to good implementations (to the author's knowledge, at present only small-scale numerical experiments serve to illustrate the method).

Following the theoretical analysis of adaptive partition strategies, in Part 3 important aspects of their implementation—including modifications and extensions—will be presented.

PART THREE

Implementation Aspects, Algorithm Modifications, and Stochastic Extensions

The most challenging problems always require more resources than can be provided by the fastest available computer...

(*Report of the Panel on Future Directions in Computational Mathematics, Algorithms, and Scientific Software.* SIAM, Philadelphia, 1985)

If mowing the lawn with the mowing machine has become too easy, you can always try doing it with the vacuum cleaner...

(Graham Hough, London Review of Books, October 17, 1985)

Chapter 3.1

Diagonally Extended Univariate Algorithms for Multidimensional Global Optimization

3.1.1 Introduction

In Part 2, a general framework was provided for analysing and constructing adaptive partition algorithms. Specific instances of the generic partition algorithm scheme (PAS, as introduced in Section 2.1.2) can be applied to solve a broad class of global optimization problems with continuous or Lipschitz-continuous structure. The case of one- and multidimensional intervals and simplices, and more general convex, star and Lipschitzian feasible sets was discussed in detail.

Obviously, the general PAS and the related theoretical convergence results allow for a broad range of additional specifications, with respect to the essential PAS (sampling, aggregation, selection and refinement) steps. In this chapter 'economic' sampling strategies, applicable on multidimensional box regions, will be studied. (As noted previously, interval feasible sets are important in many applications, since often it is easy to provide a 'safe range' of the possible decision vectors—with or without additional constraints.) Such strategies were first introduced in Pintér (1986b); the discussion below draws mainly on the (more general) exposition presented in Part 2. Specifically, applying results of the convergence analysis in Chapters 2.2–2.4, we will be able to draw straightforward theoretical conclusions, needed to establish effective rectangular partition strategies.

3.1.2 Diagonally Extended Univariate Algorithms

Let us consider again the CGO or LGO problem on n-dimensional interval regions:

$$\min f(x) \qquad (3.1.1)$$
$$a \leq x \leq b.$$

In (3.1.1), f is continuous or Lipschitzian on the finite n-interval $[a, b]$ (again, the inequalities $a \leq x \leq b$ are interpreted componentwise). As previously, X^a will denote the set of accumulation points generated by a PAS-type search method, and X^* will be the set of global solutions to (3.1.1).

In Chapter 2.4 a general rectangular PAS was introduced to solve global optimization problems on multidimensional intervals. Recalling this algorithm (Section 2.4.1), it is obvious that sampling (function evaluation) at all vertices of a partition subinterval could easily become prohibitively expensive. To see this, notice that the amount of information—vertices and function values—collected after k iterations would be of the order of $k2^n(n+1)$ real numbers, for $[a, b] \subset I\!R^n$. With this simple observation in mind, we shall introduce a straightforward extension of one-dimensional optimization strategies which makes possible much more economic sampling. Specifically, we shall define partition methods which, in essence, rely only on the 'lower left' and 'upper right' vertices xl, xu, and corresponding function values zl, zu, for each subinterval $[xl, xu]$ generated by the PAS-type method. We shall recall the general notation $P_i\{D_i, f(D_i)\}$ for the partition operator value, related to an arbitrary interval subset $D_i = [xl_i, xu_i]$ in $[a, b]$. To keep the notation simpler, we shall not use subindex i with respect to P and $[xl, xu]$, while it is understood that D_i can be an arbitrary subinterval, generated during the iterations of a PAS, such that $i = i(k)$, $k \in N$. The regularity of all partition operators discussed below will be assumed throughout, in accordance with the requirements postulated in Section 2.2.1.

Definition 3.1.1. P is a diagonal partition operator, if it has the special structure

$$P\{D_i, f(D_i)\} := P\{[xl, xu], f([xl, xu])\} = P\{xl, xu, zl, zu\}. \qquad (3.1.2)$$

That is, the value of P depends only on the vertices determining the subinterval $[xl, xu]$, and on $zl = f(xl)$, $zu = f(xu)$.

Remark 3.1.1. P in (3.1.2) will typically be dependent also on several problem and algorithm-specific parameters (such as, for example, the Lipschitz-constant or an initial lower estimate of the optimum, etc.), but this fact will only be reflected by the notation when necessary. This comment refers also to the two subsequent definitions.

Definition 3.1.2. P is a decomposable multiplicative diagonal partition operator, if it has the special form

$$P\{[xl, xu], f([xl, xu])\} = P1\{xl, xu\} \times P2\{zl, zu\}. \tag{3.1.3}$$

That is, P can be expressed as the product of two factors: the first one depends only on the vertices determining the subinterval $[xl, xu]$, and the second factor depends only on $zl = f(xl)$ and $zu = f(xu)$.

Definition 3.1.3. P is a decomposable additive diagonal partition operator, if it has the special form

$$P\{[xl, xu], f([xl, xu])\} = P1\{xl, xu\} + P2\{zl, zu\}. \tag{3.1.4}$$

That is, P can be expressed as the sum of two terms: the first one depends only on the vertices determining the subinterval $[xl, xu]$, and the second term depends only on $zl = f(xl)$ and $zu = f(xu)$.

Remark 3.1.2. Recalling the examples in Chapter 2.3.1, it is clear by simple examination that—for instance—the one-dimensional partition operators of Piyavskii and Kushner meet, respectively, the requirements of Definitions 3.1.2 and 3.1.3. Observe also that, evidently, the other cited operators of Strongin, Mockus–Tiesis–Žilinskas, Žilinskas, and Boender are univariate diagonal partition operators (cf. Remark 3.1.1, and the fact that the additional parameters of the operators listed could be set externally, so that they would not depend on the sequence of search points generated).

Let us observe now that the general convergence qualification results of Chapter 2.2, related to regular partition operators (as summarized by Theorem 2.4.1), can be applied. Consequently—in complete analogy with Theorems 2.4.3 and 2.4.4—one can immediately deduce the following assertions.

Theorem 3.1.1. *Assume that the decomposable multiplicative diagonal partition operator $P\{[xl, xu], f([xl, xu])\} = P1\{xl, xu\} \times P2\{zl, zu\}$ has the following property: for arbitrary feasible x, the limit relations*

$$\lim_k xl = \lim_k xu = x, \quad xl \le x \le xu, \quad xl = xl(k), \quad xu = xu(k), \quad k \in N,$$

imply

$$\lim_k P1\{xl, xu\} = 0, \quad \text{or} \quad \lim_k P2\{zl, zu\} = 0. \tag{3.1.5}$$

Then—applying the corresponding RPAS-type algorithm to solve (3.1.1) for a continuous objective function—an everywhere dense set of points will be generated on $[a, b]$; additionally, all accumulation points of the minimizer point sequence will belong to $X^\star$. ■

Theorem 3.1.2. *Assume that the decomposable additive diagonal partition operator $P\{[xl, xu], f([xl, xu])\} = P1\{xl, xu\} + P2\{zl, zu\}$ has the following property: for arbitrary feasible x, the limit relations*

$$\lim_k xl = \lim_k xu = x, \quad xl \le x \le xu, \quad xl = xl(k), \quad xu = xu(k), \quad k \in N$$

imply

$$\lim_k P1\{xl, xu\} = 0, \quad \text{and} \quad \lim_k P2\{zl, zu\} = P2(z), \tag{3.1.6}$$

in which $z = f(x)$, and $P2$ is a strictly monotonically decreasing function of z. Then—applying the corresponding RPAS-type algorithm to solve (3.1.1) for a Lipschitz-continuous objective function—the set of accumulation points X^a will be identical to $X^\star$, the set of globally optimal solutions. ■

We can also immediately formulate analogous results related to (general) diagonal partition operators.

Theorem 3.1.3. *Let us assume that the diagonal partition operator $P\{[xl, xu], f([xl, xu])\} = P\{xl, xu, zl, zu\}$ has the following property: the limit relations*

$$\lim_k xl = \lim_k xu = x, \quad xl \le x \le xu, \quad xl = xl(k), \quad xu = xu(k), \quad k \in N$$

imply

$$\lim P\{xl, xu, zl, zu\} = c \tag{3.1.7}$$

for arbitrary feasible x, and some constant value c. Then—applying the corresponding RPAS-type algorithm to solve (3.1.1) for a continuous objective function f—an everywhere dense set of points will be generated on $[a, b]$;

additionally, all accumulation points of the minimizer point sequence will belong to $X^\star$. ■

Theorem 3.1.4. *Let us assume that the diagonal partition operator $P\{[xl, xu], f([xl, xu])\} = P\{xl, xu, zl, zu\}$ has the following property: for arbitrary feasible x, the limit relations*

$$\lim_k xl = \lim_k xu = x, \quad xl \le x \le xu, \quad xl = xl(k), \quad xu = xu(k), \quad k \in N$$

lead to

$$P\{x, x, z, z\} = \lim_k P\{xl, xu, zl, zu\}, \qquad z = f(x); \tag{3.1.8}$$

this limit value is translation-invariant with respect to x, and strictly monotonically decreasing with respect to z. Then—applying the corresponding RPAS-type algorithm to solve (3.1.1) for a Lipschitz-continuous objective function f—the set of accumulation points X^a will be identical to $X^\star$.

■

Remark 3.1.3. One can observe the dimension-free character of the above theoretical convergence statements, so that they readily apply for arbitrary n-dimensional global optimization problems on interval feasible sets. The computational advantages of applying n-dimensional ($n \ge 1$) diagonal partition operators—at least in terms of their relative simplicity—can be significant, since they replace the possibly $O(k2^n(n+1))$ storage requirements of general RPAS operators by $O(k2(n+1))$. In other words, the information to be stored when applying diagonal operators is (merely) a linear function of both the number of PAS-type iterations k, and of the problem dimensionality n. Note immediately that, of course, even this storage demand can become quite significant for problems of higher dimensionality which typically necessitate a large number of PAS iterations, in order to arrive at a reasonably accurate approximation of the global solution. Nevertheless, using diagonal partition operators, a component of the immediate exponential complexity growth—typical in many areas of global optimization—can be avoided.

3.1.3 Examples

To illustrate the concepts introduced, several examples of multidimensional diagonally extended partition operators will be presented below. It can be instructive to compare these to the partition operators discussed earlier: see, in particular, Sections 2.3.1 (univariate operators) and 2.4.4 (decomposable operators).

Diagonally Extended Kushner Partition Operator

We shall define

$$P\{xl, xu, zl, zu\} = \exp\left\{\frac{-2(f_k^\star - \varepsilon - zu)(f_k^\star - \varepsilon - zl)}{\sigma^2 \|xu - xl\|}\right\}. \tag{3.1.9}$$

In (3.1.9), $\varepsilon > 0$, $\sigma > 0$ are fixed (function model dependent) parameters, $f_k^\star$ is the current sample minimum, and $\|\cdot\|$ denotes the Euclidean norm in the n-dimensional real space $\mathbb{R}^n$. It is simple to verify that (3.1.9) is a regular, decomposable multiplicative diagonal partition operator, if $f_k^\star - \varepsilon < z^\star$, $z^\star$ being the global optimum value of f on $[a, b]$. Note that the latter condition is sufficient to assure the monotonicity property P3 of the operator (3.1.9), since $P\{xl, xu, zl, zu\} > 0$ for all nondegenerate intervals, and

$$P\{x, x, z, z\} = \lim_k P\{xl, xu, zl, zu\} = 0. \tag{3.1.10}$$

Since $P\{xl, xu, zl, zu\}$ is a decomposable multiplicative operator, Theorem 3.1.1 implies that the diagonally extended Kushner partition operator will generate an everywhere dense sample point set in $[a, b]$; and that the sample minimum sequence provides an estimate of the global optimum of a (merely) continuous objective function f on $[a, b]$.

Diagonally Extended Piyavskii Partition Operator

Define now

$$P\{xl, xu, zl, zu\} = L\|xu - xl\| - \frac{1}{2}(zl + zu). \tag{3.1.11}$$

In (3.1.11), $L > 0$ is a strict overestimate of the Lipschitz constant of f on $[a, b]$. Note that any such L will guarantee the fulfilment of property P3 (monotonicity) needed for the regularity of operator (3.1.11). To see this, observe that for $\widehat{z} = f(\widehat{x}) := \min f(x)$, $xl \leq x \leq xu$, one can write

$$zl - \widehat{z} \leq L\|xl - \widehat{x}\| \quad \text{and} \quad zu - \widehat{z} \leq L\|xu - \widehat{x}\|. \tag{3.1.12}$$

Considering now that $P\{x,x,z,z\} = -z$, and exploiting the two inequalities in (3.1.12), one can derive the chain of relations

$$
\begin{aligned}
-z = P\{x,x,z,z\} \leq -\widehat{z} = P\{\widehat{x},\widehat{x},\widehat{z},\widehat{z}\} \leq \frac{1}{2}\{-zl + L\|xl - \widehat{x}\|\} \\
+ \frac{1}{2}\{-zu + L\|xu - \widehat{x}\|\} < -\frac{1}{2}(zl+zu) + L\|xu - xl\|,
\end{aligned}
\tag{3.1.13}
$$

which shows the monotonicity of the operator (3.1.11). The other regularity properties—continuity and boundedness—are evidently met, and perfection is guaranteed by using an appropriate refinement rule (such as,for example, halving bisection).

Now, evidently, (3.1.11) is a decomposable additive diagonal partition operator which meets the conditions of Theorem 3.1.2. Consequently, applying the diagonally extended n-dimensional Piyavskii partition operator in solving the LGOP (3.1.1), it will generate a sample point set in $[a,b]$, with the convergence property $X^a = X^\star$.

Diagonally Extended Strongin Partition Operator

As a third example, define

$$
P\{xl,xu,zl,zu\} = K\|xu - xl\| + \frac{(zu-zl)^2}{K\|xu-xl\|} - 2(zl+zu). \tag{3.1.14}
$$

In (3.1.14), $K > 4L$ is required (L being the Lipschitz constant of f on $[a,b]$). Note that any such K will guarantee the fulfilment of the monotonicity property $P3$, as can be seen by the following chain of relations (cf. (3.1.12)–(3.1.13)):

$$
\begin{aligned}
-4z = {} & P\{x,x,z,z\} \leq -4\widehat{z} \\
& = P\{\widehat{x},\widehat{x},\widehat{z},\widehat{z}\} \leq 2\{-zl + L\|xl - \widehat{x}\|\} + 2\{-zu + L\|xu - \widehat{x}\|\} \\
& \quad < -2(zl+zu) + 4L\|xu - xl\| < K\|xu - xl\| \\
& \quad + \frac{(zu-zl)^2}{K\|xu-xl\|} - 2(zl+zu).
\end{aligned}
\tag{3.1.15}
$$

(Again, all other regularity properties are also met.) Observe now that (3.1.14) defines a diagonal partition operator $P\{xl, xu, zl, zu\}$ which meets the conditions of Theorem 3.1.4, since

$$P\{x, x, z, z\} = \lim_k P\{xl, xu, zl, zu\} = -4z, \qquad z = f(x),$$

is evidently translation-invariant with respect to x, and strictly monotonically decreasing with respect to z. Consequently, applying the extended Strongin partition operator in solving the LGOP (3.1.1), the set of accumulation points X^a of the search point sequence generated will be identical to $X^\star$.

Concluding this discussion, note that—inter alia—similar multidimensional extensions of the univariate partition operators of Mockus–Tiesis–Žilinskas, Žilinskas, and Boender can also be provided, in a fairly straightforward fashion (keeping in mind the appropriate parameterization of these extensions). In the implementations discussed later, the extended Piyavskii and Strongin operators have been built in (with added options). Illustrative numerical results applying the extended Strongin operator in solving one-, two- and four-dimensional global optimization test problems were reported in Pintér (1986b); numerous examples of applications will be discussed in Part 4.

Chapter 3.2

Estimation of Lipschitzian Problem Characteristics in Global Optimization

3.2.1 Introduction

We shall consider again the Lipschitz global optimization problem on a finite n-interval $[a, b]$:

$$\min f(x)$$
$$a \leq x \leq b, \qquad a, x, b \in I\!R^n, \tag{3.2.1}$$

under the analytical conditions stated in Chapter 2.4; in particular, f is assumed to be Lipschitz-continuous with some constant L. As previously, the—not necessarily unique—optimal solution of this LGOP will be denoted by $x^\star \in X^\star$, and $f^\star = f(x^\star)$. Additionally, the set of accumulation points generated by a PAS-type globally convergent adaptive partition algorithm will be denoted again by X^a.

A predominant numerical issue in applying Lipschitz optimization to solve (3.2.1) is the fact that the minimal suitable Lipschitz-constant $L = L(f, [a, b])$ is—as a rule—unknown, and in great many cases there is no easy way to find it. Theoretically correct overestimates of L can, of course, be provided in most practical cases, but very crude estimates may lead to unacceptable numerical performance: recall, for example, the L-dependent efficiency estimates from Chapter 2.4. In practice—as, for instance, in Strongin's (1973, 1978) algorithm (see formula (2.3.13))—L is frequently estimated from the sample points and function values. Hansen, Jaumard and

Lu (1992c) point out the potential 'dangers' of such approaches: they construct an example of a univariate LGOP in which the global optimum will necessarily be missed, when using Strongin's original algorithm. Note in this context that in the Hansen–Jaumard–Lu counterexample local (non-global) convergence occurs, since L is unknown and a 'false' procedure parameter m is determined on the basis of a collection of sample points generated; with proper parameterization ($m > 2L$), the algorithm is globally convergent, as was discussed in Chapter 2.3.

Knowledge of a valid Lipschitz-constant estimate makes possible to provide appropriate bounds of objective function values, leading to corresponding partition subset selection/deletion, and algorithm termination rules: recall, for example, the case of general Lipschitz optimization, as discussed in Chapter 2.7. Horst, Nast and Thoai (1994) provide a discussion of various Lipschitzian bounding procedures. These methods are categorized into two classes. Approaches belonging to the first category are based on several direct extensions and generalizations of Piyavskii's algorithm to the multivariate case. Their implementation typically involves nontrivial auxiliary calculations which appear to limit their practical applicability to fairly small (often only $n \leq 2$ or 3) dimensional LGOPs. (See, for instance, Mayne and Polak, 1984; Wood, 1985, 1991, 1992; Mladineo, 1986, 1992; Meewella and Mayne, 1988; Baritompa, 1993, 1994; Breiman and Cutler, 1993; Zhang, Wood and Baritompa, 1993; Baritompa and Cutler, 1994; Sergeyev, 1994.)

Methods of the second category reviewed by Horst, Nast and Thoai use 'inexpensive' bounds which are computationally easy to generate; the bounds derived, however, are typically also quite 'weak'. (For examples of such approaches, see, Galperin, 1985, 1988; Pintér, 1986b, 1988; Horst and Tuy, 1987; Horst, 1988; Tuy and Horst, 1988; Hendrix and Pintér, 1991.) Horst, Nast and Thoai present a new approach based on providing bounds via solving an auxiliary linear programming problem, at each iteration (on each active subset which is possibly to be partitioned). This procedure yields more precise bounds than the ones used previously (in methods of the second category), at the cost of additional computational effort. Note, however, that all bounding approaches listed above need a correct estimate of L.

This brief exposition indicates that relatively 'inexpensive' and valid—or, at least, established in some well-defined sense—estimates of L are of apparent significance, in terms of effective numerical implementation of Lipschitzian optimization procedures. In the present chapter straightforward deterministic and statistical methods will be discussed, for estimating L and related problem characteristics on the basis of certain sample points (generated within the framework of a PAS-type algorithm). First, we shall provide simple deterministic bounds—assuming L is given; this will be followed by a brief discussion of using statistical inference to estimate L (and, possibly, also other problem characteristics).

3.2.2 Subset-Specific Estimates of Lipschitzian Characteristics

We shall make two additional comments, before going into some technical details. The first—simple, but computationally very relevant—observation to make is that the general convergence theory of adaptive partition methods presented in Part 2 does not rely per se on the 'true' Lipschitz constant which is valid for the whole search region D (in the present case, interval $[a, b]$. To assure convergence in the strong sense of the identity $X^a = X^\star$, it is sufficient to maintain regularity of the partition operator, and to apply correct bounding procedures on the partition subsets generated. (To verify this, recall the related discussion of PASs in Chapters 2.2, 2.4, 2.7, and check the sufficient conditions of the $X^a = X^\star$ type convergence results). For this reason, it is completely adequate to determine valid subset-specific Lipschitz-constant estimates, and then use this information in the aggregation, selection, and partition refinement steps of the corresponding PAS-type optimization method. (Without going into details, note that analogous statement applies with respect to the subset deletion rules in PAS strategies, discussed in Chapter 2.7.) Obviously, the numerical advantages of this approach can be very significant, since instead of working with the 'worst possible' overall value $L = L[a, b]$, often much smaller local values L can be used. To illustrate this point, note that taking an arbitrary smooth function f in (3.2.1), its local Lipschitz-constant will be small in a suitable neighbourhood of all interior stationary points, including local/global extrema. Of course, the estimation of subset-specific values L is—as a rule—theoretically not easier, than the estimation of L[a,b], but the overall efficiency of the calculations can be enhanced using appropriate local (over)estimates.

It should be also noted at this point that—although it is certainly a serious theoretical flaw to use improper ('too small') subset-specific estimates of L—from the viewpoint of global convergence it only matters to provide proper estimates for that sequence of partition subsets which converges to

$x^\star$. (In the case of multiple solutions, the same property is required for all partition set sequences which converge to points of $X^\star$.) For instance, 'ignoring' or deleting—incorrectly—a partition subset which does not contain points of $X^\star$ can, in fact, accelerate the numerical search procedure. Of course, this is not defence for incorrect estimation procedures, but still a point to keep in mind, since in many practical cases we are unable to provide and use—either global, or subset-specific—guaranteed Lipschitz parameters which would be of any practical value: hence, we simply must rely on their estimates.

3.2.3 Bounding Procedures on the Basis of Sample Points

We shall denote by $[xl, xu]$ an arbitrary subinterval of $[a, b]$. Let x_j, $j = 1, ..., J$, be a finite set of sample points selected in $[xl, xu]$; the corresponding function values are z_j, $j = 1, ..., J$. In accordance with the remarks made above, our primary objective is to find valid and possibly tight bounds for f on $[xl, xu]$. In the following discussion we shall suppose first that a correct (over)estimate of $L = L(f, [xl, xu])$ is available. For simplicity of notation, we shall write only L, unless we need to emphasize dependence on the subset in question. (In fact, the essence of the discussion could be presented using $[a, b]$ itself, as a possible instance of $[xl, xu]$.)

Upper bound

Let $x^\star[xl, xu]$ and $f^\star[xl, xu]$ denote the minimizer vector and minimum of f on $[xl, xu]$. Evaluating f at all sample points leads to the elementary estimate

$$f^\star[xl, xu] \le \min_{j=1,\ldots,J} z_j. \tag{3.2.2}$$

Lower bound

By the Lipschitz property of f, for an arbitrary sample point x_j, we have

$$f^\star[xl, xu] \ge z_j - L[xl, xu]\|x_j - x^\star[xl, xu]\|; \tag{3.2.3}$$

consequently,

$$f^\star[xl, xu] \ge \max_{j=1,\ldots,J} \{z_j - L[xl, xu]\|x_j - x^\star[xl, xu]\|\}. \tag{3.2.4}$$

The location of $x^\star[xl, xu] >$ is, of course, typically unknown; nevertheless, it is easy to provide estimates for $||x_j - x^\star[xl, xu]||$. For instance, the main diagonal length $||xu - xl||$ is a simple, frequently used bound: this leads to the estimate

$$f^\star[xl, xu] \geq \max_{j=1,\ldots,J} z_j - L[xl, xu]||xu - xl||. \tag{3.2.5}$$

For another example, one can select the centre $xc = (xl+xu)/2$ of $[xl, xu]$ as a (possibly singular) sample point: in that special case the simple bound

$$f^\star[xl, xu] \geq f(xc) - \frac{1}{2}L[xl, xu]||xu - xl|| \tag{3.2.6}$$

can be derived.

A generally somewhat sharper bound than (3.2.5) or (3.2.6) can be derived on the basis of the following observation. Since the norm functional $N(x) = ||x - x_j||$ is convex, its maximum is attained at one of the vertices of $[xl, xu]$: this is evidently that vertex which is farthest from x_j. Let $k = 1, \ldots, n$ denote the components of arbitrary n-vectors x in $[a, b]$. For a given x_j, define

$$d_{jk} = \max(x_{jk} - xl_k, xu_k - x_{jk}), \quad k = 1, \ldots, n; \quad \text{and let } \Delta_j = \left(\sum_{k=1}^{n} d_{jk}^2\right)^{\frac{1}{2}}.$$

Then, since $||x_j - x^\star[xl, xu]|| \leq \Delta_j$, we arrive at the bound

$$f^\star[xl, xu] \geq \max_{j=1,\ldots,J} \{z_j - L[xl, xu]\Delta_j\}. \tag{3.2.7}$$

Note that, as the bounds (3.2.5) to (3.2.7) are all valid, their maximum is also an obvious lower bound.

Summing up the above brief discussion, we can use, for example, the following directly computable lower an upper bounds of the minimum of f on $[xl, xu]$:

$$\underline{f}[xl, xu] \leq f^\star[xl, xu] \leq \overline{f}[xl, xu], \quad \text{in which} \tag{3.2.8}$$
$$\underline{f}[xl, xu] := \max_{j=1,\ldots,J} \{z_j - L[xl, xu]\Delta_j\}, \quad \text{and} \quad \overline{f}[xl, xu] := \min_{j=1,\ldots,J} z_j.$$

The bounds $\underline{f}[xl, xu]$ and $\overline{f}[xl, xu]$ can be used in PAS-type search procedures to accomplish subset selection, deletion, and bound refinement steps following, for example, the general algorithm-scheme discussed in Chapter 2.7. Note finally that the subset specific bounds immediately provide the following aggregated lower and upper bounds for the optimum value $f^\star$ in (3.2.1):

$$\min_{[xl,xu]\in I[a,b]} \underline{f}[xl,xu] \le f^\star \le \min_{[xl,xu]\in I[a,b]} \overline{f}[xl,xu]. \qquad (3.2.9)$$

In (3.2.9) the minimum is taken over an arbitrary selection of active subintervals $[xl,xu] \in I[a,b]$ which jointly cover $[a,b]$ (or, more precisely, cover that subset of $[a,b]$ which—after possible valid elimination steps—contains $X^\star$).

The lower estimates $\underline{f}[xl,xu]$, of course, still depend on the subset-specific—typically unknown—value $L[xl,xu]$. In the following section this issue will be resolved, at least in a statistical sense.

3.2.4 Lipschitz-Constant Estimation, Using Results from Extreme Order Statistics

For the sake of simplifying the notation—and also because the discussion is valid in a more general framework than the previous sections of this chapter—below we shall omit explicit reference to the subset on which the Lipschitz-constant is estimated. It could be, for instance, the initial embedding interval $[a,b]$, its arbitrary subinterval $[xl,xu]$, or in fact, any robust partition subset of the initial feasible set. We shall denote this subset by S.

For an arbitrary pair of sample points x_i, x_j, $i,j = 1,...,J$, from S, and $z_i = f(x_i)$, $z_j = f(x_j)$, define

$$L_{ij} = \frac{|z_i - z_j|}{||x_i - x_j||} \quad \text{for} \quad i = 1,2,\ldots,J, \quad j = 1,2,\ldots,J, \quad i \neq j. \qquad (3.2.10)$$

Then an immediately available lower bound of the Lipschitz-constant $L = L(f,S)$ is

$$\underline{L} = \max_{i,j=1,\ldots,J} L_{ij} \le L. \qquad (3.2.11)$$

As noted previously, practically meaningful deterministic upper bounds of L are often difficult—sometimes impossible—to find. (For instance, this is the case in 'black box' models, leading to sample values of f.) For this reason, statistical estimation techniques will be discussed briefly below. During this discussion, we shall assume that the sample points x_j, $i = 1, \ldots, J$, on which these estimates are based, are drawn randomly from a uniform probability distribution on the set S: this fact will be denoted by $x_j \sim U\{S\}$. Moreover, the symbols $z_1, z_2, z_3, \ldots$ will have a more general interpretation than above, although incorporating also their usual (function value) meaning; other notation may also differ slightly from that previously used.

There exists an elegant, comprehensive theory of extreme order statistics dealing with the limit distribution of sample extrema; see Gnedenko (1943), Gumbel (1958), De Haan (1970), Galambos (1978). Let $z_1, \ldots, z_N, \ldots$ be independent, identically distributed (i.i.d. random variables (r.v.); their common probability distribution function (p.d.f.) is denoted by $F(z) = P(z_1 < z)$. Define now

$$Z_N = \max\{z_1, \ldots, z_N\}, \qquad W_N = \min\{z_1, \ldots, z_N\}. \tag{3.2.12}$$

Then the p.d.f. of Z_N and W_N can be simply expressed by the p.d.f. $F(x)$, as

$$H_N(z) = P(Z_N < z) = \prod_{i=1}^{N} P(z_i < z) = (F(z))^N = F^N(z) \tag{3.2.13}$$

$$\begin{aligned} L_N(z) &= P(W_N < z) = 1 - P(W_N \geq z) \\ &= 1 - \prod_{i=1}^{N} P(z_i \geq z) = 1 - (1 - F(z))^N \end{aligned} \tag{3.2.14}$$

Since $\min(z_1, \ldots, z_N) = -\max(-z_1, \ldots, -z_N)$, for continuous and strictly monotonical p.d.f. F, it suffices to examine, for instance, the p.d.f $H_N(z)$ of the maximum r.v. Z_N. Let us note that, in general, $F^N(z)$ does not lend itself to straightforward analysis and numerical application. For this reason, approximations of $H_N(z)$ (and $L_N(z)$)—valid for sufficiently large sample size N—are of significant theoretical and practical interest.

Definition 3.2.1. The p.d.f. $F(z)$ belongs to the domain of attraction of a certain p.d.f. $H(z)$, if there exist real sequences $\{a_N\}$ and $\{b_N\}$, $b_N > 0$, such that the limit relation

$$\lim_{N \to \infty} H_N(a_N + b_N z) = H(z) \tag{3.2.15}$$

holds for all real values z. Observe that if (3.2.15) is valid, then—for sufficiently large N—one can apply the approximation

$$P(Z_N < z) = H_N(z) \approx H((z - a_N)/b_N) \tag{3.2.16}$$

to replace the—exact, but analytically non-tractable—p.d.f. $F^N(z)$. Consequently, if a_N and b_N are estimated from the sample $z_1, \ldots, z_N$, then one can approximate the p.d.f. of Z_N, providing, for example, confidence interval type estimates for its value. The significance of such estimates is evident, not only in terms of estimating the Lipschitz-constant, but also other characteristics of optimization problems: we shall briefly return to this point somewhat later.

Let us define $R(F) = \sup(z : F(z) < 1)$. Obviously, $R(F) < \infty$ corresponds to the case (most relevant in optimization), when the realizations of the r.v.'s $z_1, z_2, z_3, \ldots$ are bounded from above. For instance, this is the case if the (existing) Lipschitz-constant or maximum of a certain function f on the set S is estimated on the basis of a random sample of points $x_1, \ldots, x_N, \ldots$, from S.

The basic results of extreme order statistics, relevant for the present discussion, can be summarized by the following four theorems. For proofs and other details, see Gnedenko (1943) and earlier references therein; a more recent general exposition of the theory can be found in Galambos (1978).

Theorem 3.2.1. *Let $R(F) = \infty$, and assume that there is a constant $c > 0$ such that for all $z > 0$ we have*

$$\lim_{t\to\infty} \frac{1 - F(tz)}{1 - F(t)} = z^{-c}. \tag{3.2.17}$$

Then there exists a positive sequence $\{b_N\}$, which leads to the limit relation

$$\lim_{N\to\infty} H_N(b_N z) = H_{1,c}(z) = \begin{cases} 0 & \text{if } z \le 0 \\ \exp(-z^{-c}) & \text{if } z > 0. \end{cases} \tag{3.2.18}$$

A possible choice of $\{b_N\}$ is $b_N := \inf\{r : 1 - F(r) \le 1/N\}$. ■

Theorem 3.2.2. *Let $R(F) < \infty$, and assume that for all $z > 0$ the p.d.f.*

$$F^\star(z) = F(R(F) - 1/z)$$

satisfies the relation (3.2.17). Then there exist sequences $\{a_N\}$ and $\{b_N\}$, $b_N > 0$, which lead to the limit relation

$$\lim_{N\to\infty} H_N(a_N + b_N z) = H_{2,c}(z) = \begin{cases} \exp(-(-z)^c) & \text{if } z < 0 \\ 1 & \text{if } z \geq 0. \end{cases} \tag{3.2.19}$$

A possible choice of the normalizing sequences is $a_N := R(F)$, and $b_N := R(F) - \inf\{r : 1 - F(r) \leq 1/N\}$. ■

Theorem 3.2.3. *Assume that for some finite value a, we have*

$$I(a) = \int_a^{R(F)} (1 - F(z))dz < \infty, \tag{3.2.20}$$

and define $K(t) = I(t)/(1 - F(t))$ for $t > a$. If, for all real z, the limit relation

$$\lim_{t\to R(F)} \frac{1 - F(t + zK(t))}{1 - F(t)} = e^{-z} \tag{3.2.21}$$

is valid, then there exist sequences $\{a_N\}$ and $\{b_N\}$, $b_N > 0$, such that the following limit relation holds:

$$\lim_{N\to\infty} H_N(a_N + b_N z) = H_{3,0}(z) = \exp(-e^{-z}) \qquad -\infty < z < \infty. \tag{3.2.22}$$

A possible choice of the normalizing sequences is

$$a_N : \inf\{r : 1 - F(r) \leq 1/n\} \quad \text{and} \quad b_N := K(a_N).$$ ■

Theorem 3.2.4. *There are only three types of limit p.d.f.'s $H(z)$ for which the relation (3.2.15) can hold: these are $H_{1,c}$, $H_{2,c}$ and $H_{3,0}$ defined above. An arbitrary p.d.f. $F(z)$ belongs to the domain of attraction of the corresponding p.d.f. H exactly in the case, when, respectively, the assumptions of Theorems 3.2.1, 3.2.2, or 3.2.3 are satisfied.* ■

The general results of probability theory quoted have obvious relevance in estimating extrema in optimization problems, including—as a special case—the subset-specific estimation of Lipschitz-constant values. There has been considerable interest in applying extreme order statistics to 'hard' optimization problems in which the optimum needs to be estimated—global optimization being one obvious target area. For examples, one can refer to

Clough (1969), Archetti, Betrò, and Steffe (1975), Pintér (1978), Golden and Alt (1979), De Haan (1981), Makarov and Radashevich (1981), Zanakis and Evans (1981), Boender, Rinnooy Kan, Stougie and Timmer (1982), Ariyawansa and Templeton (1983), C. Fodor (1983), Patel and Smith (1983), Smith (1987), C. Fodor and Pintér (1988); general surveys and discussions—related to the field of global optimization—are provided, for example, by Törn and Žilinskas (1989), Zhigljavsky (1991, 1993). The following brief discussion is restricted to the case of Theorem 3.2.2. which deals with the most relevant case, in the present context.

Theorem 3.2.2 states that—if the range of the p.d.f. $F(z)$ is bounded from above by some value $M := R(F)$ (with probability 1)—then the sample maximum p.d.f. can be asymptotically approximated by

$$\lim_{N\to\infty} H_N(M + b_N z) = H_{2,c}(z) = \begin{cases} \exp(-(-z)^c) & \text{if } z < 0 \\ 1 & \text{if } z \geq 0, \end{cases} \tag{3.2.23}$$

and b_N can be chosen as M minus the $(1 - 1/N)$-quantile of $F(z)$.

In (3.2.23), $H_{2,c}(z)$ is called the Weibull distribution with tail index (shape parameter) c. Note that, in particular, (3.2.23) holds, if $F(z)$ can be represented in the form

$$F(z) = 1 - d(M - z)^c + o((M - z)^c). \tag{3.2.24}$$

In (3.2.24), $d > 0$ is a suitable function model parameter, which, in turn, determines a model of the sample value distribution.

An essential problem in applying (3.2.23) or (3.2.24) is that the function model parameters—including, in particular, the shape parameter c—are unknown in many cases of practical relevance. (A notable exception is discussed by Patel and Smith (1983): they show that in concave programming under linear constraints the tail index equals n, the dimension of the constraint space.) Hence, naturally, in constructing estimation procedures one typically needs to estimate also the shift parameter M and the scale parameter b_N.

Zhigljavsky (1991, 1993) presents a detailed account of various statistical techniques for estimating M for both cases, when the value c is known or unknown. For instance, in case of known c, the following one-sided confidence interval estimate can be proposed. Let us denote by $z_N^\star \geq z_{N-1}^\star, \ldots, \geq z_{N-k}^\star$

the first $N - k + 1$ elements of an ordered random sample; additionally, for $0 < \delta < 1$ define

$$r(k, \delta) = 1/\{(1 - \delta^{1/k})^{-1/c} - 1\}. \tag{3.2.25}$$

Then, under appropriate analytical conditions ($N \to \infty$, $k \to \infty$, $k/N \to 0$)

$$[z_N^\star, z_N^\star + r(k, \delta)(z_N^\star - z_{N-k}^\star)]$$

is a $1 - \delta$ confidence interval for M, in the asymptotic sense ($N \to \infty$). A serious practical drawback is that this (or other known) confidence interval results typically lead to very broad ranges which therefore are of little practical value.

For the case of unknown c, the general maximum likelihood estimator (or other advanced statistical methodology) can be applied. Note, however, that these approaches can lead to other complex numerical tasks—including cases of induced (embedded) global optimization problems: see Zanakis and Kyparisis (1986), Zhigljavsky (1991, 1993), Gourdin, Hansen and Jaumard (1994), and further references therein. For example, the latter authors develop a global optimization method to estimate the Weibull p.d.f. parameters: this approach is based on first order optimality conditions and solution space projection to reduce the initial three-parameter estimation problem to a univariate problem (the solution of the latter needs nontrivial embedded calculations). Observe that, in the present context, such auxiliary calculations would be needed within each PAS-type iteration cycle: this would lead to a significant—hardly justifiable—increase of the computational demand.

In summary, there exist statistically well-established—albeit numerically demanding—techniques which allow the estimation of sample extrema. As special cases, both the Lipschitz-constant and the sample minimum can be estimated in a subset-specific context, on the basis of search points randomly drawn from S.

3.2.5 Numerical Comments and Conclusions

Let us recall now the discussion from Sections 3.2.1–3.2.3, hence putting the issue of statistical inference into perspective with respect to PAS-type algorithms. Since such inference is to be made in each iteration cycle of a—theoretically infinite—algorithm, therefore it is evident that one should stay with relatively simple estimation techniques which are still acceptable, also from a statistical viewpoint. For this 'pragmatic' reason, on the basis of the results summarized by Zhigljavsky (1991, Chs. 4 and 7, with appropriate further references therein), straightforward linear estimates of the form

$$M \approx \sum_{j=0}^{k} a_{N-j} z_{n-j}^{\star} \quad (a_{N-j},\ j = 0, 1, \ldots, k, \ \text{are given real coefficients}) \tag{3.2.26}$$

are included in the Lipschitzian global optimization algorithm implemented (which will be discussed in Chapter 3.4). Let us note again that these estimates may be 'vague' (leading to low numerical efficiency), or—on the other hand—can lead to incorrect statistical inference (which may or may not be a problem, in terms of global convergence of the PAS-type algorithm). Since the global optimization problems to be solved may exhibit quite different and possibly 'unexpected' Lipschitzian behaviour, the user has the possibility of parameterizing externally the built-in relation (3.2.25) by using a 'safety factor': the larger this factor is, the more 'cautious' subset-specific Lipschitz-constant estimate is applied (and often the less numerical efficiency can be expected). There is apparently ample room for research on this area, to derive theoretically correct and practically usable statistical estimation techniques.

It is obvious from the above discussion that randomly drawn sample points are necessary, in order to establish statistical estimates of subset-specific Lipschitzian characteristics. At the same time, the multidimensional diagonal extension of univariate algorithms (presented in Chapter 3.1) calls for a deterministic point selection strategy which depends on the univariate method generalized. One can add at this point in the discussion that in many well-posed LGOPs the objective function f shows a 'quadratic-at-large' behaviour (at least in the sense that approaching the boundary of the embedding interval $[a, b]$, the function values become much larger than in the interior, around local/global optima). For the latter reason, simple quadratic approximations of f on the main diagonal of partition subintervals often yield good search points, especially in the initial phase of global search.

On the basis of these observations, in the implementation discussed in Chapter 3.4 both deterministic and random sample points are chosen, in each adaptively generated subset. The deterministic points are based on the diagonal generalization of a one-dimensional method (such as, for example,

Piyavskii's algorithm), and on a quadratic function model based approximation: the number of these sample points is fixed (within a given algorithm). The additional random sample points serve to estimate the subset-specific Lipschitz-constant, contribute to providing statistically based lower bounds of the (local) optimum, and also may improve the global optimum estimate. It is difficult to see in advance the relative usefulness of deterministic versus randomly chosen sample points, therefore the user has the possibility of selecting the number of the latter. Numerical tests and practical experience show that—aside from the theoretical need for (at least a few) random sample points—they may improve the search efficiency, especially in case of global optimization problems which have a complicated 'response surface', with a large number of 'scattered' local/global optima. Examples of such applications will be discussed in Part 4.

Chapter 3.3

General Lipschitz Optimization Applying Penalty Multipliers

3.3.1 Introduction

Consider now the Lipschitzian global optimization problem in the following general form:

$$\min f_0(x) \tag{3.3.1}$$
$$x \in D = \{a \leq x \leq b : f_i(x) \leq 0,\ i = 1, \ldots, I\} \subset \mathbb{R}^n.$$

In accordance with the discussion in Chapter 2.7, we shall assume that D is the closure of a nonempty, bounded, open set in the real n-dimensional space $\mathbb{R}^n$, and that the constraint functions f_i, $i = 0, 1, \ldots, I$, are all Lipschitz-continuous on D, with corresponding Lipschitz-constants $L_i = L_i(D, f_i)$, $i = 0, 1, \ldots, I$. In other words, the inequalities

$$|f_i(x) - f_i(y)| \leq L_i \|x - y\| \tag{3.3.2}$$

are assumed to hold for all pairs of x, y from D.

We shall also suppose that the objective function f_0 is (possibly) multiextremal on the feasible set D, and has a finite set of globally optimal solutions $X^\star$. Consequently, we are interested in numerical solution procedures which find (in practice, approximate) points of $X^\star$.

Note that the problem statement (3.3.1) is typical, for instance, in a number of engineering applications, in which explicit bounds are given for

x, the n-vector of decision variables, and the functions f_i, $i = 1,\ldots,I$, express—possibly complicated—relationships among the components of x. In particular, some of the constraints f_i can be evaluated via computationally intensive, 'black box' operations, such as, for example, running an embedded computer program. We shall illustrate the relevance of this case, in relation to several applications discussed in Part 4.

3.3.2 Solution Approach

In order to approximate the numerical solution of (3.3.1), we shall apply a classical penalty function transformation (consult Zangwill, 1967; Fiacco and McCormick, 1968; or the review of Fletcher, 1983). Following this methodology, the 'complicated' constraints f_i, $i = 1,\ldots,I$, are included into an augmented objective function which may have, for example, the following form:

$$f(x,c) := c_0 f_0(x) + \sum_{i=1}^{I} c_i \max\{f_i(x), 0\}. \tag{3.3.3}$$

In the definition of the composite function f, $c_i > 0$, $i = 0,1,\ldots,I$, are suitably chosen penalty multiplier parameters, $c = (c_0,\ldots,c_n)$. Furthermore, (3.3.1) is replaced by the parameterized Lipschitzian global optimization problem (LGOP)

$$\begin{aligned} &\min f(x,c) \\ &a \le x \le b. \end{aligned} \tag{3.3.4}$$

As can be seen, the function f attempts to balance the aims of reducing the objective function, and satisfying the implicit constraints. Evidently, the greater the multiplier c_i is, the greater importance is associated with f_i in the composite objective function f. Therefore, adaptively changing the coefficients c_i, the (approximate) feasibility and optimality of the sequentially generated solutions can be enforced.

A variety of theoretical results relating to the convergence of the sequential penalty approach is known: several basic results will be summarized below, omitting their proof. (For a detailed discussion of these and other related results, reference is made again to Fletcher (1983), and the works cited therein.)

Theorem 3.3.1. *If, for a certain parameterization c, $x^\star$ solves (3.3.4) while it satisfies all constraints f_i, $i = 1, \ldots, I$, then $x^\star$ solves (3.3.1). Conversely, if $x^\star$ solves (3.3.1), then it is (one of) the solution(s) of (3.3.4), for arbitrary parameterization c.* ■

Note that this assertion naturally leads to (exact or approximate) criteria for stopping the sequential search procedure, based on adaptively generated parameterizations of (3.3.4).

Theorem 3.3.2. *For fixed c_0 and sufficiently large values c_i, $i = 1, \ldots, I$, the optimal solution of (3.3.4) will be feasible: that is, it will meet the constraints f_i, $i = 1, \ldots, I$ in (3.3.1).* ■

This assertion implies that, by appropriately increasing the constraint multipliers, feasibility can be guaranteed.

Theorem 3.3.3. *If, for each selected c, the exact global solution is determined in (3.3.4) and the multipliers c_i, $i = 1, \ldots, I$, tend to infinity (that is, they can all be chosen arbitrarily large), then the corresponding sequence of solution points converges to the globally optimal solution of (3.3.1).* ■

The last statement shows the theoretical convergence property of the search procedure which is based on solving an adaptively chosen sequence of LGOPs, in the form (3.3.4). Note immediately that there are also other ways to assure theoretical global convergence of the penalty approach (Fletcher, 1983), and that their numerical suitability is quite often problem-dependent.

Let us recall (Chapter 1.1) at this point that—as a direct consequence of the Lipschitz property of the functions f_i, $i = 0, 1, \ldots, I$—function f is also Lipschitzian. Therefore one can apply adaptive partition strategies to solve instances of problem (3.3.4), as discussed in Chapters 2.4 and 3.1. Note in this context that the realization of adaptive rectangular subdivision strategies can, typically, be much easier than that of a direct search on the—implicitly defined and possibly quite complicated—feasible region of (3.3.1) (recall Chapter 2.7). This fact is the principal motivation for combining the penalty function approach with Lipschitzian global optimization on an n-interval region. The corresponding sequential optimization procedure has the following basic structure.

Optimization Algorithm Scheme for Solving Constrained LGOPs

Step 0. (Initialization) Set $k := 1$ (major iteration count). Choose a componentwise positive vector of initial multipliers $c^{(k)} = (c_0^{(k)}, c_1^{(k)}, \ldots, c_I^{(k)})$.

Step 1. (Embedded solution procedure) Applying the penalty multiplier $c^{(k)}$, solve the corresponding kth optimization problem, denoted by $(3.3.4)^{(k)}$. Denote its (approximate) optimal solution by $x^{(k)} = x^{(k)}(c^{(k)})$.

Step 2. (Verification) If $x^{(k)}$ is a feasible solution of (3.3.1), then it (approximately) solves it: the search procedure can be terminated. In the opposite case, Step 3 follows.

Step 3. (Multiplier update) Increase all multipliers c_i for which $f_i(x^{(k)}) > 0$, defining $c_i^{(k+1)} := a_i^{(k)} c_i^{(k)}$ ($a_i^{(k)} > 1$ multiplier factor). Set $k := k+1$ and return to Step 1 of the next major iteration cycle.

Concluding this brief description, it is worth observing that the usual nondifferentiability and/or nonconvexity issues associated with the penalty method are now irrelevant, as the present approach is gradient-free and exploits only Lipschitz-continuity. Nevertheless, the relative magnitude of the multipliers c_j may nonnegligibly influence the numerical properties of the objective function in (3.3.4) and, hence, the solution efficiency in the embedded problems. The adaptive choice of the multiplicator vector c was studied by numerous authors; in the context of global optimization, see, for example, Timmer (1984). Examples of applying the penalty function transformation approach will be discussed in Part 4.

Chapter 3.4

An Implementation of a Lipschitzian Global Optimization Procedure

3.4.1 The LGO Program System

Scope of Application

The general framework and underlying convergence theory of adaptive partition algorithms in global optimization leave ample room for constructing robust and effective, computationally viable algorithm-instances. The program system described below has been implemented (in gradually developed versions), extensively tested, and applied by this author in the past decade. This, of course, does not imply any notion of 'perfection'—in fact, the program modules have often been (and surely will be) modified and improved, not infrequently in connection with new applications.

The LGO program system serves for finding—that is, numerically approximating—the best solution or, theoretically, all solutions of the (possibly) multiextremal optimization problem in its standardized form

$$\min f(x), \qquad a \le x \le b. \tag{3.4.1}$$

In (3.4.1), a, x, b are finite real n-vectors, and f is Lipschitz-continuous on $[a, b]$. As discussed in Chapter 3.3, a more general class of Lipschitz optimization problems can be approximated in the form (3.4.1), possibly following a penalty transformation based incorporation of some constraints into the objective function. The penalty multipliers can be adaptively changed, if necessary, to enforce (approximate) feasibility and optimality of the solution found.

Problem Scaling

The LGO program system performs self-scaling with respect to the search n-interval $[a, b]$ which is automatically transformed into $[0, 1] \subset \mathbb{R}^n$. This transformation helps a more 'uniform', scale-free screening of the feasible set. At the same time—depending on the original problem structure—it often may lead to very large 'jumps' of the objective function (on $[0, 1]$), and hence, to quite huge Lipschitz-constants. (Note that, of course, the initial problem may already display such characteristics.) To avoid related numerical problems during the solution procedure, it is advisable to scale also the objective function. This can be easily done, applying a suitable multiplier which transforms the function values, say, into the range $[-100, 100]$. Some experimentation (argument and function value sampling) may be advisable in connection with complicated problems, since appropriate scaling may indeed lead to improved numerical performance.

Solution Methodology

The current program system includes, in an integrated fashion, the following optimization procedure options:

- Lipschitzian global optimization on the interval $[a, b]$
- global optimization by pure random search on the interval $[a, b]$
- local optimization launched from the current estimate of the global optimum (Fletcher–Reeves–Polak–Ribiere method, FRPR)
- local optimization launched from the current estimate of the global optimum (Powell-Brent method, PB).

A few comments related to these solvers are in order. As discussed earlier, pure random search is a very simple, 'folklore' approach to global optimization, which has been used extensively. In the present context, it can be applied, for example, by novice users who want to 'explore' a bit the feasible set, and/or as a preliminary search phase for possibly reducing the initially chosen search region. The Lipschitzian solver is much more sophisticated, and—enabled with proper parameterization—it should serve to generate a reasonable approximation of the global optimizer point(s), before the algorithm 'switches' to local search. Among the two local methods included,

FRPR relies on gradient information—related to the objective function f—and, for sufficiently 'smooth' problems, it often assures a rapid termination of the search. PB is gradient-free, and is somewhat more 'cautious', but—according to computational experience—is also more robust, than FRPR.

Note that both of the local search algorithms FRPR (Fletcher and Reeves, 1964; Polak and Ribiere, 1969; Polak, 1971) and PB (Powell, 1964; Brent, 1973) are discussed, for example, by Press, Teukolsky, Vetterling and Flannery (1992). For purposes of the present implementation, FRPR and PB were both modified and adapted, to assure their proper connection to the global optimization search phase. Note additionally that the implementation of pure random search in LGO is also based on the (fairly sophisticated) random number generator described in Press, Teukolsky, Vetterling and Flannery (1992): of course, this search phase also had to be connected to the local optimizers FRPR and PB.

The listed solver options can be activated either in an interactive fashion (user control mode), or can be launched by the LGO system (automatic mode). In the first case, the user can specify on-line the computational effort to be spent by the chosen option; observing the results obtained, one can switch to another solver option, or terminate the run. In automatically controlled runs, a Lipschitzian global optimization phase is followed by a PB refinement. The development of additional options to solve (3.4.1) will be the subject of further research.

Structure of LGO Application Programs

The following abbreviations will be used (they reflect a Fortran development environment):

NAME	–	name of application
NAME-DR.FOR	–	source code of the main driver, written/adapted by user
NAME-OF.FOR	–	objective function file, written/adapted by user
NAME.IN	–	input parameter file, written/adapted by user
USERSRC.FOR	–	additional source code provided by user (optional)
LGOSBRS.OBJ	–	object file of LGO program system routines
NAME.OUT	–	detailed output file generated by LGO
NAME.SUM	–	summary output file generated by LGO.

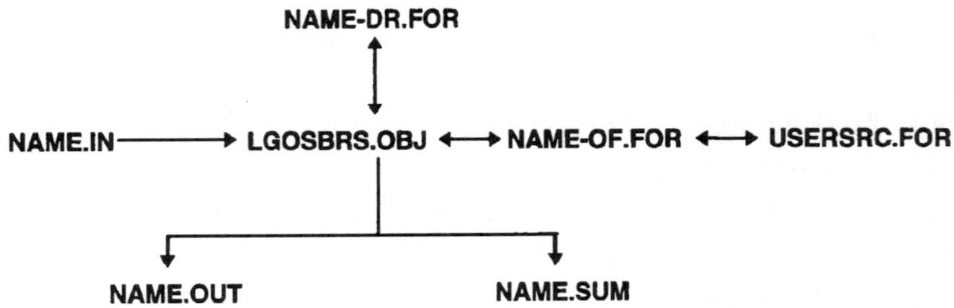

Figure 3.4.1
Basic structure of LGO application program systems

Commented templates of the files NAME-DR.FOR, NAME-OF.FOR, and NAME.IN are provided for assisting the user: this way, the setup of these application-dependent files is quite straightforward, and does not require special skills or intimate knowledge of the LGO system. (Of course, the parameterization of LGO in NAME.IN profits from an understanding of the basic principles of the optimizer, but sample parameter values and commentary on parameter choices are provided in the template input file.)

Fig. 3.4.1 shows the principal structure of LGO application programs.

A brief description of the program structure outlined by Fig. 3.4.1 follows.

The main driver file NAME-DR.FOR serves for the (optional) printout of application header/title information; it opens the (in general, application-specific) input parameter file NAME.IN and output files NAME.OUT and NAME.SUM, and then calls the LGO system. On return, the results stored in NAME.OUT and NAME.SUM can be (additionally) reported in an arbitrary form specified by the user, and other tasks—analysis of results, repeated program runs, etc.—can also be initiated.

LGOSBRS.OBJ is the 'heart' of the optimizer: it integrates all subroutines and functions which are necessary to operate the solver variants. Its operations are governed by the input parameter file NAME.IN. In the latter are defined the lower and upper bounds (interval $[a, b]$) of the optimization problem (3.4.1), and the parameterization of the LGO program system. Note that this structure allows for rerunning the same executable program under different feasible range and LGO system parameter specifications.

During the optimization run, LGOSBRS.OBJ iteratively calls the objective function segment NAME-OF.FOR which—for each parameter vector x

chosen by LGO—returns the corresponding function value. Frequently—for example, in solving nonlinear systems of equations by global optimization—NAME-OF.FOR is simply a corresponding programmed 'translation' of an explicitly given numerical problem. In many other cases—notably, in 'black box' system optimization—the objective function value is calculated via running an attached program system denoted here by USERSRC.FOR. For instance, USERSRC.FOR may be the descriptive model of a complex—industrial, engineering, ecological, etc.—system which is to be parameterized (calibrated) applying global optimization. Examples will be provided in Part 4 for both cases mentioned. Note that—for all program development purposes—NAME-DR.FOR and NAME-OF.FOR can be given in the same (ASCII text) file; this applies also to USERSRC.FOR when present.

Two output files are generated by LGO, in a complete optimization run. NAME.OUT is a more detailed result file: the level of detail is driven by parameter choice in the input file NAME.IN. Typically, NAME.OUT can be most useful during the development of an application in which the problem statement is verified and the optimization process is traced. Another application can be related to very difficult or poorly defined problems, in which it is informative to inspect a variety of feasible solutions.

The file NAME.SUM contains the summary results of the optimization run. It is useful in 'routine' LGO runs, since it provides concise information, without the need of browsing a possibly large amount of additional data.

3.4.2 Current System Requirements and Problem Size Limitations

Hardware

In order to run LGO with desirable efficiency, the following basic hardware components are recommended, in a PC environment. (Only the 'minimal' requirements are indicated; more or higher is typically better, in terms of solvable problem-sizes and execution speed.)

- IBM PS/2 compatible, Intel 386&387DX, 486DX or 586DX (Pentium) processor based machine; the numerical processor is essential, in order to use the program with reasonable efficiency (some of the software components used may also require one of the above processors)
- 4–8 Mbyte RAM

- 4–12 Mbyte available hard drive space
- EGA, VGA, or SVGA monitor (especially in case of using graphic representation options).

  Note that implementations of LGO on other platforms—for example, workstations or mainframes—can also be made available. The expectable speed increase may be quite significant, in case of computationally intensive optimization problems. (We have both related workstation and mainframe experience.)

Software

Additional to the hardware needs outlined, an appropriate software environment is a prerequisite. Again, minimal recommended version numbers are provided as examples; higher versions are expected to function properly. Note also that not all software components listed will 'automatically' work together, so that proper consideration is to be given to the selection of the software platform.

- Up-to-date PC operating system; examples:
  MS DOS version 6.21 © Microsoft Corp.
  Windows 3.1 © Microsoft Corp.

  Note that the LGO implementation discussed here is DOS-based, and it does not need per se an environment like Windows: the system, however, can be connected to and run under Windows. Note additionally that a (DOS- or Windows-based) graphical interface—and other graphics capabilities—can also be of use in connection with a number of complex applications: for instance, when ex post graphical analysis or the running of several other program systems is required in conjunction with the use of LGO.

- Appropriate memory management software; examples:
  Windows 3.1 © Microsoft Corp.
  QEMM Memory Manager 4.2 © Quarterdeck Office Systems
  Ergo OS/386 Operating System 2.00 © Ergo Computing, Inc.
  Phar Lap TNT DOS-Extender © Phar Lap Software, Inc.

- Up-to-date, professional level Fortran 77 or 90 development environment; examples:
  Lahey Fortran F77L-EM/32 3.01 © Lahey Computer Systems, Inc.
  Lahey Fortran LF90 1.10b © Lahey Computer Systems, Inc.
  Salford FTN77/486 © Polyhedron Software Ltd.
  Interacter 3.0 © Interactive Software Services Ltd.
  Tecplot 6.0 © Amtec Engineering, Inc.

Other platforms can, of course, also be considered.

Problem Size Limitations

For reasons of keeping the program storage demands (and runtimes) within reasonable bounds—in a PC environment—the following limitations were introduced:

- The present (explicitly declared) array structure in LGO supports the formulation and solution of problems up to 50 variables.
- Maximum 10,000 (currently active) partition subintervals can be simultaneously present in the Lipschitz global optimization mode.

According to our experience, these bounds are not too restrictive with respect to many application problems. Note also that the above limitations can be easily lifted, but the storage requirements and computational times may (will) correspondingly increase. For illustration, note that the Lahey Fortran 77 executable (.EXP) file corresponding to the above declarations is over 4 Mbyte. Consequently, either appropriate RAM space is necessary, or the program will run in—slower—virtual memory management mode.

It is difficult to estimate the computational resource demands of 'general' global optimization problems discussed here—especially in light of the rapid development of the available hardware and software tools. Several illustrative examples (test problems) will be discussed later in this chapter; various applications will be the subject of Part 4.

3.4.3 Using LGO in an Interactive Environment

The current version of the program system LGO can be activated and used in an interactive fashion, using a menu-based interface. In the following description it is assumed that the LGO system and the interface programs are all in a specified directory on the hard disk (e.g., C:\LGO), where they are run (for simplicity, under DOS).

Before going into details, let us emphasize again that the interactive use of LGO presupposes the availability of a professional Fortran environment which is installed on the PC; consult, for instance, Lahey Computer Systems (1994a,b). (In the absence of such an environment, automatically running demonstration versions can be made available.)

Typing 'LGO %1' (%1 stands for the application filename given by the user) at the C:\LGO prompt, a menu shell is activated, which shows the following screen:

LIPSCHITZIAN GLOBAL OPTIMIZATION
PROGRAM SYSTEM
Version 3.0

© Janos D. Pinter, 1986...1995

Application Menu (Options)	**Option Symbols**
Introduction (General Information)	**I**
Demonstration Program Information	**D**
Edit Optimization Input File(s)	**E %1**
Compile Optimization Program Files	**C %1**
Run Optimization Program System	**R %1**
Analyse Optimization Output File(s)	**A %1**
Graphical Analysis of Results	**G %1**
Quit (Return to MS-DOS)	**Q**

Please type in the symbol of a valid option, and press Enter.
(If you are not familiar with LGO, choose first I, and then D.)

For instance, typing 'LGO TEST2' when invoking this menu initiates an interactive session, for running a variety of two-dimensional test problems (given in the file TEST2.FOR) which can be modified and adapted by the user. In the following a brief description of the listed menu options follows.

Introduction (General Information)

Selecting 'I' in the main menu, a brief information session is launched. This provides a concise description related to multivariate global optimization of continuous and Lipschitz functions on interval regions (for novice users). The

CGO/LGO problem class is defined; the graph of a univariate multiextremal function f is shown, indicating local and global optima. This is followed by a summary of the solution approach applied in LGO, and a list of several existing and prospective applications. Finally, basic information related to running the LGO program system is provided. With respect to program names, the following conventions are applied (see Section 3.4.2, and observe some slight modifications):

filename	–	name of application, given by user;
filename.FOR	–	source code of the main driver and objective function file, defined by user;
filename.IN	–	input parameter file (definition of bounds and parameterization of program system), this file is also defined by the user;
filename.OUT	–	detailed output file generated by the LGO program system;
filename.SUM	–	summary output file generated by LGO;

Templates and sample programs are provided for the .FOR and .IN files, under the names TEST2.FOR and TEST2.IN.

Upon completing the information session, the program returns to the main (applications) menu.

Demonstration Program Information

Selecting 'D' in the main menu, information related to running TEST2 is provided. In any single run, the TEST2 program can be executed, to minimize one of a given set of two-variate multiextremal test functions. The form of these functions can be modified by the user in the file TEST2.FOR; the parameterization of the optimizer system can also be modified, in the file TEST2.IN. Both TEST2.FOR and TEST2.IN are provided; they can be edited, the program can be compiled and run, and the output can be analysed, as indicated by the main menu. For instance, typing 'A TEST2'—after running the TEST2 program—allows the analysis of the results.

Note that user applications can also be written and then run, making use of the templates provided by the files TEST2.FOR and TEST2.IN.

At the end of the demo program information session, the main menu is displayed again.

Edit Optimization Input Files

Selecting 'E filename' in the main menu, the optimization input (filename. FOR and filename.IN) files specified by the user are invoked into a text editor. (In the LGO system the IBM editor PE is called, but other simple ASCII editors—such as the EDIT command in MS DOS—could also be used.) At this point, the user can check and modify the optimization input files: as mentioned above, the sample files TEST2.FOR and TEST2.IN can serve as a basis for preparing new files. After the editing session, the program returns to the main menu.

Compile Optimization Program Files

Selecting 'C filename' in the main menu, compilation of the user-dependent source code (filename.FOR file) follows. The actual realization of this step obviously depends on the programming environment available to the user (that is, it needs to be executed, depending on the available Fortran compiler). In case of compilation and/or linking errors, corresponding error messages appear, and the menu program prompts the user to return to the editing option. If the program compilation was successful, then—communicating this message to the user—control is returned again to the main menu level.

Run Optimization Program System

Having a successfully compiled and linked code, one can select 'R filename' in the main menu. This invokes the runtime environment, and the program will be executed (again, according to the development system available to the user). After the optimization run is terminated (successfully, or with runtime error messages), the system returns again to the main menu: repeated editing and/or analysis of results may then follow.

Analyse Optimization Output Files

Following a (formally) successful optimization run, one can invoke the result analysis mode, typing 'A filename'. The optimization output (filename.SUM and filename.OUT) files are called into the built-in ASCII editor, and they can be viewed or possibly edited (for purposes of subsequent use). Following the analysis of output files, control is returned to the main menu level.

Graphical Analysis of Results

Following an optimization run—again depending on the available environment—a graphical summary of the results can be activated by typing 'G filename'. This step typically needs the (automatic) loading of the output file(s) filename.OUT and filename.SUM. For instance, the sample points generated by the optimizer (their two-dimensional projection into a specified subspace), or a graph of the best function value found versus actual number of steps, etc., may follow. After inspecting these graphs, control is returned again to the main menu.

Quit

Typing 'Q' at the main menu, the LGO program system application is terminated, and the program returns to the operating system (DOS) from which LGO was called.

3.4.4 Illustrative Test Results

To illustrate the use of the LGO program system, the solution of two 50-variable test problems of the form (3.4.1) will be discussed below.

Example 1

In the first example, the objective function f is defined as follows.

$$f(\boldsymbol{x}) = f_1(\boldsymbol{x}) + f_2(\boldsymbol{x}) + f_3(\boldsymbol{x}), \quad \text{in which} \tag{3.4.2}$$

$$f_1(\boldsymbol{x}) = \sum_{i=1}^{n} i x_i^2$$

$$f_2(\boldsymbol{x}) = \sum_{i=1}^{n} (x_{i-1} + 5 \sin x_i + x_{i+1}^2)^2$$

$$f_3(\boldsymbol{x}) = \sum_{i=1}^{n} \ln^2(1 + |i \sin^2 x_{i-1} + 2x_i + 3x_{i+1}|).$$

(By definition, in (3.4.2) $x_0 = x_n$ and $x_{n+1} = x_1$; $\ln(\cdot)$ denotes the natural logarithm function.)

Let us make a few comments related to the function (3.4.2). The first sum, f_1, consists of the simplest possible quadratic terms. Nevertheless—due to the multipliers i of the individual terms, in case of large n—the resulting sum will become 'badly scaled': specifically, the rate of the largest and smallest axes in the ellipsoid level sets of the function f_1 equals $\sqrt{n}$. Note also that component f_1 serves to model the situation in which the 'far side' of the 'vaguely defined' search region $[a, b]$ contains only inferior solutions: this is typical in many well-posed global optimization problems of the form (3.4.1).

The two other component functions of f (the sums f_2 and f_3) model a noise structure superimposed on f_1, leading to a multiextremal function f in n variables. The function f_2 is a combination of embedded linear, trigonometric and quadratic terms, in simple quadratic function addenda. The oscillation amplitude of the trigonometric term is magnified by its multiplier factor. Finally, the function f_3 is made up by logarithmic terms, with embedded trigonometric and linear arguments. All embedded terms in f_2 and f_3 depend on three subsequent component of the vector x. One can also observe that the noise terms become relatively more significant in a closer neighbourhood of the optimal solution. In summary, although function f defined by (3.4.2) is not very complicated, it surely has 'somewhat tricky' nonlinearities and multiextremal structure, and in case of a large initial region of uncertainty (given by $[a, b]$) most local scope methods would probably have difficulties in finding the global solution—the zero vector.

```
a(1)= -8.8  a(11)=-7.8  a(21)=-8.5  a(31)=-3.5  a(41)=-5.5   (3.4.3)
b(1)=  1.4  b(11)= 3.2  b(21)= 7.2  b(31)= 1.9  b(41)= 2.2
a(2)= -6.2  a(12)=-5.2  a(22)=-1.2  a(32)=-6.2  a(42)=-3.2
b(2)=  0.9  b(12)= 3.9  b(22)= 4.9  b(32)= 3.9  b(42)= 4.9
a(3)= -8.7  a(13)=-6.1  a(23)=-5.7  a(33)=-7.1  a(43)=-4.3
b(3)=  1.7  b(13)= 1.8  b(23)= 3.5  b(33)= 3.5  b(43)= 7.8
a(4)= -7.7  a(14)=-2.7  a(24)=-7.7  a(34)=-7.7  a(44)=-4.7
b(4)=  0.8  b(14)= 4.2  b(24)= 1.5  b(34)= 4.8  b(44)= 2.5
a(5)= -3.2  a(15)=-5.6  a(25)=-8.6  a(35)=-5.6  a(45)=-3.6
b(5)=  5.3  b(15)= 3.3  b(25)= 5.3  b(35)= 2.3  b(45)= 8.3
a(6)= -3.5  a(16)=-7.1  a(26)=-9.5  a(36)=-6.5  a(46)=-4.5
b(6)=  7.9  b(16)= 2.9  b(26)= 6.8  b(36)= 2.8  b(46)= 1.9
a(7)= -5.1  a(17)=-2.1  a(27)=-5.1  a(37)=-5.1  a(47)=-4.1
b(7)=  8.7  b(17)= 6.7  b(27)= 3.7  b(37)= 8.7  b(47)= 1.7
a(8)= -2.2  a(18)=-5.2  a(28)=-6.7  a(38)=-3.2  a(48)=-7.2
b(8)=  4.7  b(18)= 3.7  b(28)= 1.7  b(38)= 1.7  b(48)= 3.2
a(9)= -9.1  a(19)=-4.1  a(29)=-4.1  a(39)=-5.1  a(49)=-4.1
b(9)=  3.8  b(19)= 2.8  b(29)= 1.8  b(39)= 1.8  b(49)= 1.8
a(10)=-6.3  a(20)=-7.3  a(30)=-4.3  a(40)=-3.3  a(50)=-5.3
b(10)= 1.7  b(20)= 4.7  b(30)= 6.7  b(40)= 7.7  b(50)= 1.3
```

In the test example, we chose $n = 50$. The components of the lower and upper bound vectors a and b are given in (3.4.3). (The numerical values and results are directly cited from the output files LGOTEST.OUT and LGOTEST.SUM, except for minimal necessary text editing.)

The value of the objective function (3.4.2) at the lower and the upper bound vectors equals $f(a) = 106613.50$, and $f(b) = 80112.91$. These values indicate that the Lipschitz constant will be quite large, when $[a, b]$ is scaled—automatically, by the optimization algorithm—into the 50-dimensional unit interval.

To minimize function (3.4.2) on $[a, b]$ of (3.4.3) by the LGO program system, the maximal number of global and local phase search steps was chosen as $\max_g = 2 \cdot 10^5$ and $\max_l = 10^5$, respectively. To put these 'large' values into proper perspective, observe that a 50-dimensional interval has, for example, $2^{50} > 10^{15}$ vertices: hence, the above values $\max_g$ and $\max_l$ can be considered quite modest. The other LGO parameters were chosen in such a manner that $\max_g$ and $\max_l$, in effect, served to control the program execution. It should be noted at this point that—since these stopping criteria are not checked at every single function evaluation—the actual number of global/local search steps may slightly exceed the stated bounds $\max_g$ / $\max_l$.

The results of the optimization run are summarized below.

Global Search Phase

Number of function evaluations: 200082.

Optimum estimate found:

Function value: 2860.00

Estimated optimal solution vector (the components x_1 to x_{50} are listed sequentially, by rows):

```
-0.582359E+00  -0.480671E+00  -0.323575E+00  -0.856682E+00  -0.552137E+00
 0.556333E-01  -0.782664E+00  -0.312315E-01   0.131451E+01  -0.182171E+01
-0.164235E+01  -0.105945E+00  -0.167769E+01   0.116252E+01  -0.617902E+00
-0.150214E+01   0.282612E+01  -0.217902E+00  -0.237475E+00  -0.582565E+00
 0.288644E+00   0.221470E+01  -0.549966E+00  -0.254997E+01  -0.818971E+00
-0.375484E+00  -0.173881E+00  -0.199780E+01  -0.797261E+00   0.185765E+01
-0.477154E+00  -0.546158E+00  -0.116627E+01  -0.702672E+00  -0.117769E+01
-0.129399E+01   0.262505E+01  -0.457047E+00  -0.123747E+01   0.285765E+01
-0.118965E+01   0.133427E+01   0.247341E+01  -0.669539E+00   0.306146E+01
-0.917368E+00  -0.853239E+00  -0.137822E+01  -0.797261E+00  -0.160541E+01
```

Although this vector is, obviously, not a very precise approximation of the optimum—considering the initial large search interval—it is not too far from it. (Increasing max_g, improved global phase optimum estimates can be provided.) The subsequent local search, automatically launched from this point essentially finds the global optimum, as shown by the results below. (Note that, in fact, the local search phase was automatically terminated by LGO—in the absence of appreciable improvement of the objective function value—before the preset maximal number of steps would have been attained.)

Local Search Phase

Number of additional function evaluations: 50262.

Estimated optimum value: 0.235088E-04.

Estimated optimal solution vector (the components x(1) to x(50) are listed sequentially, by rows):

0.169440E-03	0.162151E-03	-0.272220E-03	-0.115536E-03	0.525944E-03
0.355519E-03	0.460719E-03	0.303982E-03	-0.247225E-03	0.193031E-04
0.138270E-03	0.138394E-04	-0.491507E-04	0.161346E-03	0.834514E-04
0.256642E-03	0.895554E-04	0.767251E-04	-0.388216E-04	-0.397363E-04
-0.137227E-04	-0.137235E-03	-0.415906E-04	0.989787E-04	-0.825170E-04
-0.384140E-04	-0.253712E-03	0.181291E-04	-0.158941E-03	0.935380E-04
-0.204893E-03	0.757348E-04	0.123985E-03	-0.754476E-04	0.155654E-03
0.548079E-05	0.286140E-03	0.866373E-04	-0.107684E-03	0.120136E-03
0.924189E-04	0.176875E-04	0.365546E-04	0.183546E-03	0.125534E-03
0.965736E-04	0.410046E-04	-0.621115E-04	-0.803805E-04	-0.169211E-03

Considering the fact that the LGO solver does not rely on gradient information, the solution found is considered as an acceptable approximation of the zero vector.

The total computational effort—in terms of function evaluations—was 250344. According to the built-in PC timer, the described LGO test run was started at 12:28:35.10, and terminated at 12:56:42.79. The total runtime, hence, was about 28 min (on a PS/2 Intel 386&387DX, 8 Mbyte RAM PC, in a Lahey Fortran 77 runtime environment.) Note finally that this example was 'not too difficult' for LGO, but the global search phase was essential. To illustrate this point, note that launching a local search phase from the best solution found among 100 uniformly generated random sample points leads to a suboptimal solution in some 44,650 additional steps, when the local search is automatically terminated. (The corresponding objective function value equals 35.04.)

Example 2

In the second example, the objective function f is defined as follows.

$$f(x) = f_1(x) + f_2(x) + f_3(x), \qquad \text{in which} \tag{3.4.4}$$
$$f_1(x) = \sum_{i=1}^{n} i x_i^2$$
$$f_2(x) = \sum_{i=1}^{n} i \sin^2(x_{i-1} \sin x_i - x_i + \sin x_{i+1})$$
$$f_3(x) = \sum_{i=1}^{n} i \ln(1 + i(x_{i-1}^2 - 2x_i + 3x_{i+1} - \cos x_i + 1)^2).$$

Again, by definition, in (3.4.4), $x_0 = x_n$ and $x_{n+1} = x_1$. (The unique global minimum of (3.4.4) is the zero vector.)

Similar observations to those in connection with function (3.4.2) can be made. We have again the—rather badly scaled—quadratic function f_1, to which badly scaled noise terms are added this time. Although the individual terms in functions f_2 and f_3 can be relatively small, the ith terms are multiplied by the same factor i, as in f_1. Consequently, the relative 'noise effect' referred to can become quite substantial, especially for certain vectors x in the neighbourhood of the optimum. In summary, we have a similar test problem to that of Example 1, but the noise terms are probably somewhat more 'tricky'. In the present test example, we choose again $n = 50$; the components of the lower and upper bound vectors a and b are also given by (3.4.3), as in Example 1.

To minimize function (3.4.4) on $[a, b]$ of (3.4.3) by the LGO program system, the maximal number of global and local phase search steps is chosen now as $max_g = 10^6$ and $\max_l = 10^5$, respectively. Although there are other built-in termination parameters, in the test example these values serve again to control the execution of LGO.

The results of the optimization run are summarized below; for additional related comments, recall the discussion of Example 1.

The value of the objective function (3.4.4) at the lower and the upper bound vectors equals $f_a = 51386.43$, and $f_b = 37781.05$—implying again a quite significant overall Lipschitz constant value.

Global Search Phase

Number of function evaluations: 1000405.

Optimum estimate found:

Function value: 11021.52

Estimated optimal solution vector (the components x(1) to x(50) are listed sequentially, by rows:

-0.872199E+00	-0.261361E+00	0.958211E+00	0.355886E+00	0.977738E+00
0.255926E+01	0.195261E+01	0.629367E-01	-0.221148E+01	-0.304696E+00
-0.174208E+01	-0.188451E+00	-0.174932E+01	0.109997E+01	-0.698596E+00
-0.159280E+01	0.274633E+01	-0.298595E+00	-0.300035E+00	-0.691365E+00
0.146298E+00	0.215939E+01	-0.633379E+00	-0.263338E+01	-0.944997E+00
-0.523271E+00	-0.253667E+00	-0.207396E+01	-0.850754E+00	0.175792E+01
-0.526114E+00	-0.637731E+00	-0.126237E+01	-0.816005E+00	-0.124932E+01
-0.137831E+01	0.249993E+01	-0.501474E+00	-0.130003E+01	0.275792E+01
-0.125946E+01	0.126083E+01	0.236371E+01	-0.734819E+00	0.295356E+01
-0.975394E+00	-0.905826E+00	-0.147252E+01	-0.850754E+00	-0.166525E+01

One should recall that the objective function value is the sum of 150 terms, several of them being highly 'amplified'. This explains the large objective function value, found in a relatively close vicinity of the global optimum. Again, the subsequent local search phase automatically refines this solution.

Local Search Phase

Number of additional function evaluations: 64776.

Estimated optimum value: 0.116325E-02.

Estimated optimal solution vector (the components x(1) to x(50) are listed sequentially, by rows):

0.246135E-04	-0.496429E-02	-0.256523E-02	-0.948861E-03	0.523864E-04
0.546945E-03	0.743775E-03	0.942086E-03	0.904422E-03	0.825253E-03
0.707861E-03	0.607685E-03	0.559501E-03	0.493742E-03	0.403791E-03
0.312093E-03	0.200017E-03	0.940176E-04	0.361621E-04	-0.168630E-04
-0.556142E-04	-0.722397E-04	-0.671281E-04	-0.372672E-04	-0.188569E-04
-0.130694E-04	-0.354038E-04	-0.456414E-04	-0.669419E-04	-0.683881E-04
-0.887829E-04	-0.146445E-03	-0.142459E-03	-0.137900E-03	-0.103197E-03
-0.680875E-04	-0.669518E-04	-0.721837E-04	-0.702192E-04	-0.619646E-04
-0.476193E-04	-0.441888E-04	-0.424293E-04	-0.609721E-04	-0.608845E-04
-0.573608E-04	-0.468054E-04	-0.331098E-04	-0.107399E-04	0.107129E-04

Again—since LGO is a derivative-free solver—the accuracy obtained seems to be acceptable.

The total number of objective function evaluations in Example 2 is 1065181. The LGO run was started at 09:25:01.22, and was automatically terminated at 13:37:22.03—the runtime is over four hours(!). This can be said to be quite significant, but problem (3.4.4) indeed is a bit 'tricky', and—at least in the field of numerical global optimization—the number of variables is not too small. On faster personal computers, workstations, mainframes, supercomputers, parallel computer architectures, etc. (available today), the runtime could be reduced by, say, a factor of 10-500. The essential numerical message, however, is that—for all speed computers of today and tomorrow are able to provide—it is easy to find (test and real) global optimization problems which would immediately call for more power.

A variety of further numerical examples, related to diverse applications, will be discussed in Part 4. Since in many of these applications—and also in other areas of multiextremal optimization—uncertainties and statistical fluctuations will play a significant role, in the remainder of Part 3 stochastic decision models and solution techniques will be discussed. Within this framework, we shall discuss a general class of stochastic solution strategies that can be applied to solve—deterministic or stochastic—global optimization problems, under mild analytical assumptions. This will provide another generic methodology for continuous and Lipschitzian global optimization (and, in fact, will be applied in relation to several applications in Part 4).

Chapter 3.5

Decision Making under Uncertainty: Stochastic Model Forms

3.5.1 Introduction

In deterministic management models it is assumed that all relevant problem characteristics are known with certainty (or that with respect to these characteristics sufficiently exact point estimates can be provided). Although this assumption can be essentially valid in connection with many engineering problems, it certainly does not hold automatically, for instance, in the context of environmental modelling and decision making. To illustrate this point, note that—even in reasonably 'well-defined' engineering systems—the expert estimates of system failure probabilities and consequences may differ by several orders of magnitude (especially in the case of 'rare' events with unforeseeable serious consequences).

In many decision problems of practical relevance, essential components of the problem can be regarded as partially unknown, uncertain, or statistically fluctuating. Stochastic optimization models explicitly take this fact into account, handling model uncertainties as random variables. This implies that 'rigid' deterministic criteria are to be replaced by suitable stochastic 'model building blocks'. In the following discussion a number of such model blocks will be presented briefly. For more detailed expositions on the subject of stochastic modelling and optimization, see, for example, Ermoliev (1976), Kall (1976), Yudin (1979), Dempster (1980), Kall and Prékopa (1980), Prékopa (1980), Pintér (1984b), Berger (1985), French (1986), Klein

Haneveld (1986), Hillier and Lieberman (1986), Kumar and Varaiya (1986), Prékopa and Wets (1986), Ermoliev and Wets (1988), Pintér (1990d), Frauendorfer (1992), Kushner and Dupuis (1992), Wets (1994).

The basic components of stochastic programming models considered below are the following:

- a decision vector x belonging to the feasible set $D \subset \mathbb{R}^n$
- a vector random variable ξ, $\xi(w) \in \mathbb{R}^m$
- an evaluation procedure ('rule', function(s)) of the decision x, which depends also on the (possible, random) realizations of ξ.

For illustrative purposes, consider one of the usual environmental management paradigms, expressed verbally as

$$\begin{aligned}&\{\text{obtain 'best possible' environmental state}\} \\ &\{\text{all feasible decision options have to meet given resource constraints}\}.\end{aligned} \tag{3.5.1}$$

Now define $C(x)$, a real function which expresses the resource ('cost') implications, and $U(x)$, the (positive, scalar) environmental effect ('utility') of decision x: then, for example,

$$\begin{aligned}&\max U(x) \\ &C(x) \leq C_{\max}\end{aligned} \tag{3.5.2}$$

can be interpreted as a (very simple) valid deterministic mathematical programming formulation of the problem statement (3.5.1). For example—in the context of a groundwater quality management problem—x may denote locational, injection, and/or pumping rate decisions related to reclamation wells; $C(x)$ may then express the total investment plus discounted operational costs for the time horizon considered; and $U(x)$ may be the groundwater quality attained. As this example indicates, the dependence on the decision x may not be obvious, and often it is, indeed, quite complicated.

For explicitly including stochasticity aspects into our model, functions U and C in (3.5.2) can be formally extended as

$$\begin{aligned}&\text{'}\max U(x,\xi)\text{'} \\ &\text{'}C(x,\xi) \leq C_{\max}\text{'}.\end{aligned} \tag{3.5.3}$$

The symbols '. . .' in the formulation (3.5.3) indicate that—the values of the functions U and C themselves being now random variables—the direct model extension (3.5.3) has no rigorous mathematical meaning. Therefore, applying the concepts and tools of probability theory and statistics, mathematically sound optimization model forms should replace (3.5.3).

Before proceeding further, let us note that in the discussion below it will be tacitly assumed that all uncertainties, fluctuations, etc., can be adequately characterized using probabilistic concepts and tools: naturally, this assumption is valid only via certain abstractions. In particular, we shall suppose that the necessary statistical characteristics of the modelled random variables (such as their mean, variance, covariance, or possibly their probability distribution function, p.d.f.) can be sufficiently well estimated, when this is (implicitly) required by the suggested model form.

3.5.2 Model Variants and Solution Approaches

Statistical Performance Analysis

First, two fairly straightforward, but—as a rule—computationally intensive methods are described: both of them may serve to analyse the influence of the stochasticity in the decision problem. Assume first that a feasible decision is selected; then one might be interested in the statistical performance characteristics of that decision. A conceptually simple, 'universal' tool is to apply Monte Carlo (stochastic) simulation (see, for example, Fishman, 1978; Rubenstein, 1981; or Bratley, Fox and Schrage, 1983). Generating (typically within a computer based model) a sufficiently large number of realizations of all important random variables, one can estimate, for example, the—scalar or vector—probability distribution function of the system behaviour, or some more aggregated characteristics like its expected (average) performance. Recalling the groundwater quality management example: if a well location and operational scheme are chosen, then by Monte Carlo simulation one can derive a collection of random realizations (trajectories) of the resulting water quality in space and time. Note that the 'sufficient' number of Monte Carlo simulation cycles can be determined, applying classical results of probability theory; see, for example, Feller (1971): this issue will be discussed later in Chapter 3.7.

Applying the outlined Monte Carlo method to several selected ('promising') decision variants, one can obtain a detailed statistical evaluation of those alternatives. Note that this approach does not necessitate any sort of formal optimization, although the options investigated can be derived, say, by solving a corresponding (probably simplified) optimization model of the problem analysed.

A second direct approach, closely related to the above, is based on a 'reverse' philosophy. In contrast to the prior selection of decision variants and the posterior elicitation of random model components, it may also be instructive to analyse variations/changes in the 'optimal' solution, as a function of the occurring random elements in the scenarios (this issue is called the distribution problem in stochastic programming). In other words, generating joint realizations of all stochastic model elements, one can determine the corresponding 'best' strategy, by solving each of the resulting deterministic problems. Evidently, the computational burden of this approach can be much greater than that of the previously considered method.

The directly applicable techniques outlined may lead to two opposite conclusions. If the performance statistics of the decisions investigated—in terms of U and C—vary only to a negligible (acceptable) extent, then the analyst may conclude that the random factors considered have no significant influence on the results. In this case an appropriately chosen aggregated description, say, taking temporal and/or spatial averages of the random factors, seems to be justified. In the opposite case—experiencing significant variations in the performance of the analysed decisions or in the optimal decision, depending on the occurring particular realizations of the random circumstances—there is a strong quantitative rationale to replace the basic deterministic model by its suitable stochastic extension. The numerical solution of such stochastic models frequently necessitates the proper combination of optimization strategies with Monte Carlo simulation.

In the following discussion several approaches to stochastic modelling will be highlighted. Note that, generally speaking, the model forms presented below are non-equivalent: hence, there remains the important issue of selecting such formulation(s) that correspond 'best' to the problem at hand.

Expected Value Models

Taking the expected (mean) value of all involved random variables in the 'improper' model form (3.5.3), one obtains the deterministic optimization problem

$$\begin{aligned} &\max U(x, E\xi) \\ &C(x, E\xi) \leq C_{\max}. \end{aligned} \tag{3.5.4}$$

$E\xi$ (or $E\{\xi\}$) denotes the expected value of the argument random variable ξ. Note that if U and C are nonlinear functions of the random realizations of ξ, then (3.5.4)—as a rule—is different from the model

$$\max E\{U(x,\xi)\} \qquad (3.5.5)$$
$$E\{C(x,\xi)\} \le C_{\max}.$$

The latter formulation can also be considered as a 'natural' expected value model based on (3.5.3).

Evidently, both (3.5.4) and (3.5.5) are aimed—though probably in a different form—at the maximization of the expected utility of the decision x, under resource constraints which are also interpreted in a mean value sense. In certain cases (for example, when U and C are linear functions of ξ), the explicit form of dependence on the decision variable x can be easily derived. In general, however, the evaluation of the expected value operators may need the application of auxiliary numerical techniques (such as, for example, multidimensional integration via Monte Carlo simulation).

Both (3.5.4) and (3.5.5) are often applied, and in many cases they actually serve as the implicit motivation for deriving deterministic models. However, if the underlying 'expected value philosophy' does not seem to be acceptable—that is, the random variables are not sufficiently well characterized by their mean value, but, say, their variance/covariance structure is also important—then the familiar 'averaging' description is obviously of limited value. More precisely, the consideration of additional statistical constraints on the acceptable performance of decisions will generally restrict the set of feasible alternatives: thus, ignoring them one may arrive at decisions which have questionable performance.

Expected Value–Standard Deviation (E-D) Models

Consider now the following model form:

$$\max E\{U(x,\xi)\} - r_1 D\{U(x,\xi)\} \qquad (3.5.6)$$

$$E\{C(x,\xi)\} + r_2 D\{C(x,\xi)\} \le C_{\max}.$$

In (3.5.6), $D\{\cdot\}$ denotes the standard deviation of the embedded random variable, that is, $V(\xi) := E\{(\xi - E\xi)^2\}$, $D(\xi) = V^{1/2}\{\xi\}$; r_1 and r_2 are given positive parameters of the model.

As can be seen, the objective function in (3.5.6) is again aimed partly at the maximization of the expected benefits of decision x. This time, however, the variance of these benefits is also taken into account, expressing the fact that decisions leading to largely fluctuating utility outcomes are less favourable. Hence, this model reflects a 'risk averse' attitude to decision making: it can happen that decisions with 'smaller, but safer' return are preferred to 'more profitable, but riskier' ones (one may think of financial applications of this principle).

Analogous comments are valid with respect to the constraint (set) of (3.5.6): it is required that not only the expected value of the resource demands, but its sum with an additional parameterized standard deviation term, should remain below the 'rigid' bound $C_{\max}$. Note that increasing the values of r_1 and r_2 yields gradually more 'safe' decisions, and (typically) decreasing expected benefits. More precisely, if r_1 and r_2 tend to zero, then (3.5.6) approximates the model (3.5.5); furthermore, if r_1 tends to infinity, then the minimum variance solution is sought, and if r_2 tends to infinity, then the problem gradually becomes infeasible. Having these observations in mind, it may be instructive to analyse the 'menu' of solutions $x(r_1, r_2)$, derived under various parameterizations (r_1, r_2).

Let us note finally that the E-D model form introduced can be of particular relevance in those cases, when the p.d.f. of the involved random variable is completely characterized by the mean and variance values (examples being many of the two-parameter p.d.f.'s: the normal, lognormal, standard gamma p.d.f., and so on).

Chance Constrained Models

The probabilistic constrained model formulations presented below explicitly reflect the fact that—due to the random factors in the problem studied—one may have to be satisfied with decisions that are 'acceptable', at least for most realizations of the random factors considered. (For instance, water distribution or environmental quality management models may fall into this category.) This verbal statement can be mathematically formalized in different ways, leading, for example, to the model variants (3.5.7) and (3.5.8) below. The first model is

$$\max P\{U(x,\xi) \geq U_{\min}\} \tag{3.5.7}$$

$$E\{C(x,\xi)\} + rD\{C(x,\xi)\} \leq C_{\max}.$$

In (3.5.7), $P\{\cdot\}$ denotes the probability of the embedded random event, $U_{\min}$ is a given model parameter (the target utility level); in addition, as above, $r > 0$ parameterizes the risk aversion aspect in the E-D type resource constraint. The second model form, presented for illustration, is

$$\begin{aligned} &\min E\{C(x, \xi)\} \\ &P\{U(x, \xi) \geq U_{\min}\} \geq p. \end{aligned} \tag{3.5.8}$$

In (3.5.8) the parameter p expresses a prescribed reliability level that, in most practical problems, is chosen near to 1.

The two models introduced above reflect closely related approaches to handling uncertainty. In (3.5.7) 'safe' resource constraints are to be met and, under this condition, the objective is to find a feasible decision that maximizes the probability of a 'satisfactory' environmental state. In (3.5.8) one is seeking that decision which is expected to be the minimum-cost option, while it guarantees 'acceptable' environmental conditions with a sufficiently high probability. Note that the solutions of both problems are parameter-dependent: in particular, it may also happen that (3.5.7) or (3.5.8) has no feasible solutions. It is worth emphasizing at this point that such infeasibility does not necessarily mean 'bad' model parameterization: for example, if (3.5.8) has no solution, then this means that the stated probabilistic environmental quality requirement simply cannot be met—therefore either the target level $U_{\min}$ or the probability bound p on its attainment should be respectively changed (decreased). Analogous comment applies with respect to the setting of parameters r and $C_{\max}$ in model (3.5.7).

The literature discussing and applying chance constrained models is extensive; the present exposition is confined to a qualitative summary concerning some of the main theoretical and numerical tractability/solution aspects. Let us remark first that the general Monte Carlo approach can obviously be applied to check the solution variants considered: this is to be combined with optimization steps, in order to derive an improving sequence of decisions. In certain cases, there is a possibility of applying deterministic approximations of chance constraints: first, one of these will be introduced here in summary form, together with some additional notes on further approximations.

Following the classical approach of Charnes and Cooper (1962), certain separate (scalar) chance constraints can be directly approximated. Specifically, consider a scalar random variable ξ and the associated chance constraint (CC)

$$P\{\xi \le c\} \ge p \qquad (c \text{ and } 0 < p < 1 \text{ are fixed model parameters}). \quad (3.5.9)$$

Under simple analytical conditions, this CC can be equivalently transcribed as

$$P\{\xi \le c\} = P\Big(\frac{\xi - E\xi}{D\xi} \le \frac{c - E\xi}{D\xi}\Big) = H\Big(\frac{c - E\xi}{D\xi}\Big) \ge p,$$

where $H(\cdot)$ is the p.d.f. of the corresponding standardized random variable $\xi^s = (\xi - E\xi)/D\xi$. For instance, assume now that ξ is normally distributed (this assumption is frequently justified by the central limit theorem; see, for example, Feller, 1971). Then $H = \Phi$ (Φ is the standard normal p.d.f.) and (3.5.9) is equivalent to the explicit deterministic constraint

$$E\xi + \Phi^{-1}(p)D\xi \le c.$$

Note that $\Phi^{-1}(p) \ge 0$ for $p \ge 0.5$; hence, in most practically relevant cases the constraint derived is stricter than the corresponding expected value constraint $E\xi \le c$.

In general, recalling the expected value–standard deviation models, a close relationship between CCs and parametric constraints of the form

$$E\xi + rD\xi \le c, \qquad r = H^{-1}(p),$$

can be immediately recognized. Without going into further detail, unnecessary here, let us mention that a number of other—usually more complicated—explicit deterministic equivalents and approximations of chance constraints are known; see, for example, the results surveyed and extended by Pintér (1989), with further references therein. These approximations are, in general, applicable to separate chance constraints of the form (3.5.9), when limited parametric information concerning the embedded random variable ξ—such as, for example, its mean, variance, upper bound, range, etc.—is available. Note also that a number of formally more complicated functions of random variables, for example, their linear combinations, can be deduced to (3.5.9), but this type of transformation is certainly not 'universally applicable'.

Note finally that, for a fairly broad class of simultaneous (joint) chance constraints stated with respect to a vector of random variables, convexity results are known (see, for example, Prékopa, 1973; or the survey of Wets, 1983). This property implies remarkable numerical advantages concerning the optimization module applicable during the solution procedure (as in the case referred to, the set of feasible decision variants in (3.5.8) is convex). Nevertheless, the evaluation of the joint chance constraints, generally speaking, can only be accomplished via Monte Carlo simulation techniques. Therefore it is of significant practical interest to approximate joint chance constraints, say, via separate constraints (see, for example, Galambos, 1977; or Prékopa, 1986).

Bounding Unfavourable Outcomes

Although chance constrained models explicitly take into account desirable or undesirable system performance, they do not offer any quantitative guarantee with respect to the extent of possibly occurring unfavourable events. (Consult Hogan, Morris and Thompson (1981) for a related discussion, with many references.) For example, a particular decision alternative may assure 'good' groundwater quality with 98 per cent probability, but in the remaining cases water quality may become absolutely unacceptable. This deficiency can be partly rectified by introducing appropriate additional restrictions. For example, the chance constraint

$$P\{U(x,\xi) \geq U_{\min}\} \geq p$$

can be supplemented by a requirement of the form

$$E\{U(x,\xi) \mid U(x,\xi) < U_{\min}\} \geq U_{\min} - d. \tag{3.5.10}$$

The relation (3.5.10) expresses the requirement that—even if the utility level $U_{\min}$ cannot be attained—the expected outcome (the conditional expectation) of such random occurrences will not be beyond control: the parameter $d > 0$ reflects this safety requirement. Note that the feasibility of (3.5.10) is not assured automatically (recall the remarks made with respect to the parameterization of (3.5.7) and (3.5.8)).

A similar philosophy to (3.5.10) is expressed by the constraint

$$E\{\max[0, U_{\min} - U(x,\xi)]\} \leq e, \tag{3.5.11}$$

see Klein Haneveld (1986). In (3.5.11), $e > 0$ is a model parameter, by which the (unconditional) expectation of the utility $U(x, \xi)$ is bounded, in the case of unfavourable realizations of ξ.

With respect to the 'model building blocks' (3.5.10) and (3.5.11), similar comments to those made for chance constraints are valid. Note, however, that their analytical properties (smoothness, convexity, etc.) are often more favorable than those of the probabilistic constraints.

Target Level Models

Suppose now that there exists an 'ideal' target system state. In accordance with the above discussion, the corresponding utility itself can typically be interpreted as a random variable $U_{\text{ideal}}(\xi)$, probably depending also on some 'ideal' decision: $U_{\text{ideal}}(\xi) := U_{\text{ideal}}(x_{\text{ideal}}, \xi)$. (Take, for example, the groundwater quality that would result, in the absence of all controllable man-made pollution effects.) The model form presented below is aimed at the approximation of this 'ideal' state, under resource constraints (the latter is simply considered here in the expected value sense):

$$\begin{aligned} &\min \rho\{U(x, \xi), U_{\text{ideal}}(\xi)\} \\ &E\{C(x, \xi)\} \leq C_{\max}. \end{aligned} \tag{3.5.12}$$

In (3.5.12), ρ denotes an appropriate, statistically established discrepancy measure that expresses the 'distance' between the probability distribution functions of the 'ideal' and the actually realized environmental quality. Examples of ρ may then be based on statistical hypothesis tests or on some other usual deviation measures (a collection of possible functions ρ is presented later, in Chapter 4.3). As an example, let us cite here only the symmetrized form of the Fisher–von Mises statistics:

$$\begin{aligned} &\rho\{U(x, \xi), U_{\text{ideal}}(\xi)\} \\ &= \int_{-\infty}^{\infty} |F_x(z) - F_{\text{ideal}}(z)| dF_{\text{ideal}}(z) + \int_{-\infty}^{\infty} |F_x(z) - F_{\text{ideal}}(z)| dF(z). \end{aligned} \tag{3.5.13}$$

In (3.5.13), $F_x(z)$ and $F_{\text{ideal}}(z)$ denote respectively the p.d.f. of the resulting outcomes (parameterized by the decision x) and of the 'ideal' outcomes.

Recourse Models

The basic 'underlying philosophy' of the stochastic model forms introduced so far suggests that, first, a decision is taken (that cannot be changed later), and then the random factors are realized. In this way, all these models reflect a static decision-making procedure. Although this concept is realistic in many actual situations—related, for example, to the long-term design of reservoirs, wastewater treatment plants, etc.—it is not a 'true' problem representation, when dynamic system characteristics also play a role and adaptive decision making is required. This comment motivates the formulation of stochastic decision models with a consideration for possible recourse actions.

For illustration, we shall consider a two-stage problem formulation. Assume that first a decision x_1 is to be chosen; then an observation considering the realizations of the involved random variables ξ_1 (of that decision stage) is made. Following this, there is a possibility of taking another—supplementary, recourse—decision x_2 that, in general, may depend on both x_1 and ξ_1: $x_2 = x_2(x_1, \xi_1)$. In the context of the groundwater quality management example, stages may represent, for example, several subsequent years: this way, decision x_1 may be based on our prior knowledge about the environmental system considered, including its statistical behaviour, while decision x_2 may symbolize an additional adjustment based on the effect of x_1 and probably some more information related to stochastic system behaviour. Evidently, the concept outlined can be naturally generalized for handling an arbitrary (finite) number of decision-making stages: to keep the notation simple, only the two-stage case will be discussed below.

Similarly to what was said above, there are many non-equivalent possibilities of formulating dynamic models (evidently so as, inter alia, the single-stage model variants presented above can all be extended to yield corresponding multistage generalizations). Let us consider, for example, the chance constrained model (3.5.8)

$$\begin{aligned}
&\min E\{C(x,\xi)\}\\
&P\{U(x,\xi) \geq U_{\min}\} \geq p.
\end{aligned}$$

A possible two-stage extension then has the form (using notation analogous to that introduced above)

$$\begin{aligned}
&\min C_1(x_1) + E\{C_2[x_2(x_1,\xi_1)]\}\\
&P\{U_1(x_1,\xi_1) \geq U^{(1)}_{\min}\} \geq p\\
&E\{U_2(x_1,\xi_1,x_2)\} \geq U^{(2)}_{\min}.
\end{aligned} \tag{3.5.14}$$

The model (3.5.14) can be interpreted as follows. The first-stage decision x_1 is to be selected in such a way that the probabilistic constraint is satisfied (that is, the utility of the decision is bounded from below by $U_{\min}^{(1)}$, at least with probability p). After having realized x_1 and observed ξ_1, x_2 is chosen to satisfy the second model constraint (the expected two-stage utility U_2 should be above $U_{\min}^{(2)}$). Under these conditions, the summed expected costs of the two-stage decision are to be minimized. (Note that in (3.5.14) it is assumed that the cost C_1 is a deterministic function of x_1, while the embedded cost function C_2 is a random variable.)

It is usual to investigate models with complete recourse, assuming that the last (here, the second) stage decision makes possible the exact satisfaction of the corresponding constraint. Under this assumption, the third relation in (3.5.14) may have the form

$$U_2(x_1, \overline{\xi}_1, x_2)\} = U_{\min}^{(2)};$$

here $\overline{\xi}_1$ is the actually occurring realization of the random variable ξ_1.

Let us remark finally that the numerical solution of recourse models—especially in cases of complete recourse—is typically based on discretization of the probability space (that is, the discretization of the p.d.f. of ξ_1) and the construction of corresponding approximation schemes; see, for example, Kall (1976, 1987), Wets (1983, 1994), Gassmann and Ziemba (1986), Frauendorfer (1992), and references therein.

3.5.3 Conclusions

Closing the above brief discussion of diverse stochastic model variants, one can see that—given an appropriate information base and a corresponding model—these may provide valuable additional insight into the problem characteristics, when compared to more traditional deterministic modelling approaches. Of course, none of the stochastic models is 'perfect' or 'complete' either: rather, they should be conceived as possible modules ('building blocks') of a general 'conceptual toolkit' to assist decision making, in the presence of uncertainty.

With respect to numerical solution aspects, the following main points deserve attention. In certain cases, it is possible to apply diverse approximations of relevant stochastic programming model forms: in these cases deterministic optimization methodology can be directly applied to solve the models derived. Frequently, however, there is no possibility of applying such approximations, and hence the solution method is to be based on combined Monte Carlo simulation and stochastic optimization. (It is useful to recall in this context that accomplishing a sufficient number of Monte Carlo simulation cycles, a broad class of statistical requirements—interpreted as possible model constraints—can be verified simultaneously: this comment has obvious importance, considering the computationally intensive character of most stochastic programming model versions.) In the subsequent two chapters the issues of optimizing and evaluating stochastic systems will be discussed in more detail.

Chapter 3.6

Adaptive Stochastic Optimization Procedures

3.6.1 Introduction

Consider the following stochastic programming problem:

$$\begin{aligned} &\min E f_0(x,y) && \\ &E f_i(x,y) \le 0, && i=1,\dots,I, \\ &x \in D_0 \subset \mathbb{R}^n, && y(w) \in \mathbb{R}^q. \end{aligned} \tag{3.6.1}$$

In (3.6.1), x is a decision vector to be selected from a closed, bounded subset D_0 of the n-dimensional real Euclidean space $\mathbb{R}^n$; y is a q-dimensional vector valued random variable; f_i, $i = 0, 1, \dots, I$, are respectively defined measurable functions; E is the symbol of mathematical expectation (the expected values are supposed to exist). Note that (3.6.1) is a fairly general stochastic programming model form; it encompasses—under suitable transformations—the 'model blocks' and types discussed in the previous chapter.

The continuous structure and the existence of optimal solution of different stochastic optimization problems can be assured under rather weak assumptions, and essentially independently from the type of probability distribution function of the involved random variables. To illustrate this point, we shall prove the following assertion.

Theorem 3.6.1. *Assume that the following conditions are fulfilled:*

- *The set of feasible vectors in* (3.6.1) *is nonempty.*

- *The p.d.f.* $H(z) = P(y < z)$, $z \in \mathbb{R}^q$, *is absolutely continuous with respect to the Lebesgue measure* $\mu = \mu^{(q)}$; *the density function will be denoted by* $h(z) = H'(z)$.
- *There exists a support set* $S \subseteq \mathbb{R}^q$ *such that*

$$\int_S h(z)\,dz = 1.$$

- *The functions* $f_i(x, z)$ *are measurable in* $z = y(w)$ *for each* $x \in D_0$, *and* $i = 0, 1, \ldots, I$.
- *For each* $\varepsilon_i > 0$, $i = 0, 1, \ldots, I$, *a corresponding* δ_i *can be chosen, such that* $x_1, x_2 \in D_0$, $\|x_1 - x_2\| < \delta_i$ *implies*

$$\sup_{z \in S} |f_i(x_1, z) - f_i(x_2, z)| < \varepsilon_i, \qquad i = 0, 1, \ldots, I.$$

Then (3.6.1) has a globally optimal solution.

Proof. Define

$$F_i(x) = Ef_i(x, y) = \int_{\mathbb{R}^q} f_i(x, z)h(z)\,dz, \qquad i = 0, 1, \ldots, I. \tag{3.6.2}$$

We shall prove that $F_i(x)$ is continuous on D_0. Then, since the feasible set is nonempty, bounded, and (by continuity of F_i, $i = 1, \ldots, I$) closed, and F_0 is continuous, one can directly apply the theorem of Weierstrass, which guarantees the existence of a global solution.

For arbitrary $\varepsilon_i > 0$, one can write

$$\begin{aligned} |F_i(x_1) - F_i(x_2)| &\le \int_{\mathbb{R}^q} |f_i(x_1, z) - f_i(x_2, z)| h(z)\,dz \\ &\le \sup_{z \in S} |f_i(x_1, z) - f_i(x_2, z)| < \varepsilon_i, \end{aligned}$$

if $x_1, x_2 \in D_0$ and $\|x_1 - x_2\| < \delta_i$.

The continuity of F_i, $i = 0, 1, \ldots, I$ in (3.6.2) holds, leading directly to the assertion. ■

Note that, in spite of the mild continuity requirements, differentiability assumptions related to (3.6.1) may not be valid. Furthermore, although there exist results which assure convexity of certain broad classes of stochastic programming problems (see, for instance, Prékopa, 1973; or the survey of Wets, 1983), the sufficient conditions of these theorems are not always satisfied. Therefore it is important to study stochastic optimization methods which are applicable also for solving (deterministic or stochastic) optimization problems—under continuity, but even without Lipschitz-continuity assumptions.

The known iterative stochastic algorithms can basically be divided into two groups: stochastic approximation type methods and their generalizations (see, for example, Vaysbord and Yudin, 1968; Kushner and Clark, 1978; Poljak, 1978; Kumar and Varaiya, 1986; Kushner and Dupuis, 1992) and random search techniques (see, for instance, Rastrigin, 1974; Gaviano, 1975, 1978; Bogomolov and Karmanov, 1977; Denisov, 1978; Marti, 1980; Pintér, 1980c; Baba, 1981; Solis and Wets, 1981; Born and Bellman, 1982; Timmer, 1984; Patel, Smith and Zabinsky, 1988; Zhigljavsky, 1991; Zabinsky and Smith, 1992; or Boender and Romeijn, 1995; with extensive lists of references). Stochastic approximation algorithms, as a rule, can be applied only in the presence of convexity and differentiability assumptions. Pure random search methods can in principle solve optimization problems under less stringent assumptions; however, their numerical efficiency may be questioned, even for simpler deterministic problems.

The purpose of this chapter is to examine general convergence properties of certain classes of adaptive stochastic optimization strategies, without convexity and differentiability assumptions. To increase the efficiency of the basic random optimization methods studied, subsequently a general class of stochastically combined (hybrid) algorithms is introduced. Finally, some experimental results with hybrid algorithm implementations are summarized. The exposition draws mainly on Pintér (1984a,b), and on several applications and numerical studies related to the subject.

3.6.2 Convergence of Random Search Based Stochastic Optimization Methods

For the purpose of the following general discussion, we shall consider the continuous global optimization problem introduced earlier:

$$\min_{x \in D \subseteq \mathbb{R}^n} f(x) = f(x^\star), \quad x^\star \in X^\star \subset D, \quad X^\star \neq \emptyset. \tag{3.6.3}$$

To find points of $X^\star$, a stochastic sequence $\{x_k\}$ is generated according to the sequential rule formalized as

$$\begin{aligned} &x_0 \in D \quad \text{arbitrary}, \\ &x_{k+1} = S_k(u_k, \overline{\omega}_k) \in D, \qquad k = 0, 1, 2, \ldots, \\ &u_k = (x_0, f(x_0), x_1, f(x_1), \ldots, x_k, f(x_k)), \qquad \overline{\omega}_k = (\omega_0, \omega_1, \ldots, \omega_k). \end{aligned} \tag{3.6.4}$$

Expressed in words, for each index $k = 0, 1, 2, \ldots$, x_{k+1} is selected according to some conditional probability measure μ_k defined on the measurable space $\{D, \mathcal{D}\}$: for example, $\mathcal{D}$ is the σ-algebra of Borel-measurable subsets of the (Borel-measurable) set D. The decision rule S_k depends on the current search information u_k and the kth projection of the complete (infinite) sequence of random factors $\omega = (\omega_0, \omega_1, \omega_2, \ldots)$, $k = 0, 1, 2, \ldots$. Let $\omega \in \Omega$; we shall suppose that a probability measure P defined on the measurable space $(\Omega, \mathcal{A})$ corresponds to the stochastic algorithm scheme (3.6.4). Our objective is to find general sufficient conditions which assure that, for the generated sequence $\{x_k\}$, the probabilistic convergence relation

$$P(\lim_{k\to\infty} \varrho(x_k, X^\star) = 0) = 1, \qquad \varrho(x, X^\star) := \inf_{x^\star \in X^\star} \|x - x^\star\|,$$

holds.

Let $\{A_k\}$, $A_k \in \mathcal{A}$, $k = 0, 1, 2, \ldots$, be a sequence of random events. First we prove two simple but important auxiliary results. As in the sequel certain criteria are formulated in terms of conditional probabilities, where (for the stochastic algorithms in question) it is not assured a priori that the conditions have always positive probabilities, let us define for $A, B \in \mathcal{A}$,

$$P(A|B) = P(A), \quad \text{if } P(B) = 0.$$

Specifically, $B = \bigcup_{j=k}^{k-1} \overline{A}_j$ is also considered as an event of probability 0. ($\overline{A}_j$ is the complementary event to A_j.)

Lemma 3.6.1. *Assume that for every pair of nonnegative integers k, n, $k \geq n$, we have*

$$P\Big(A_k \,\Big|\, \bigcap_{j=n}^{k-1} \overline{A}_{k-j}\Big) \geq p_k, \qquad 0 \leq p_k \leq 1, \qquad k, n = 0, 1, 2, \ldots, \tag{3.6.5}$$

holds, and the sequence $\{p_k\}$ *satisfies the relation*

$$\sum_{k=0}^{\infty} p_k = \infty. \tag{3.6.6}$$

Then we have

$$P(A_\infty) = P\Big(\bigcap_{n=0}^{\infty} \bigcup_{k=n}^{\infty} A_k\Big) = 1, \tag{3.6.7}$$

i.e., among the random events $\{A_k\}$*, infinitely many occur with probability 1 (w.p.1).*

Proof. It evidently suffices to show that $P\Big(\bigcup_{k=n}^{\infty} A_k\Big) = 1$ for $n = 0, 1, 2, \ldots$. By virtue of (3.6.5), for $n = 0, 1, 2, \ldots$, we have

$$P\Big(\bigcap_{k=n}^{N} \overline{A}_k\Big) = \prod_{k=n}^{N} P\Big(\overline{A}_k \,\Big|\, \bigcap_{j=n}^{k-1} \overline{A}_j\Big) \leq \prod_{k=n}^{N} (1 - p_k) \leq \exp\Big(-\sum_{k=n}^{N} p_k\Big).$$

Therefore, applying (3.6.6), $P(\bigcap_{k=n}^{N} \overline{A}_k) \underset{N\to\infty}{\longrightarrow} 0$. The implied relation $P(\bigcup_{k=n}^{N} A_k) \underset{N\to\infty}{\longrightarrow} 1$, $n = 0, 1, 2, \ldots$, completes the proof. ■

Remark 3.6.1. The above result is an extension of the second assertion of the well-known Borel–Cantelli lemma (see, for example, Rényi, 1968) for the case of dependent event sequences: the classic lemma states that, for independent A_k's, the relation

$$\sum_{k=0}^{\infty} P(A_k) = \infty$$

implies (3.6.7). Note that, according to the first part of the Borel–Cantelli lemma, from

$$\sum_{k=0}^{\infty} P(A_k) < \infty$$

follows $P(A_\infty) = 0$. This result can be readily applied to prove the following assertion, which is formulated symmetrically to Lemma 3.6.1.

Lemma 3.6.2. *Applying the notation introduced previously, assume now that*

$$P\Big(A_k \;\Big|\; \bigcap_{j=n}^{k-1} \overline{A}_{k-j}\Big) \le p_k, \quad 0 \le p_k \le 1, \quad k \ge n, \quad k, n = 0, 1, 2, \ldots, \tag{3.6.8}$$

where the sequence $\{p_k\}$ satisfies the relation

$$\sum_{k=0}^{\infty} p_k < \infty. \tag{3.6.9}$$

Then we have

$$P(A_\infty) = 0, \tag{3.6.10}$$

that is, among the events $\{A_k\}$ only a finite number takes place, w.p.1.

Proof. As for $k = n$, (3.6.8) specifically implies that

$$P(A_k) \le p_k, \qquad k = 0, 1, 2, \ldots,$$

therefore

$$P(A_\infty) \le P\Big(\bigcup_{k=n}^{\infty} A_k\Big) \le \sum_{k=n}^{\infty} P(A_k) \le \sum_{k=n}^{\infty} p_k, \quad n = 0, 1, 2, \ldots . \tag{3.6.11}$$

Hence, by (3.6.9) and (3.6.11), (3.6.10) follows immediately. ■

Remark 3.6.2. Lemmas 3.6.1 and 3.6.2 provide general sufficient and necessary conditions for the convergence of different adaptive stochastic optimization procedures. Lemma 3.6.1 can be applied to prove convergence of a large number of stochastic algorithms based on random search, in possible combinations with known deterministic methods of nonlinear programming. For examples and related results, see Rastrigin (1974), Gaviano (1975), Bogomolov and Karmanov (1977), Denisov (1978), Devroye (1978), Marti (1980), Baba (1981), Solis and Wets (1981), Born and Bellman (1982), Patel, Smith and Zabinsky (1988), Zhigljavsky (1991), Zabinsky and Smith (1992).

Consider now the problem of minimizing a measurable function $f : D \to \mathbb{R}^1$ on the measurable set $D \subseteq \mathbb{R}^n$. For obvious reasons, we assume that $f(x) > -\infty$ $x \in D$, and $X^\star \neq 0$. In accordance with the scheme (3.6.4), the following conceptual algorithm can be applied to find points of the set $X^\star$ of optimal solutions.

Prototype Pure Adaptive Stochastic Search Algorithm

Step 0. Let $x_0 \in D$, set $k := 0$.

Step 1. Select a new point

$$y = S_k(u_k, \overline{\omega}_k) \in D.$$

Step 2. Set

$$x_{k+1} = \begin{cases} y & \text{if } f(y) < f(x_k), \\ x_k & \text{otherwise;} \end{cases}$$

then—setting $k := k+1$—return to Step 1.

For arbitrary $\delta > 0$ and $\widehat{\varepsilon} > 0$, define the sets

$$G(X^\star, \delta) = \{y \in D : \varrho(y, X^\star) < \delta\}$$
$$\widehat{D} = \widehat{D}(\delta) = D \setminus G(X^\star, \delta) = \{y \in D : \varrho(y, X^\star) \geq \delta\},$$

and the random events

$$F_k = F_k(\delta) : \{x_k \in \widehat{D}\}, \quad A_k = A_k(\widehat{\varepsilon}) : \{f(x_{k+1}) \leq f(x_k) + \widehat{\varepsilon}\},$$
$$k = 0, 1, 2, \ldots .$$

Theorem 3.6.2. Suppose that

- *for every $\delta > 0$ there exist $\widehat{\varepsilon} = \widehat{\varepsilon}(\delta) > 0$ and a sequence $\{\widehat{p}_k\}$, $\widehat{p}_k = \widehat{p}_k(\delta)$, $0 \leq \widehat{p}_k \leq 1$, such that for every pair of indices n, k, $0 \leq n \leq k$, we have*

$$P\Big\{A_k \cup \overline{F}_k \,\Big|\, \bigcap_{j=n}^{k-1} (\overline{A}_j \cap F_j)\Big\} \geq \widehat{p}_k; \tag{3.6.12}$$

- *for the sequence $\{\widehat{p}_k\}$,*

$$\sum_{k=0}^{\infty} \widehat{p}_k = \infty \tag{3.6.13}$$

holds. Then

$$P(\liminf_{k\to\infty} \varrho(x_k, X^\star) = 0) = 1.$$

That is, with probability 1, the generated sequence of points $\{x_k\}$ contains a subsequence $\{x_k^\star\}$, which converges to the set of optimal solutions $X^\star$.

Remark 3.6.3. The minimizer subsequence determined by

$$f(x_k^\star) = \min_{0 \le j \le k} f(x_j), \qquad x_j \in D$$

is an obvious choice for $\{x_k^\star\}$.

Proof. Define

$$F = F(\delta) = \{\omega : \liminf_{k\to\infty} \varrho(x_k, X^\star) \ge \delta\}.$$

It is sufficient to show that for every $\delta > 0$, $P(F(\delta)) = 0$ holds. Assume indirectly that for some $\delta > 0$, we have

$$P(F(\delta)) = p, \qquad p = p(\delta) > 0.$$

Then for $\omega \in F(\delta)$ we have $\varrho(x_k, C^\star) \ge \delta/2$, if $k \ge n = n(\delta, \omega)$, that is,

$$F(\delta) \subseteq \bigcup_{n=0}^{\infty} \bigcap_{k=n}^{\infty} F_k(\frac{\delta}{2}).$$

Let $\widehat{\delta} = \delta/2$, then by the assumptions of the theorem there exist $\widehat{\varepsilon} = \widehat{\varepsilon}(\widehat{\delta}) > 0$ and $\{\widehat{p}_k\}$, $\widehat{p}_k = \widehat{p}_k(\widehat{\delta})$, $0 \le \widehat{p}_k \le 1$, such that the relations (3.6.12) and (3.6.13) are satisfied. Therefore—analogously to the proof of Lemma 3.6.2—we can write

$$P\Big\{\Big(\bigcap_{k=n}^{N} \overline{A}_k\Big) \cap \Big(\bigcap_{k=n}^{N} F_k\Big)\Big\} = P\Big\{\bigcap_{k=n}^{N} (\overline{A}_k \cap F_k)\Big\}$$
$$= \prod_{k=n}^{N} P\Big\{\overline{A}_k \cap F_k \,\Big|\, \bigcap_{j=n}^{k-1} (\overline{A}_j \cap F_j)\Big\}$$
$$\le \prod_{k=n}^{N} (1 - \widehat{p}_k) \le \exp\Big(-\sum_{k=n}^{N} \widehat{p}_k\Big) \xrightarrow[N\to\infty]{} 0,$$

that is,

$$P\Big\{\Big(\bigcap_{k=n}^{\infty}\overline{A}_k\Big)\cap\Big(\bigcap_{k=n}^{\infty}F_k\Big)\Big\}=0,\qquad n=0,1,2,\ldots.$$

Denote $A^{(n)}=\bigcap_{k=n}^{\infty}\overline{A}_k$, $F^{(n)}=\bigcap_{k=n}^{\infty}F_k$; then $P(A^{(n)}\cap F^{(n)})=0$ implies that

$$P(A^{(m)}\cap F^{(m)})=P(A^{(n)}\cap F^{(m)})=0,\qquad m\le n.$$

Therefore

$$\begin{aligned}P(\overline{A}_\infty\cap F)&\le P\Big\{\Big(\bigcup_{n=0}^{\infty}A^{(n)}\Big)\cap\Big(\bigcup_{m=0}^{\infty}F^{(m)}\Big)\Big\}\\&\le\sum_{n=0}^{\infty}\sum_{m=0}^{\infty}P(A^{(n)}\cap F^{(m)})=0.\end{aligned}$$

Consequently, we have $P(A_\infty\mid F)=1$. Considering the positivity of $\widehat{\varepsilon}$ and Step 2 of the algorithm, hence we obtain that

$$P(\omega:\lim_{k\to\infty}f(x_k)=-\infty\mid\liminf_{k\to\infty}\varrho(x_k,X^\star)\ge\delta)=1,$$

that is,

$$P(\omega:\lim_{k\to\infty}f(x_k)=-\infty)\ge p(\delta)>0.$$

As the last relation contradicts the assumption that f is bounded from below on D, we can conclude that $p(\delta)=0$ for every $\delta>0$. This completes the proof. ∎

Corollary 3.6.1. *In addition to the assumptions of Theorem 3.6.2, suppose that*

- *f is uniformly continuous on D;*
- *for every $\delta>0$ there exists $\varepsilon=\varepsilon(\delta)>0$ such that*

$$x\in D\setminus G(X^\star,\delta)$$

implies

$$f(x)\ge f(x^\star)+\varepsilon. \tag{3.6.14}$$

Then

$$p\big(\lim_{k\to\infty} \varrho(x_k, X^\star) = 0\big) = P\big(\lim_{k\to\infty} f(x_k) = f^\star\big) = 1 \quad (f^\star = \min_{x\in D} f(x)).$$

Proof. From the uniform continuity of f and convergence of $\{f(x_k)\}$ follows

$$\begin{aligned} 1 &= P(\omega : \liminf_{k\to\infty} \varrho(x_k, X^\star) = 0) \\ &= P(\omega : \exists\{k_i(\omega)\} \subseteq \{k\} \lim_{i\to\infty} \varrho(x_{k_i}, X^\star) = 0) \\ &= P(\omega : \exists\{k_i(\omega)\} \subseteq \{k\} \lim_{i\to\infty} f(x_{k_i}) = f^\star) = P(\omega : \lim_{k\to\infty} f(x_k) = f^\star). \end{aligned} \tag{3.6.15}$$

Define now

$$K(\vartheta) = \{\omega : \limsup_{k\to\infty} \varrho(x_k, X^\star) \geq \vartheta\}.$$

It suffices to show that $P(K(\vartheta)) = 0$ for arbitrary $\vartheta > 0$. Let $\delta = \vartheta/2$, then by the definition of $K(\vartheta)$ and (3.6.14) we obtain the following chain of relations:

$$\begin{aligned} K(\vartheta) &\subseteq \bigcap_{n=0}^{\infty} \bigcup_{k=n}^{\infty} \{\omega : \varrho(x_k, X^\star) \geq \delta\} \\ &\subseteq \bigcap_{n=0}^{\infty} \bigcup_{k=n}^{\infty} \{\omega : f(x_k) \geq f^\star + \varepsilon\} \\ &\subseteq \{\omega : \lim_{k\to\infty} f(x_k) \geq f^\star + \varepsilon\}. \end{aligned} \tag{3.6.16}$$

From (3.6.15) and (3.6.16) we can conclude that $P(K(\vartheta)) = 0$ for $\vartheta > 0$, and, hence,

$$P(\lim_{k\to\infty} \varrho(x_k, X^\star) = 0) = P(\limsup_{k\to\infty} \varrho(x_k, X^\star) = 0) = 1.$$

This completes the proof. ∎

As an application of the above general convergence result, let us consider the global optimization problem

$$\min_{x \in D} f(x), \tag{3.6.17}$$

where $D \subseteq I\!R^n$ is a nonempty compact set and $f(x)$ is a continuous function. From these assumptions it follows that the set of global solutions

$$X^\star = \{x^\star \in D : f(x^\star) = \min_{x \in D} f(x)\}$$

is nonempty. Suppose that the Lebesgue-measure $\lambda^{(n)} = \lambda(I\!R^n)$ of the set

$$G(X^\star, \delta) = \{x \in D : \varrho(x, X^\star) \leq \delta\}$$

is positive for every $\delta > 0$. For solving the problem (3.6.17) a conceptual stochastic algorithm scheme can be applied, which consists of the following basic steps:

Adaptive Random Search Algorithm Scheme

Step 0. Let $x_0 \in D$, set $k := 0$.

Step 1. Generate a search direction $s_k = s_k(u_k, \overline{\omega}_k)$ according to some probability measure μ_k given in $I\!R^n$. We assume that the sequence $\{\mu_k\}$ is defined in accordance with the following requirement: if $D_1 \subseteq D$, $D_1 \in \mathcal{D}$, $\lambda^{(n)}(D_1) > 0$, then for every $\delta > 0$ there exist $\widehat{\varepsilon} = \widehat{\varepsilon}(\delta) > 0$ and $\widehat{p} = \widehat{p}(D_1, \delta) > 0$ such that

$$P\Big\{B_k \cup F_k \;\Big|\; \bigcap_{j=n}^{k-1} (\overline{A}_j \cap F_j)\Big\} \geq \widehat{p}, \qquad k, n = 0, 1, 2, \ldots, \quad k \geq n,$$

where the random events A_k, B_k, F_k are defined as follows:

$$\begin{aligned} A_k &:= A_k(\widehat{\varepsilon}) : \{f(x_{k+1}) \leq f(x_k) - \widehat{\varepsilon}\} \\ B_k &:= B_k(D_1) : \{\exists\, \gamma \geq 0 : x_k + \gamma s_k \in D_1\} \\ F_k &:= F_k(\delta) : \{\varrho(x_k, X^\star) \geq \delta\}. \end{aligned}$$

Step 2. Having selected s_k, the step-size $\gamma_k = \gamma_k(u_k, \overline{\omega}_k)$ is determined by univariate global search on D:

$$f(x_{k+1}) = f(x_k + \gamma_k s_k) = \min_{\substack{\gamma \geq 0 \\ x_k + \gamma s_k \in D}} f(x_k + \gamma s_k).$$

Then—setting $k := k + 1$—return to Step 1.

Theorem 3.6.3. *For the generated sequence $\{x_k\}$ we have*

$$P(\lim_{k\to\infty} \varrho(x_k, X^\star) = 0) = P(\lim_{k\to\infty} f(x_k) = f(x^\star)) = 1. \quad \blacksquare$$

The proof of this theorem is based on a straightforward argument which shows that the assumptions of both Theorem 3.6.2 and Corollary 3.6.1 are satisfied. Related results for establishing the convergence of different random search algorithms were presented, for example, in Gaviano (1975), Marti (1980), Baba (1981), Solis and Wets (1981).

Remark 3.6.4. The global optimization scheme described by Steps 0 to 2 can be implemented in a number of ways. For example, if s_k, $k = 0, 1, 2, \ldots$, are simply generated as uniformly distributed random vectors on the set D, then the probabilistic assumption in Step 1 is satisfied. Moreover, if, for example, f is continuous, then different known univariate global search methods can approximate Step 2 with arbitrary accuracy. (Examples of such algorithms were discussed in Part 2.)

3.6.3 Convergence of Stochastically Combined Optimization Procedures

As can be seen from the results of the previous section, for example, random search based stochastic optimization methods possess global convergence properties, under mild assumptions. At the same time, computational tests and actual applications indicate that, though usually such methods are rather efficient in the initial phase of optimization, their efficiency may sharply decrease in the later search phase. Therefore, it seems advantageous to start the optimization procedure with some fast 'crude search' method, which is then 'switched' to some locally efficient method. In many cases, the objective function in the neighbourhood of minima can be suitably approximated quadratically: in these circumstances, conjugate direction methods (see, for example, the monograph of Hestenes, 1980) can be efficiently used.

Semi-heuristic combinations of different techniques have been extensively applied to solve global optimization problems: see examples and further references in Törn (1978), Devroye (1978), Pintér, 1980a,b). Note, however, that the global convergence of combined methods cannot be assured by their local subalgorithms, even for deterministic and smooth optimization problems. At the same time, stochastic search components can assure global convergence of properly combined methods, even in cases of nonconvex, nondifferentiable, and stochastic problems.

As in Section 3.6.2, let $D \subset I\!R^n$ be the feasible set, and $f : I\!R^n \to I\!R^1$ the objective function; a sequence $\{x_k\}$ is sought which converges w.p.1 to $X^\star$, the nonempty optimal subset of D. To generate the sequence $\{x_k\}$, the following stochastically combined (hybrid) method can be applied. Let $x_0 \in D$ be arbitrary. The sequence of search points $\{x_k\}$ is defined by the relations

$$\begin{aligned} x_{k+1} &= H_k(u_k, \overline{\omega}_k) \qquad\qquad k = 0, 1, 2, \ldots \qquad (3.6.18)\\ u_k &= (x_0, f(x_0), x_1, f(x_1), \ldots, x_k, f(x_k))\\ \overline{\omega}_k &= (\omega_0, \omega_1, \ldots, \omega_k)\\ H_k &: \alpha_k G_k(u_k, \overline{\omega}_k) + (1 - \alpha_k) L_k(u_k, \overline{\omega}_k), \qquad 0 \le \alpha_k \le 1. \end{aligned}$$

Applying similar notation to (3.6.4), formulae (3.6.18) symbolize that at step k—independently from the results of the previous iterations—either the algorithmically defined subprocedure G_k is accomplished with probability α_k, or subprocedure L_k is accomplished with probability $(1 - \alpha_k)$. Both G_k and L_k can depend on the search information u_k and also on random factors represented by $\overline{\omega}_k$. Subalgorithms G_k and L_k typically correspond to generating the next search point according to a step (or several steps) of a global, respectively local, optimization algorithm. As a simple example, G_k can be the choice of one or several new search points by a random search algorithm, and L_k can be a conjugate directional search cycle accomplished starting from the best point found so far.

Applying again the notation $\omega = (\omega_0, \omega_1, \omega_1, \ldots)$, $\omega \in \Omega$, we shall suppose that there exists a probability measure P, defined on the measurable space $(\Omega, \mathcal{A})$, which corresponds to the stochastic algorithm scheme (3.6.18).

In connection with the convergence properties of hybrid optimization methods, we shall prove first a simple extension of Lemma 3.6.1. Let $\{A_k\}$, $A_k \in \mathcal{A}$, $k = 0, 1, 2, \ldots$, be a sequence of random events, where A_k is defined by $\overline{\omega}_k$, the kth projection of ω. Let us denote by $P(A|B; G_k)$ (respectively $P(A|B; L_k)$) the conditional probability of the event A on the condition B, if subprocedure G_k (respectively L_k) is applied. (As before, $P(A|B)$, by definition, is equal to $P(A)$, if $P(B) = 0$.)

Lemma 3.6.3. *If the application of subalgorithm G_k in (3.6.18) assures that for every pair of nonnegative integers k, n, $k \ge n$, we have*

$$P\Big(A_k \,\Big|\, \bigcap_{j=n}^{k-1} \overline{A}_j; G_k\Big) \geq p_k, \quad 0 \leq p_k \leq 1, \quad k \geq n, \quad k, n = 0, 1, 2, \ldots, \tag{3.6.19}$$

and the sequences $\{p_k\}$, $\{\alpha_k\}$ satisfy the relation

$$\sum_{k=0}^{\infty} \alpha_k p_k = \infty, \tag{3.6.20}$$

then

$$P(A_\infty) = P\Big(\bigcap_{n=0}^{\infty} \bigcup_{k=n}^{\infty} A_k\Big) = 1, \tag{3.6.21}$$

that is, among the events $\{A_k\}$, infinitely many occur, w.p.1.

Proof. Again, it suffices to show that $P\big(\bigcup_{k=n}^{\infty} A_k\big) = 1$ for $n = 0, 1, 2, \ldots$.

By (3.6.18), (3.6.19), and $A_k = A_k(\overline{\omega}_k)$, we can write

$$\begin{aligned} P\Big(A_k \,\Big|\, \bigcap_{j=n}^{k-1} \overline{A}_j\Big) &= \alpha_k P\Big(A_k \,\Big|\, \bigcap_{j=n}^{k-1} \overline{A}_j; G_k\Big) + (1 - \alpha_k) P\Big(A_k \,\Big|\, \bigcap_{j=n}^{k-1} \overline{A}_j; L_k\Big) \\ &\geq \alpha_k p_k. \end{aligned}$$

Therefore

$$\begin{aligned} P\Big(\bigcup_{k=n}^{N} A_k\Big) &= 1 - P\Big(\bigcap_{k=n}^{N} \overline{A}_k\Big) = 1 - \prod_{k=n}^{N} P\Big(\overline{A}_k \,\Big|\, \bigcap_{j=n}^{k-1} \overline{A}_k\Big) \\ &\geq 1 - \prod_{k=n}^{N} (1 - \alpha_k p_k) \geq 1 - \exp\Big(-\sum_{k=n}^{N} \alpha_k p_k\Big). \end{aligned} \tag{3.6.22}$$

From (3.6.22), exploiting (3.6.20), immediately follows $P\big(\bigcup_{k=n}^{N} A_k\big) \underset{N\to\infty}{\longrightarrow} 1$ for $n = 0, 1, 2, \ldots$. The lemma is proved. ■

Note (without unnecessary repetitions) that Lemma 3.6.2 can be applied also in connection with hybrid algorithms.

On the basis of the results of Section 3.6.2 and Lemma 3.6.3, we can immediately investigate now the convergence properties of hybrid optimization procedures. For a unified and generalized presentation of these results, the following stochastic optimization problem will be studied. Let us suppose that instead of the 'embedded' deterministic problem

$$\min_{x \in D} f(x) \tag{3.6.23}$$

(in which $D \subseteq I\!R^n$, and $f : D_0 \to I\!R^1$ is defined on a suitable open set D_0, $D \subset D_0$) only its 'noisy' estimates are available. In other words, for the sequentially generated search points x_k, $k = 0, 1, 2, \ldots$, we have estimates of the objective function and the feasible set in the form

$$\begin{aligned} &f(x_k) + \xi(x_k), \\ &D + \eta(x_k). \end{aligned} \tag{3.6.24}$$

With respect to the noise structure in (3.6.24)—which may model, for example, system evaluations by Monte Carlo simulation—we shall make the following assumptions:

1. $\xi(x_k) = \xi(x_k, \overline{\omega}_k)$ is a random variable depending on the decision $x_k \in I\!R^n$, i.e., $\xi : \Omega \to I\!R^1$ is a P-measurable function for any fixed x_k. In addition, there exists a constant $0 < \sigma < \infty$ such that for every sequence $\{x_k\}$ generated by the conceptual algorithm scheme (3.6.18) we have

$$P\left(\left|\sum_{k=1}^{\infty} \xi(x_k)\right| \le \delta\right) = 1; \tag{3.6.25}$$

   and for every $\mu > 0$ there exists an index $k(\mu)$ such that $n \ge k(\mu)$ implies

$$P\left(\left|\sum_{k=n}^{\infty} \xi(x_k)\right| \le \mu\right) = 1. \tag{3.6.26}$$

2. $\eta(x_k) = \eta(x_k, \overline{\omega}_k)$ is a point-to-set P-measurable random mapping depending on the parameter $x_k \in I\!R^n$, that is, for any fixed x_k, $\eta : \Omega \to \mathcal{B}^n$ ($\mathcal{B}^n$ is the σ-algebra of the Borel-subsets in $I\!R^n$). Define the norm of $\eta(x_k, \overline{\omega}_k)$ by the essential supremum of the norm of its realizations:

$$\|\eta(x_k, \overline{\omega}_k)\| = \operatorname*{vrai\,sup}_{z \in \eta(x_k, \overline{\omega}_k)} \|z\|.$$

Then, by assumption, there exists a constant $0 < \tau < \infty$, such that for every sequence $\{x_k\}$ generated by the conceptual algorithm (3.6.18) we have

$$P\Big(\sum_{k=0}^{\infty} \|\eta(x_k)\| \leq \tau\Big) = 1; \tag{3.6.27}$$

and for every $\nu > 0$ there exists an index $k(\nu)$ such that $n \geq k(\nu)$ implies

$$P\Big(\Big\|\sum_{k=n}^{\infty} \eta(x_k)\Big\| \leq \nu\Big) = 1. \tag{3.6.28}$$

Observe that the technical Assumptions 1 and 2 model an optimization procedure in which f and D are determined with increasing accuracy.

The global convergence properties of hybrid optimization methods will now be investigated under the following additional assumptions.

3. D is the closure of an open, bounded set in $I\!R^n$; D is contained by an open set D_0.

4. $f : \overline{D}_0 \to I\!R^1$ is a continuous function, $\overline{D}_0$ is the closure of D_0. From Assumptions 3 and 4 it follows that the set

$$X^\star = \{x^\star \in \overline{D}_0 : f(x^\star) = f^\star = \min_{x \in \overline{D}_0} f(x)\}$$

is nonempty; we will suppose that $X^\star \subset D$.

5. The sequence $\{x_k\}$ is generated according to the following conceptual algorithm scheme. Let $x_0 \in D_0$, set $k := 0$, then for $k = 0, 1, 2, \ldots$, select x_{k+1} by the hybrid rule (recall (3.6.18))

$$x_{k+1} : \alpha_k G_k(u_k, \overline{\omega}_k) + (1 - \alpha_k) L_k(u_k, \overline{\omega}_k), \quad 0 \leq \alpha_k \leq 1, \quad \sum_{k=0}^{\infty} \alpha_k = \infty,$$

$$x_{k+1} \in D + \eta(x_k).$$

We shall suppose that the subprocedures G_k have the following property. For arbitrary $\varepsilon > 0$, define the set

$$D_k = \{x \in D + \eta(x_k) : f(x) \leq f^\star + \varepsilon\}$$

and the random event

$$A_k = A_k(\varepsilon) : \{x_{k+1} \in D_k\}.$$

Then we shall require, for $k, n = 0, 1, 2, \ldots$, $k \geq n$, the relation

$$P\Big(A_k \,\Big|\, \bigcap_{j=n}^{k-1} \overline{A}_j; G_k\Big) \geq p_k, \quad p_k = p_k(\varepsilon), \quad 0 < p_k \leq 1, \tag{3.6.29}$$

whenever $G_k = G_k(\varepsilon_k)$ is accomplished with sufficient accuracy, that is, $\varepsilon_k < \varepsilon$. The accuracy of subprocedures $\{G_k\}$ is parameterized by a number sequence $\{\varepsilon_k\}$, which satisfies the relations

$$\varepsilon_k > 0, \quad k = 0, 1, 2, \ldots, \quad \lim_{k\to\infty} \varepsilon_k = 0. \tag{3.6.30}$$

Finally, we assume that $p_k(\varepsilon)$ in (3.6.29) is a monotonically increasing function of ε and, for arbitrary $\varepsilon > 0$, we have

$$p(\varepsilon) = \liminf_{k\to\infty} p_k(\varepsilon) > 0. \tag{3.6.31}$$

The subprocedures $\{L_k\}$ can be chosen arbitrarily. We shall define the kth estimate of $f^\star$ and $x^\star \in X^\star$ by the relations

$$f_{k,\varepsilon_k} = f(\overline{x}_{k,\varepsilon_k}) + \xi(\overline{x}_{k,\varepsilon_k}) = \min_{\substack{0 \leq j \leq k \\ |\xi(x_j)| \leq \varepsilon_k}} [f(x_j) + \xi(x_j)], \tag{3.6.32}$$

that is, in (3.6.32), f_{k,ε_k} and $\overline{x}_{k,\varepsilon_k}$ are selected on the basis of the observations of $\{f(x_j) + \xi(x_j)\}$ with no more absolute error than ε_k. Here we suppose that at least the actual estimate of the optimum $f^\star$ is evaluated with the desired accuracy, thus the set of indices $\{0 \leq j \leq k : |\xi(x_j)| \leq \varepsilon_k\}$ is nonempty. Note that, selecting $\overline{x}_{k,\varepsilon_k}$ as the point x_j with the smallest index j satisfying (3.6.32), the sequence $\{\overline{x}_{k,\varepsilon_k}\}$ is defined unambiguously.

Theorem 3.6.4. *Suppose that assumptions 1 to 5 detailed above are satisfied. Then $\{x_k\}$ is bounded w.p.1,*

$$P(\lim_{k\to\infty} f_{k,\varepsilon_k} = f^\star) = 1, \tag{3.6.33}$$

and

$$P(\lim_{k\to\infty} \varrho(\overline{x}_{k,\varepsilon_k}, X^\star) = 0) = P(\liminf_{k\to\infty} \varrho(x_k, X^\star) = 0) = 1. \tag{3.6.34}$$

Proof. As $x_{k+1} \in D + \eta(x_k)$, D is bounded and $\|\eta(x_k)\| \leq \tau$ w.p.1, it evidently follows that $\{x_k\}$ is bounded w.p.1. First we shall prove (3.6.33). According to (3.6.32) we have w.p.1

$$f_{k,\varepsilon_k} \geq f^\star - \varepsilon_k, \qquad k = 0, 1, 2, \ldots . \tag{3.6.35}$$

Next, we will show that for arbitrary $\varepsilon > 0$ there exist $p_0 = p_0(\varepsilon) > 0$ and $k_0 = k_0(\varepsilon)$, such that for $k > k_0$, we have

$$P(f_{k,\varepsilon_k} > f^\star + 2\varepsilon) \leq \exp\Big(-p_0 \sum_{j=k_0}^{k-1} \alpha_j\Big). \tag{3.6.36}$$

Then from (3.6.35) and (3.6.36) it is easy to obtain (3.6.33). First, by (3.6.30), $k \geq k_1(\varepsilon)$ implies that $\varepsilon_k < \varepsilon$, thus for the subprocedures $G_k(\varepsilon_k)$, $k \geq k_1$, we have (3.6.29). Second, by (3.6.31), $k \geq k_2 = k_2(\varepsilon)$ implies that $p_k(\varepsilon) \geq p_0 = (p(\varepsilon)/2) > 0$. Finally, by (3.6.26), $k \geq k_3 = k_3(\varepsilon)$ implies that $|\xi(x_k)| < \varepsilon$ w.p.1. Let $k_0 = \max\{k_1, k_2, k_3\}$, then by (3.6.29), following the proof of Lemma 3.6.3, we can write, for $k > k_0$,

$$\begin{aligned}
&P(f_{k,\varepsilon_k} > f^\star + 2\varepsilon) \\
&\quad = P\Big(\min_{\substack{0 \leq j \leq k \\ |\xi(x_j)| \leq \varepsilon_k}} [f(x_j) + \xi(x_j)] > f^\star + 2\varepsilon\Big) \leq P\Big(\bigcap_{j=k_0}^{k-1} \overline{A}_j\Big) \\
&\quad = \prod_{j=k_0}^{k-1} P\Big(\overline{A}_j \,\Big|\, \bigcap_{i=k_0}^{j-1} \overline{A}_i\Big) = \prod_{j=k_0}^{k-1} \Big[1 - P\Big(A_j \,\Big|\, \bigcap_{i=k_0}^{j-1} \overline{A}_i\Big)\Big] \\
&\quad \leq \prod_{j=k_0}^{k-1} \Big[1 - \alpha_j P\Big(A_j \,\Big|\, \bigcap_{i=k_0}^{j-1} \overline{A}_i; G_i\Big)\Big] \leq \prod_{j=k_0}^{k-1} [1 - \alpha_j p_j(\varepsilon)] \\
&\quad \leq \prod_{j=k_0}^{k-1} [1 - p_0 \alpha_j] \leq \exp\Big(-p_0 \sum_{j=k_0}^{k-1} \alpha_j\Big).
\end{aligned} \tag{3.6.37}$$

As by (3.6.26) and (3.6.32) there holds w.p.1

$$f^\star - \varepsilon < f_{i,\varepsilon_i} < f_{k,\varepsilon_k} + 2\varepsilon, \qquad i \geq k \geq k_0,$$

from (3.6.35) and (3.6.37), by virtue of $\sum_{k=0}^{\infty} \alpha_k = \infty$, we can deduce that

$$P(\sup_{i \geq k} |f_{i,\varepsilon_i} - f^\star| > 4\varepsilon) \leq P(|f_{k,\varepsilon_k} - f^\star| > 2\varepsilon) \underset{k\to\infty}{\longrightarrow} 0,$$

that is, we obtain (3.6.33).

Let us turn now to the proof of (3.6.34). Evidently, it is sufficient to show that for arbitrary $\delta > 0$ we have $P(F(\delta)) = 0$, where

$$F(\delta) = \{\omega : \limsup_{k\to\infty} \varrho(\overline{x}_{k,\varepsilon_k}, X^\star) \geq \delta\}.$$

For $\widehat{\delta} > 0$, define

$$G(X^\star, \widehat{\delta}) = \{y \in D_0 : \rho(y, X^\star) < \delta\},$$
$$\widehat{D} = \widehat{D}(\delta) = \overline{D}_0 \setminus G(X^\star, \delta).$$

As $\widehat{D} \subset I\!R^n$ is compact, there exists $\widehat{f} = \min_{x\in\widehat{D}} f(x)$. Thus, by virtue of $X^\star \subset D$, we have $2\widehat{\varepsilon} = \widehat{f} - f^\star > 0$, that is, for $x \in \widehat{D}$ there holds $f(x) \geq f^\star + 2\widehat{\varepsilon}$, $\widehat{\varepsilon} = \widehat{\varepsilon}(\widehat{\delta})$. Let us select now an index k_0 such that $k \geq k_0$ implies the relations

$$D + \eta(\overline{x}_{k,\varepsilon_k}) \subset D_0, \qquad |\xi(\overline{x}_{k,\varepsilon_k})| < \widehat{\varepsilon}$$

(see (3.6.26), (3.6.28)); then for $\widehat{\delta} = \delta/2$ we can write

$$F(\delta) \subseteq \bigcap_{n=k_0}^{\infty} \bigcup_{k=n}^{\infty} \{\varrho(\overline{x}_{k,\varepsilon_k}, X^\star) \geq \widehat{\delta}\} \subseteq \bigcap_{n=k_0}^{\infty} \bigcup_{k=n}^{\infty} \{f(\overline{x}_{k,\varepsilon_k}) \geq f^\star + 2\widehat{\varepsilon}\}$$
$$\subseteq \bigcap_{n=k_0}^{\infty} \bigcup_{k=n}^{\infty} \{f_{k,\varepsilon_k} \geq f^\star + \widehat{\varepsilon}\} \subseteq \{\limsup_{k\to\infty} f_{k,\varepsilon_k} \geq f^\star + \widehat{\varepsilon}\}.$$

Hence, we obtain that

$$P(F(\delta)) \leq P(\limsup_{k\to\infty} f_{k,\varepsilon_k} \geq f^\star + \widehat{\varepsilon}) = 0 \quad \text{for any} \quad \delta > 0,$$

that is, the relations (3.6.34) also hold. ∎

Remark 3.6.4. In consequence of the stochastic noise $\{\xi(x_k)\}$, generally speaking the sequence $\{f(x_k) + \xi(x_k)\}$ is not monotonically decreasing. Therefore—in contrast with the deterministic case (Corollary 3.6.1, Theorem 3.6.3)—the relations

$$P(\lim_{k\to\infty} f(x_k) = f^\star) = P(\lim_{k\to\infty} \varrho(x_k, X^\star) = 0) = 1,$$

as a rule, do not hold under the given general assumptions.

Remark 3.6.5. From the proof of Theorem 3.6.4, it can be seen that the usual centeredness and independence assumptions concerning the noise sequences $\{\xi(x_k)\}$, $\{\eta(x_k)\}$, and $\{\varepsilon(x_k)\}$ are not necessary. At the same time, uniform (in ω) convergence of the series

$$\sum_{k=1}^{\infty} \xi(x_k), \qquad \sum_{k=0}^{\infty} \|\eta(x_k)\|, \qquad \sum_{k=0}^{\infty} \varepsilon(x_k)$$

is assumed. In many stochastic programming problems the objective and constraint functions are evaluated by independent experiments, Monte Carlo simulation, etc.: thus the accuracy of these calculations is controllable. The accuracy of function evaluations may be governed, for example, in accordance with the Chebyshev–Bernstein inequality, or the law of iterated logarithm (consult Rényi, 1968; or Feller, 1971). We shall return to the discussion of this point in Chapter 3.7.

Remark 3.6.6. The fairly general assumptions of Theorem 3.6.4 indicate that there exists a wide variety of hybrid optimization procedures which are globally convergent on a broad class of (stochastic) optimization problems. For instance, the conceptual algorithm described in connection with Theorem 3.6.3 can certainly possess the prescribed properties of the subprocedures $\{G_k\}$.

Remark 3.6.7. In early numerical experiments (Pintér, 1981) a globally convergent hybrid method was implemented as the stochastic combination of a global (G_k) and a local (L_k) optimization procedure for solving test problems. At each step k of this method, with probability α_k, G_k, or with probability $(1 - \alpha_k)$, L_k is executed. The subalgorithm G_k first randomly chooses from the vectors, uniformly distributed on the surface of the n-dimensional unit sphere; then, on the feasible section of the selected direction (half line) an adapted version of Strongin's (1978) global optimization method is performed. The subalgorithm L_k accomplishes a single cycle of the finite difference approximation of the Fletcher–Reeves (1964) conjugate directions method, starting from the actual estimate x_k of the global solution.

The procedure outlined was tested on a number of standard global optimization problems, which were investigated in Dixon and Szegö (1975,

1978). The introductory paper of Dixon and Szegö (1978) presents summarized results with different types of global optimization methods. The computational experiments described in Pintér (1981) show that the efficiency of the hybrid procedure (starting from randomly generated initial points, on average) compares favourably to these results. In addition, the optimization procedure proved to be fairly stable, when the test functions were perturbed with gradually decreasing, independent, normally distributed noise terms, according to the model

$$F(x) \approx f(x)(1+r), \quad r \sim \mathcal{N}(0, \sigma_k), \quad \sigma_k = \frac{\sigma_0}{2^k}, \quad k = 0, 1, 2, \ldots, \quad \sigma_0 > 0.$$

Remark 3.6.8. Without going into details, note that locally convergent pure and hybrid stochastic algorithms can be investigated along analogous lines (Pintér, 1982, 1984a,b). Related numerical experiments—applying adaptive random search to minimize randomly generated unconstrained convex objective functions, and a set of nonlinear programming test problems (Himmelblau, 1972b)—are summarized in Pintér (1980c), with further references.

In conclusion, one can make the following general observations:

- Appropriate stochastic optimization strategies possess strong (w.p.1) global convergence properties, without the usual convexity and differentiability assumptions related to the optimization problem.
- The efficiency of such strategies can be enhanced introducing stochastically combined (hybrid) optimization methods, in which a globally convergent stochastic search component and one (or several) locally convergent algorithm component(s) are embedded.

The experimental results mentioned above, and numerous other related studies (see, for instance, Boender, Rinnooy Kan, Stougie and Timmer, 1982; Rinnooy Kan and Timmer, 1987a,b; Törn and Žilinskas, 1989; Zhigljavsky, 1991; Törn and Viitanen, 1992, 1994; Romeijn and Smith, 1994) indicate that adaptive stochastic algorithms are among the most efficient practical strategies for global optimization under very mild analytical assumptions. In several applications, discussed later (in Part 4), such stochastic solution strategies were applied.

Chapter 3.7

Estimation of Noise-Perturbed Function Values

3.7.1 Introduction

As discussed in the previous two chapters, in stochastic optimization problems some or all of the defining functions (that is, objective and constraints in the model (3.6.1)) depend not only on the decision variables, but also on certain random factors. In such cases the optimization procedure needs to be combined with some—usually time-consuming—function value estimation method, such as experiments or Monte Carlo simulation. Stochastic programming formulations of various engineering or economic design problems (Prékopa, 1980; Ermoliev and Wets, 1988, etc.) may serve as examples of this situation. Obviously, in such cases it is desirable to decrease the number of necessary realizations of the involved random factors as much as possible, while prescribing accuracy and reliability levels (confidence intervals) of the estimated function values: this issue will be addressed below.

For an arbitrary random variable ξ with expectation $E\xi$ and finite variance $D^2\xi$, by the Bienaymé–Chebyshev inequality (Rényi, 1968; Feller, 1971), we have

$$P(|\xi - E\xi| \geq \lambda D\xi) \leq \frac{1}{\lambda^2}, \qquad \lambda > 1 \text{ arbitrary value.} \tag{3.7.1}$$

From (3.7.1) it follows that, for independent, identically distributed (i.i.d.) random variables $\xi_1, \ldots, \xi_n, \ldots$, $E\xi_k = E_1$, $D\xi_k = D_1$, $k = 1, 2, 3, \ldots$, the relation

$$P\Big(\Big|\frac{1}{n}\sum_{k=1}^{n}\xi_k - E_1\Big| \geq \varepsilon\Big) \leq \frac{D_1^2}{n\varepsilon^2}, \qquad \varepsilon > 0, \tag{3.7.2}$$

is valid. Relation (3.7.2) makes it possible to provide an (upper) estimate for n, the necessary number of realizations of the i.i.d. random variables ξ_k, $k = 1, 2, 3, \ldots$, in order to estimate their mean value E_1 with accuracy ε and reliability $1 - \delta$. Namely,

$$n \geq \frac{D_1^2}{\delta\varepsilon^2} \text{ implies } P\Big(\Big|\frac{1}{n}\sum_{k=1}^{n}\xi_k - E_1\Big| \geq \varepsilon\Big) \leq \delta, \quad \text{for } \varepsilon > 0, \quad 0 < \delta < 1. \tag{3.7.3}$$

Based on the observation that for the random variable $\zeta = e^{v(\xi - E\xi)} (v > 0)$, the Markov inequality (Rényi, 1968) yields

$$P(e^{v(\xi - E\xi)} \geq e^t E(e^{v(\xi - E\xi)})) \leq e^{-t}, \qquad t > 0,$$

that is,

$$P\Big(\xi \geq E\xi + \frac{t + \ln E(e^{v(\xi - E\xi)})}{v}\Big) \leq e^{-t}, \tag{3.7.4}$$

it was Bernstein who first proved that, for bounded summands ξ_k, $k = 1, 2, 3, \ldots$, inequality (3.7.2) can be significantly sharpened. Hoeffding (1963) presents a review of different probability inequalities for sums of bounded random variables, including several improvements of the Bienaymé-Chebyshev inequality; extensions are discussed in Pintér (1989).

In this chapter a Bernstein-type improvement of the Bienaymé-Chebyshev inequality is derived for estimating the value of 'noisy' functions. The discussion draws mainly on Pintér (1985).

3.7.2 Estimation of Noisy Function Values

Let $h : \mathbb{R}^n \times \mathbb{R}^q \to \mathbb{R}^1$ be a bounded Borel-measurable function. Assume that the values of the (bounded) function

$$H(x) = E(h(x, \eta)) \tag{3.7.5}$$

are not analytically computable, but are to be estimated (for different arguments $x \in \mathbb{R}^n$) on the basis of realizations $y_k \in \mathbb{R}^q$ of the i.i.d. to η random variables η_k, $k = 1, 2, 3, \ldots$. Our objective is to find a possibly tight bound of the minimal number of realizations n, for which the following confidence interval estimate is valid:

$$P\Big(\Big|\frac{1}{n}\sum_{k=1}^{n} h(x, \eta_k) - H(x)\Big| \geq \varepsilon\Big) \leq \delta, \quad \varepsilon > 0, \quad 0 < \delta < 1. \tag{3.7.6}$$

We shall define the following random variables and probability distribution functions (p.d.f.'s):

$$\xi_k^\star(x) = h(x, \eta_k), \quad G_x^\star(z) = P(\xi_k^\star(x) < z), \qquad k = 1, 2, 3, \ldots, \tag{3.7.7}$$

$$\xi_k(x) = \xi_k^\star(x) - E(\xi_k^\star(x)) = \xi_k^\star(x) - H(x), \quad G_x(z) = P(\xi_k(x) < z),$$
$$k = 1, 2, 3, \ldots,$$

$$\xi^{\star(n)}(x) = \sum_{k=1}^{n} \xi_k^\star(x), \quad \xi^{(n)}(x) = \sum_{k=1}^{n} \xi_k(x), \quad G_x^{(n)}(z) = P(\xi^{(n)}(x) < z),$$
$$n = 1, 2, 3, \ldots.$$

In order to simplify notation, the parameter x of the random variables and p.d.f.'s introduced will be omitted, if it does not lead to ambiguity.

Lemma 3.7.1. *Suppose that $\xi^{(n)}$ has a symmetrical to zero probability distribution for $n = 1, 2, 3, \ldots$, and for every $0 < v \leq V$ there exists $E(e^{v\xi_k^\star})$. Then for arbitrary $0 < v \leq V$ and $\varepsilon > 0$ the relation*

$$P\Big(\Big|\frac{\xi^{\star(n)}}{n} - H(x)\Big| \geq \varepsilon\Big) \leq 2\exp\{-n[v(\varepsilon + H(x)) - \ln E(e^{v\xi_k^\star})]\} \tag{3.7.8}$$

holds.

Proof. By symmetry of the distribution of $\xi^{(n)}$ in (3.7.7) we have

$$P\Big(\Big|\frac{\xi^{\star(n)}}{n} - H(x)\Big| \geq \varepsilon\Big) = P(|\xi^{(n)}| \geq n\varepsilon)$$
$$\leq P(\xi^{(n)} \geq n\varepsilon) + P(-\xi^{(n)} \geq n\varepsilon) = 2P(\xi^{(n)} \geq n\varepsilon).$$

Hence, applying the notation

$$\varepsilon = \frac{t + \ln E(e^{v\xi^{(n)}})}{nv} > 0,$$

analogously to (3.7.4) we obtain (3.7.8), as

$$
\begin{aligned}
P(\xi^{(n)} \geq n\varepsilon) = P\Big(\xi^{(n)} \geq \frac{t + \ln E(e^{v\xi^{(n)}})}{v}\Big) \leq e^{-t} \\
= \exp\{-nv\varepsilon + \ln E(e^{v\xi^{(n)}})\} = \exp\{-n[v\varepsilon - \ln E(e^{v\xi_k})]\} \\
= \exp\{-n[v(\varepsilon + H(x)) - \ln E(e^{v\xi_k^*})]\}.
\end{aligned}
$$

■

Remark 3.7.1. The symmetry of the distribution of $\xi^{(n)}(x)$ is satisfied in the practically important case, when the distribution of $\xi_k^*(x)$, $k = 1, 2, 3, \ldots$, is symmetrical (to the mean value $H(x)$). Obviously, instead of that requirement it is sufficient to assume that

$$
P(\xi^{(n)} \geq z) \geq P(-\xi^{(n)} \geq z), \qquad z \in \mathbb{R}^1, \quad n = 1, 2, 3, \ldots,
$$

holds—as in this case the proof of Lemma 3.7.1 can be directly applied.

In the following we want to make inequality (3.7.8) tight. For fixed $\varepsilon > 0$ and $H(x)$, define

$$
F(v) = v(\varepsilon + H(x)) - \ln E(e^{v\xi_k^*}).
$$

We shall be interested only in values of v, for which $F(v)$ is positive: this is natural, as the result of Lemma 3.7.1 is meaningless (trivial), if $F(v) \leq 0$. Considering the fact that $F(0) = 0$, $F'(0) = \varepsilon > 0$, and—as it is not difficult to check—$F(v)$ is concave, the positivity subinterval of $(0, V)$ always has nonzero length.

Now, if $F(v)$ has a unique maximum on $(0, V)$, then it can be found as the solution of the equation

$$
\frac{\partial F(v)}{\partial v} = \varepsilon + H(x) - \frac{\partial}{\partial v} E(e^{v\xi_k^*}) \cdot \frac{1}{E(e^{v\xi_k^*})} = 0. \tag{3.7.9}
$$

Because the integrand in $E(e^{v\xi_k^*})$ can be differentiated inside the integral operator, (3.7.9) implies the following result.

Lemma 3.7.2. *If there exists $0 < v_* \leq V$ such that the function*

$$
K(v) = \frac{E(e^{v\xi_k^*} \cdot \xi_k^*)}{E(e^{v\xi_k^*})},
$$

at argument $v_\star$ satisfies the equation

$$K(v_\star) = \varepsilon + H(x), \tag{3.7.10}$$

then

$$F(v_\star) = \max_{0 \le v \le V} F(v). \quad \blacksquare \tag{3.7.11}$$

By virtue of (3.7.7)–(3.7.11), the following assertion is valid:

Theorem 3.7.1. *Suppose that the assumptions of Lemmas 3.7.1 and 3.7.2 hold true. Then, for the necessary number of function evaluations in assuring (3.7.6), we have the following estimate:*

$$n \ge n_\star(H(x), \varepsilon, \delta) = \ln \frac{2}{\delta} \cdot \frac{1}{v_\star(\varepsilon + H(x)) - \ln E(e^{v_\star \xi_k^\star(x)})}. \tag{3.7.12}$$

In other words, (3.7.12) *implies* (3.7.6), *when the value of $H(x)$ in (3.7.5) is estimated on the basis of n independent realizations of the random variables $\xi_k^\star(x) = h(x, \eta_k)$; the value $v_\star$ is defined by* (3.7.10). $\blacksquare$

Remark 3.7.2. It is easy to verify that the function $K(v)$ is strictly monotonically increasing: consequently, $v_\star$ is unambiguously defined, if it exists. (Note that the concavity of $F(v)$ also follows by the properties of $K(v)$.)

Remark 3.7.3. The value of n in (3.7.6) can be estimated also in such cases when $v_\star$ is unknown, but, for example, for the given x and a suitable $v > 0$, a positive lower bound $C(x, \varepsilon, v)$ for $F(v)$ can be calculated. In this case

$$n \ge \ln \frac{2}{\delta} \cdot \frac{1}{C(x, \varepsilon, v)} \tag{3.7.13}$$

evidently implies (3.7.6). Assume for example, that

$$E(e^{v\xi_k(x)}) \le e^{\frac{1}{2}B(x)v^2}, \qquad 0 \le v \le V_1(x) \tag{3.7.14}$$

holds; here $B(x)$ and $V_1(x)$ are positive constants depending on x. Suppose additionally that $0 < \varepsilon \le B(x)V_1(x)$. Now we have

$$\begin{aligned} F(v) &= v(\varepsilon + H(x)) - \ln E(e^{v\xi_k^\star(x)}) = v\varepsilon - \ln E(e^{v\xi_k(x)}) \\ &\ge v\varepsilon - \frac{1}{2}B(x)v^2 = \widehat{F}(v), \qquad 0 \le v \le V_1(x). \end{aligned}$$

Therefore, choosing

$$\hat{v} = \underset{0 \le v \le V_1(x)}{\arg\max}\ \widehat{F}(v) = \frac{\varepsilon}{B(x)} \quad \left(F(\hat{v}) = \frac{\varepsilon^2}{2B(x)}\right),$$

we obtain that from

$$n \ge \ln\frac{2}{\delta}\cdot\frac{1}{\widehat{F}(\hat{v})} = \ln\frac{2}{\delta}\cdot\frac{2B(x)}{\varepsilon^2}, \quad 0 < \varepsilon \le B(x)\cdot V_1(x), \quad 0 < \delta < 1, \tag{3.7.15}$$

there follows (3.7.6). Note additionally that it is not difficult to see that (3.7.14) is equivalent to the assumption

$$E(e^{v\xi_k(x)}) < \infty, \qquad 0 \le v \le V_2,$$

where $V_2 = V_2(x) > 0$, and $B(x)$ can be any positive number which exceeds $E(\xi_k^2(x))$.

3.7.3 Estimation of Probabilities

To illustrate an application of the general framework presented, we shall consider now the special case where the unknown probability $0 < p < 1$ of a random event A is to be estimated, on the basis of its relative frequency $\xi^{\star(n)}/n$, where $\xi^{\star(n)}$ is the number of occurrences of A among n independent trials with outcomes A and $\overline{A}$. Recall the discussion of chance constrained models (Chapter 3.5), in which such estimates are typically needed.

Applying first the Bienaymé–Chebyshev inequality,

$$n \ge n_C(p,\varepsilon,\delta) = \frac{pq}{\delta\varepsilon^2} \qquad 0 < \varepsilon \le pq, \ \ q = 1-p, \ \ 0 < \delta < 1 \tag{3.7.16}$$

implies the relation

$$P\left(\left|\frac{\xi^{\star(n)}}{n} - p\right| \ge \varepsilon\right) \le \delta. \tag{3.7.17}$$

Second, according to Bernstein's approach (applying (3.7.4)), from

$$n \ge n_B(p,\varepsilon,\delta) = \ln\frac{2}{\delta}\cdot\frac{2pq\left(1+\frac{\varepsilon}{2pq}\right)^2}{\varepsilon^2} \tag{3.7.18}$$

also follows (3.7.17). Note (see, for example, Rényi, 1968) that in most practical cases (that is, for sufficiently small values of δ),

$$n_B(p,\varepsilon,\delta) < n_C(p,\varepsilon,\delta)$$

holds, and the relative improvement n_B/n_C can be quite significant.

Applying now the technique presented above, one can easily obtain that in this special case Lemma 3.7.1 yields the inequality

$$P\Big(\Big|\frac{\xi^{\star(n)}}{n} - p\Big| \geq \varepsilon\Big) \leq 2\exp\{-n[v(\varepsilon+p) - \ln(1-p+pe^{v})]\}. \qquad (3.7.19)$$

Now, maximizing the function

$$F(v,\varepsilon,p) = v(\varepsilon+p) - \ln(1-p+pe^{v}), \qquad v > 0$$

(with respect to v, while ε and p are held fixed), Theorem 3.7.1 readily implies the following result.

Corollary 3.7.1. *For arbitrary* $0 < p < 1/2$ *and* $0 < \varepsilon \leq pq$,

$$P\Big(\Big|\frac{\xi^{(n)}}{n} - p\Big| \geq \varepsilon\Big) < 2\Big[\Big(\frac{\varepsilon-p}{p}\Big)^{\varepsilon+p}\Big(\frac{q}{q-\varepsilon}\Big)^{\varepsilon-q}\Big]^{-n} \qquad (3.7.20)$$

holds. In other words,

$$n \geq n_\star(p,\varepsilon,\delta) = \ln\frac{2}{\delta}\cdot\frac{1}{\ln\Big[\Big(\frac{\varepsilon+p}{p}\Big)^{\varepsilon+p}\Big(\frac{q}{q-\varepsilon}\Big)^{\varepsilon-q}\Big]} \qquad (3.7.21)$$

implies the relation (3.7.17). ■

The result (3.7.20) was essentially obtained by Okamoto (1958). Note also that for independent random variables ξ_k, $k = 1,2,3,\ldots$, $0 \leq \xi_k \leq 1$, the respective generalization of (3.7.20) also holds, as can be shown, for example, applying Lemma 3.7.1, and Theorem 1 of Hoeffding (1963).

By Taylor series expansion of $n_\star(p,\varepsilon,\delta)$ in (3.7.21), it can be shown that

$$n_\star(p,\varepsilon,\delta) < n_B(p,\varepsilon,\delta), \qquad 0 < \varepsilon \leq pq, \quad 0 < p < 1,$$

that is, the method presented improves Bernstein's bound in the special case considered (for more details see Pintér, 1983a). Numerical examples

Table 3.7.1
Ratios of the number of necessary realizations to estimate p as a function of ε

p	ε	$n_\star/n_B$
0.01	0.001	0.937
	0.002	0.878
	0.005	0.735
	0.009	0.597
0.05	0.005	0.931
	0.010	0.868
	0.025	0.718
	0.045	0.575
0.10	0.010	0.924
	0.020	0.856
	0.050	0.695
	0.090	0.546
0.20	0.020	0.907
	0.040	0.827
	0.100	0.644
	0.180	0.517
0.50	0.010	0.980
	0.020	0.924
	0.050	0.825
	0.100	0.690
	0.250	0.425

showing the ratio of improvement of $n_\star(p,\varepsilon,\delta)$ over $n_B(p,\varepsilon,\delta)$ are displayed in Table 3.7.1: one can see that the improvement becomes more significant if ε and/or p increases.

Without going into further detail, unnecessary in the context of this book, note that a more general framework for deriving deterministic approximations of probability inequalities is discussed in Pintér (1989), with additional references therein. The main point here is that solving stochastic global (or convex) optimization problems via combined optimization–simulation techniques implies a nontrivial computational burden—adding to the burden of deterministic global search. Recall in this context the (theoretical) convergence requirements of such combined solution strategies, in which increasing evaluation accuracy is required (especially Theorem 3.6.4, and the subsequent remarks).

Following the preceding exposition on implementing continuous and Lipschitzian optimization procedures, in Part 4 a variety of applications will be presented. In these case studies, the methodology discussed in Part 3 will be applied throughout, in diverse implementations.

PART FOUR

Applications

When one tugs at a single thing in nature, he finds it attached to everything else in the world.

(John Muir)

The interlocking of the world's economy and the Earth's ecology is the new reality of the century...

(J. MacNeill, P. Winsemius, and T. Yokoshiji, *Beyond Interdependence.* Oxford University Press, 1991.)

Recommending ill-fitting computer systems is the same as a custom tailor designing a business suit that is three sizes large...

(Alan R. Simon, *How To Be A Successful Computer Consultant.* McGraw-Hill, New York, 1994.)

...(But) we live in times when it is the natural scientist who can clearly see certain dangers. Therefore it is his duty to speak.

(Konrad Lorenz)

Introductory Notes

In Part 1, a variety of mathematical problems were reviewed which can be modelled and solved, applying diverse global optimization strategies. Following the detailed treatise in Parts 2 and 3 of Lipschitz and continuous GO algorithms and their implementations, in Part 4 a special emphasis will be placed on applications of LGO and CGO methodology. (Several of these applications were highlighted in Pintér, 1992b.)

The problem-classes discussed in Part 4 originate from the following broad areas:

- (General) nonlinear approximation
- Systems of nonlinear equations and inequalities
- Data classification
- Facility location
- Aggregation of expert opinions
- Product (mixture) design
- Globally optimized calibration of complex system models
- Industrial wastewater treatment system design and operation
- Multiple point source river pollution management
- Lake eutrophication management
- Risk analysis and management of accidental toxic river pollution

All examples and illustrative case studies presented here are based on joint research work of this author and colleagues (referred, when appropriate). Since these studies were conducted during the past decade, they represent several different stages of global optimization algorithm and software development. In particular, some of the numerical results (and the computational environment) reported may reflect somewhat earlier implementations of the related LGO/CGO software. For this reason (especially the earlier) computational information—including the runtimes cited—is of purely illustrative character.

As seen, the applications listed above are collected from several fields—primarily from some areas of general interest in numerical mathematics, and from the increasingly important field of environmental system modelling and management. Part 4 will be concluded by a review of numerous other fields in which global optimization can be applied.

In each chapter, notation characteristic of that area of application is typically followed: hence, the notation applied will vary somewhat in the individual chapters, and it may differ from the notation used earlier.

Chapter 4.1

Nonlinear Approximation: Systems of Equations and Inequalities

4.1.1 Introduction

The best approximation of a function h by means of a (parameterized) family of functions H has a great many applications in numerical analysis, natural sciences, engineering, and economics. Below we shall consider this problem in a fairly general setting (Pintér, 1992b), and then show the relevance of Lipschitzian global optimization for its solution.

Let $S \subset \mathbb{R}^m$ denote the set on which the function $h_0 = h$, $h : \mathbb{R}^m \to \mathbb{R}^1$ is to be approximated. We shall assume that S is nonempty, closed and bounded, and that it is defined by a finite set of functions

$$S = \{y \in \mathbb{R}^m : h_k(y) \leq 0,\ k = 1, \ldots, K\}, \tag{4.1.1}$$

where all h_k are Lipschitzian in some neighbourhood of S. The approximation

$$H(x, y) \cong h(y), \qquad H : \mathbb{R}^{n+m} \to \mathbb{R}^1, \tag{4.1.2}$$

is sought in a Lipschitzian n-parameter family of functions: the parameter-vector $x \in D \subset \mathbb{R}^n$ is subject to optimization. We shall assume that D is bounded, robust, and is defined by a finite number of Lipschitz-continuous functions.

The discrepancy between H and h will be expressed in L_p-norm (for some fixed $1 \le p \le \infty$). The nonlinear approximation problem—with respect to h, H, S, D, and p—then can be stated as

$$\min_{x \in D} L_{p,S}\{H(x,y), h(y)\}, \tag{4.1.3}$$

where

$$L_{p,S}\{H(x,y), h(y)\} = \Big(\int_S |H(x,y) - h(y)|^p dy\Big)^{1/p}.$$

Note that well-known special cases of the global optimization problem (4.1.3) include, for instance, Chebyshev-approximation

$$\min_{x \in D} \max_{y \in S} |H(x,y) - h(y)|$$

(corresponding to $p = \infty$), and least squares approximation ($p = 2$).

It is easy to verify that the objective function in (4.1.3) is Lipschitz-continuous with respect to x, as for $x_1, x_2 \in D$ one has the estimate

$$\begin{aligned} &|L_{p,S}\{H(x_1,y), h(y)\} - L_{p,S}\{H(x_2,y), h(y)\}| \\ &\quad = \Big|\Big(\int_S |H(x_1,y) - h(y)|^p dy\Big)^{1/p} - \Big(\int_S |H(x_2,y) - h(y)|^p dy\Big)^{1/p}\Big| \le \end{aligned}$$

(by virtue of the Minkowski-inequality)

$$\Big(\int_S |H(x_1,y) - H(x_2,y)|^p dy\Big)^{1/p} \le \operatorname{vol}(S)^{1/p} \max_{y \in S} |H(x_1,y) - H(x_2,y)| \le$$

(by compactness of S and Lipschitz-continuity of H with Lipschitz-constant L_H)

$$\operatorname{vol}(S)^{1/p} L_H \|x_1 - x_2\|.$$

Note, at the same time, that uniqueness of the solution to (4.1.3), or convexity of its objective (error) function, cannot be postulated. In fact, nonlinear approximation problems frequently have multiple, nonequivalent (local and global) solutions: see, for example, the related general exposition in Jongen, Jonker and Twilt (1983, 1986), Braess (1986) or, specifically, in the context of L_2-approximation, Diener (1987a). Consequently, Lipschitzian global optimization is a valid generic approach to solving various instances of the problem-class (4.1.3), without further structural requirements. Some important special cases of nonlinear approximation problems will be discussed later, in this and subsequent chapters.

4.1.2 Systems of Nonlinear Equations

Solving a system of nonlinear equations of the general form

$$f_j(x) = 0, \quad j = 1, \ldots, m, \qquad x \in D \subset I\!R^n, \tag{4.1.4}$$

is one of the most important problems in numerical analysis with a broad range of applications. Note that in (4.1.4) the relation $m = n$ and/or the existence and uniqueness of the solution to (4.1.4) is not postulated. Thus, systems of equations which do not have an exact solution or may have several different solutions are not excluded from the present investigations.

The literature focused on problem (4.1.4)—in most cases under additional specifications—is extensive, see, for example, the monographs and surveys by Ortega and Rheinboldt (1970), Rabinowitz (1970), Wacker (1978), Forster (1980), Robinson (1980); Garcia and Zangwill (1981), Allgower and Georg (1983), Dennis and Schnabel (1983). From the discussion in these works, it is apparent that most of the 'classical' approaches—for example, standard one-dimensional root finding methods, fixed point search procedures, continuation (homotopy, embedding) algorithms, quasi-Newton methods and their generalizations, convex programming techniques, etc.—cannot guarantee the solution of the general problem (4.1.4). The primary reason for this is the (mostly) local scope of the approaches listed, namely, the local conditions on their applicability which, in general, may or may not be satisfied. For example, lack of good starting points may induce convergence to locally best approximate 'solutions' (at which the system of equations (4.1.4) has nonzero residuals) or even nonconvergence. In addition, singularities, ill-conditionedness, lack of necessary smoothness properties or of higher order information may lead to poor numerical performance, even in the case of theoretically convergent methods. While it is widely accepted that the solution of general nonlinear systems of equations will remain an inherently complicated numerical problem, the postulated global scope of multiextremal optimization essentially reduces or eliminates some of the theoretical difficulties listed above.

The main subject of this chapter is to show the application of Lipschitz global optimization for solving nonlinear systems of equations (the exposition draws on Pintér, 1990b). Related—global optimization based—investigations are described, for example, in Strongin (1978), Forster (1980), Vysotskaya and Strongin (1983), Diener (1987b), Horst and Thoai (1988), Forster (1992, 1995), and Thach (1993).

4.1.3 Nonlinear Equations and Equivalent Global Optimization Problems

We shall suppose the Lipschitz continuity of all functions f_j on the feasible set D:

$$|f_j(x_1) - f_j(x_2)| \le L_j \|x_1 - x_2\|, \quad x_1, x_2 \in D, \quad L_j > 0, \quad j = 1, \ldots, m, \tag{4.1.5}$$

($\|\cdot\|$ is the Euclidean norm in $\mathbb{R}^n$). Recall that, although the minimal values L_j are rarely known in practice, the Lipschitzian assumption (4.1.5) is fulfilled for a broad class of problems; this includes, for example, systems of equations given by continuously differentiable functions f_j defined on closed, bounded sets D. To simplify the present discussion, D will be taken as a (finite) n-dimensional interval; recall that more general (convex, star and Lipschitzian) constraint sets were handled in the general framework in Part 2, Chapters 2.6–2.7.

The approach followed here is based on transforming the system of nonlinear equations (4.1.4) into an equivalent optimization problem which subsequently can be solved, in the global sense, by some appropriate method. To this end, let us introduce the notation

$$F(x) = (f_1(x), \ldots, f_m(x))^T, \qquad F : \mathbb{R}^n \to \mathbb{R}^m, \tag{4.1.6}$$

(^{T} denotes transpose of the row vector $(f_1(x), \ldots, f_m(x))$. Then (4.1.4) can be written as

$$F(x) = 0 \quad (\text{here } 0 \in \mathbb{R}^m), \qquad x \in D. \tag{4.1.7}$$

The essential requirement with respect to all transformations of (4.1.4) or (4.1.7) is that the transformed problem should have exactly the same solution set X^* as (4.1.4). In this case the transformation will be called 'proper'. A simple way of defining norm-induced proper transformations is the following. Let $N(\cdot)$ be an arbitrary norm on $\mathbb{R}^m$ and let us define

$$f(x) = N(F(x)). \tag{4.1.8}$$

Then, obviously, N induces a proper transformation of (4.1.4) into the optimization problem

$$\min f(x), \qquad x \in D, \tag{4.1.9}$$

since for all global solutions of the system (4.1.4)—and only for these—the solution of (4.1.9) has zero optimum value.

Remark 4.1.1. The above observation implies that, whenever (4.1.4) has solution(s), all norm-induced transformations (4.1.9) of the system (4.1.4) are equivalent, in a theoretical sense. However, if (4.1.4) has no solutions, then the generalized solution of (4.1.4) is naturally defined as its best approximate solution, with respect to the particular norm functional (4.1.8) used. Hence, as a rule, the latter solutions will be norm-dependent. Note that, in the general case, distinction should be made between locally best and global generalized solutions, defined as local and global minimizers of f on D. We shall suppose in the sequel that the norm-dependent, global generalized solution set, denoted by $X^\star$, is, at most, countable. In many cases of practical relevance, $X^\star$ consists of a unique point $x^\star$, or is finite (multiple solutions).

The following simple proposition, needed in order to justify the approach taken, shows the Lipschitzian property of f.

Lemma 4.1.1. *If the functions f_j, $j = 1, \ldots, m$, are Lipschitz-continuous, then f is also Lipschitzian with a constant which depends on the vector function F, the norm N, and the set D.*

Proof. Since all norms on $\mathbb{R}^n$ are topologically equivalent, the inequality

$$N(u) \le c \cdot \sum_{j=1}^{m} |u_j|, \qquad u = (u_1, \ldots, u_m) \tag{4.1.10}$$

holds, with an appropriate constant $c = c(N) > 0$. In the present context, one can consider $u_j = f_j(x)$ for each selected x. According to this interpretation, the constant $c = c(N)$ also depends (implicitly) on F and the set

D: $c = c(N, F, D)$. Now, by the relations

$$|N(u) - N(v)| \leq N(u - v) \leq c \cdot \sum_{j=1}^{m} |u_j - v_j|$$

and the Lipschitzian property of the functions f_j, we obtain for all pairs $x_1, x_2 \in D$, the chain of relations

$$|f(x_1) - f(x_2)| = |N(F(x_1)) - N(F(x_2))| \leq c \cdot \sum_{j=1}^{m} |f_j(x_1) - f_j(x_2)|$$
$$\leq c \cdot \sum_{j=1}^{m} L_j \|x_1 - x_2\| = \left(c \cdot \sum_{j=1}^{m} L_j\right) \cdot \|x_1 - x_2\|.$$

This shows that f is also Lipschitz-continuous with (at most) $L = c \cdot \sum_{j=1}^{m} L_j$ as an appropriate Lipschitz constant. ∎

In view of Lemma 4.1.1, an arbitrarily chosen realization of the general Lipschitzian approach discussed in Part 2 can be applied to solve (4.1.9). More precisely, if the global optimization method selected satisfies the analytical requirements detailed in Chapter 2.4, then the limit point set of the generated sample (search) point sequence is identical to the (nonempty) global solution set X^*.

As an example of applying this general approach, in Pintér (1990b) a diagonal type multivariate extension of Piyavskii's univariate algorithm was used, with additional 'tuning' (recall Chapter 3.1). The method generates an adaptive rectangular subdivision of the initially given interval region D. This basic global optimization strategy has been supplemented with refinements, including a conjugate directions type local search. The basic structure of this algorithm is summarized below.

Diagonal Type Extended Piyavskii Algorithm for Solving Nonlinear Systems of Equations

Step 0. (Initialization) Set $l := 1$ (iteration count); $s := 1$ (actual number of partition subintervals); $k := 2$ (actual number of search points selected). The initial search information is represented by the vertices xl_1 and xu_1 of

$D = [xl_1, xu_1] = \{x \in \mathbb{R}^n : xl_1 \leq x \leq xu_1\}$, and corresponding function values $zl_1 = f(xl_1)$ and $zu_1 = f(xu_1)$ where f is defined by applying the l_1-norm (recall (4.1.8)).

Step 1. (Interval selection) Choose a subinterval $[xl_i, xu_i]$, $i = 1, \ldots, s$, for which the function value

$$L\|xu_i - xl_i\|_2 - \frac{1}{2}(zl_i + zu_i)$$

is maximal; denote the index of this subinterval by t. (L is assumed to be a valid overestimate of the Lipschitz constant of f.)

Step 2. (Point selection and subinterval bisection) Choose a new point from $[xl_t, xu_t]$ according to

$$x\text{new}_t = \frac{1}{2}(xl_t + xu_t) - \frac{(zu_t - zl_t)}{2L\|xu_t - xl_t\|_2}(xu_t - xl_t);$$

set $k := k + 1$.

The point $x\text{new}_t$ defines a bisection of subinterval t by that hyperplane which contains $x\text{new}_t$ and is orthogonal to (one of) the longest edge(s) of $[xl_t, xu_t]$. Replace subinterval t by the two intervals generated in this way; set $l := l + 1$, $s := s + 1$, $k := k + 2$. Evaluate f at the points generated in this step, namely, $x\text{new}_t$ and the two new subinterval-defining vertices.

Step 3. (Stopping criteria of global search phase) Stop global search and go to Step 4, if one of the following conditions holds:

i) sufficiently good global optimum estimate is obtained: $f(x_g^{\min}) < \alpha$ where $x_g^{\min}$ is the best argument found so far, and α is a given tolerance level;

or

ii) subinterval t is sufficiently small: $\|xu_t - xl_t\|_1 < \delta$ (δ is a positive tolerance parameter). If none of the above criteria is met, then return to Step 1.

Step 4. (Solution refinement in local search phase) Initiate local search from $x_g^{\min}$ in order to find an improved solution estimate. In order to possibly enhance the search efficiency, in this phase the objective function is defined by the l_2-norm of the vector function F.

Before turning to an illustrative numerical performance study, let us present an example of solving a three-variable test problem. This problem consists of finding a vector $x = (x_1, x_2, x_3)$ on the interval

$$D = \{1.0 \le x_1 \le 2.5;\ 0.2 \le x_2 \le 2.0;\ 0.1 \le x_3 \le 3.0\} \tag{4.1.11}$$

which (approximately) solves the following system of equations:

$$\begin{aligned} f_1(x) &= 0.1x_1 + \cos(2x_2) + 0.09240 = 0 \qquad (4.1.12)\\ f_2(x) &= \sin(10x_1/3) + \ln(2x_2) - 2.52x_3 + \sin(3x_3) + 0.08805 = 0\\ f_3(x) &= 2(x_1 - 0.75)^2 + \sin(16\pi x_2 - \pi/2) - 3.26815 = 0. \end{aligned}$$

Note that (4.1.12) has a number of 'locally best' generalized solutions on D with significant nonzero residuals: thus, the application of some standard (local) solution method would typically lead to one of these. The approximate global solution is

$$x_i^\star \approx 1.85210, \quad x_2^\star \approx 0.92605, \quad x_3^\star \approx 0.61737, \quad f(x^\star) = \sum_{j=1}^{3} |f_j(x^\star)| < 10^{-5}.$$

Solving the optimization problem, induced by (4.1.11)–(4.1.12), in the form

$$\min\left\{\sum_{j=1}^{m} |f_j(x)| : x \in D\right\},$$

and applying the diagonal extension of Piyavskii's algorithm, one can follow the optimum estimates produced by the outlined algorithm in its global search phase: several steps are shown in Table 4.1.1. (The global search phase was restricted to 150 steps in this illustrative numerical example.)

The best approximation of the solution, found in step 96, is fairly close to $x^\star$ but, because of 'abrupt' changes of the objective function, the corresponding residual is still significant. The local refinement phase needs an additional 180 steps (function and gradient evaluations) to reach the optimal solution from its approximation found in step 96: this indicates the relative difficulty of problem (4.1.11)–(4.1.12).

Table 4.1.1
Solution of problem (4.1.11)–(4.1.12):
several global search phase approximations

No. of global search phase steps	Approximate solution found	Function value
14	(1.8465, 1.2159, 0.7363)	1.4030
29	(1.7500, 0.7079, 0.6637)	1.9309
34	(1.9044, 1.6886, 0.7797)	1.2089
64	(1.7500, 1.2159, 0.6478)	1.7735
83	(1.9044, 1.2159, 0.6637)	1.2156
96	(1.7500, 0.9251, 0.6478)	0.8698
148	(2.2022, 1.6079, 0.9455)	1.0251

4.1.4 Test Results on Randomly Generated System of Equations

In order to obtain a more realistic and detailed assessment concerning robustness and efficiency of the global optimization approach outlined, two classes of pseudo-random test problems are chosen for further analysis. Randomized problem classes are advantageous when compared to individually selected problems, for the following reasons:

- they make possible a statistically sound evaluation of the method in question (with respect to the problem-class studied) under mild and general probabilistic assumptions;
- they are easy to generate sequentially;
- they are unambiguously reproducible, thus facilitating more objective evaluation and comparison of different methods.

For instance, random two-variate test problems were considered in Strongin (1978), for comparing several global optimization algorithms.

Trigonometric Systems of Equations

In the following set of numerical experiments, our objective is to find the solution of the system of equations

$$f_i(x) = \sum_{j=1}^{n} \{A_{ij} \sin^2\{B_{ij}(x_j - x_j^\star)\} + C_{ij} \sin^2\{D_{ij}(x_j - x_j^\star)\}\} \\ + (x_i - x_i^\star)^2 = 0, \qquad i = 1, \ldots, n, \tag{4.1.13}$$

$$x \in D = \{x = (x_1, \ldots, x_n) : -\pi \leq x_j \leq \pi,\ j = 1, \ldots, n\}. \tag{4.1.14}$$

The parameters in (4.1.13) follow randomized generation rules as described below:

- the components of the solution vector, $x_j^\star$, $j = 1, \ldots, n$, are independent identically distributed (i.i.d.), uniform random variables on the interval $[-\pi, \pi]$;
- the multipliers A_{ij}, B_{ij}, C_{ij}, D_{ij} $(i, j = 1, \ldots, k)$ are i.i.d. uniform random variables on the interval $[1, 6]$;
- these parameters are generated sequentially in the following order: $x^\star$ componentwise, the matrices A, B, C, D row by row;
- for generating these values, the simple pseudo-random number generator

$$r_1 = \sin(0.86), \quad r_2 = \sin(0.23), \quad r_{k+2} = r_k + r_{k+1} \pmod 1, \\ (k = 1, 2, 3, \ldots)$$

is applied. (This random number generator is not very sophisticated, but it is very easy to (re)produce, and is adequate for the purposes of the present numerical example.)

For illustration, we shall cite the first pseudo-random problem generated:

$$n = 2, \qquad x_1^\star = 3.052498, \qquad x_2^\star = -1.798262;$$

$$A = [A_{ij}] = \begin{bmatrix} 1.998089 & 3.067078 \\ 4.065167 & 1.132245 \end{bmatrix}; \quad B = [B_{ij}] = \begin{bmatrix} 4.197412 & 4.329657 \\ 2.527069 & 5.856725 \end{bmatrix};$$

$$C = [C_{ij}] = \begin{bmatrix} 2.384794 & 2.240519 \\ 3.624313 & 4.864832 \end{bmatrix}; \quad D = [D_{ij}] = \begin{bmatrix} 2.489146 & 1.353978 \\ 2.843124 & 3.197102 \end{bmatrix}.$$

A few comments related to the problem class (4.1.13)–(4.1.14) are in order. Besides the unique global optimum $x^\star$, several locally best 'solutions'

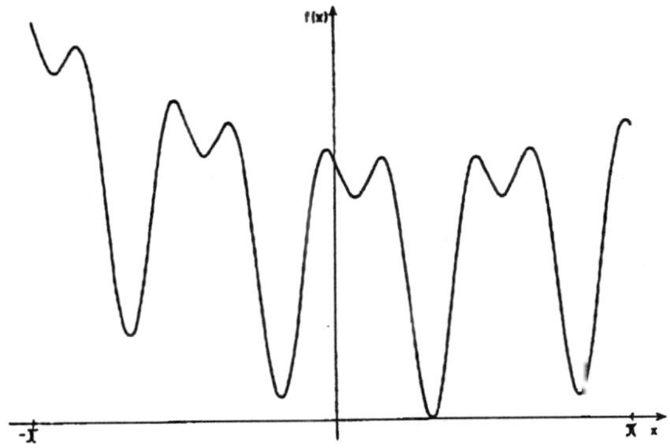

$$f(x) = 5\sin^2\{2(x-1)\} + 3\sin^2\{4(x-1)\} + 0.2(x-1)^2$$
$$-\pi \le x \le \pi, \quad (x^\star = 1,\ f(x^\star) = 0)$$

Figure 4.1.1
A univariate trigonometric equation with several local 'solutions' (minima)

exist with nonzero residuals. The actual number of these 'pseudo-solutions' strongly depends on the random realization of problem parameters. Note also that the fluctuating trigonometric terms may cause significant 'jumps' of the derived objective function

$$\sum_{i=1}^{n} |f_i(x)|, \qquad x \in D. \tag{4.1.15}$$

For illustration, see Fig. 4.1.1, which shows a function closely related to the class of functions (4.1.13). (For a more pronounced graphical effect of the trigonometric terms, the quadratic term is 'scaled down.')

It is obvious that only the additional quadratic terms $(x_i - x_i^\star)^2$ provide an 'explicit' orientation towards finding the global solution in each function f_i, $i = 1, \ldots, n$.

The structure of (4.1.13)–(4.1.14) would not permit the application of a standard equation solver, unless a close local estimate of $x^\star$ can be provided. For example, several locally best generalized (approximate) solutions of the equation

$$5\sin^2\{2(x-1)\} + 3\sin^2\{4(x-1)\} + 0.2(x-1)^2 = 0, \qquad -\pi \le x \le \pi$$

are (see Fig. 4.1.1):

$$x_1 = -2.898 \quad f_1 = 8.062; \quad x_2 = -0.566 \quad f_2 = 0.492;$$

$$x_3 = 1.780 \quad f_3 = 5.123; \quad x_4 = 2.566 \quad f_4 = 0.492;$$

while, obviously, $x^* = 1$ and $f(x^*) = 0$.

A number of instances of the problem-class (4.1.13)–(4.1.14) were solved by the global optimization procedure under different algorithm specifications. These specifications are related primarily to the stopping criteria applied and the method by which the Lipschitz constant of the objective function (4.1.15) is (over)estimated.

Recall that the algorithm outlined earlier has two termination criteria, as follows:

- The approximate solution found in the global search phase is 'acceptable' (as expressed by parameter α). Note that this criterion is specific to the present problem-class, and is based on the assumption that we are interested only in the (by supposition, existing) solution. Although this reflects a somewhat 'ideal' problem setting, for testing purposes and by its practical relevance its use is justified.
- The subinterval selected for further partition is 'sufficiently small' (as expressed by parameter δ). (This criterion is based on the convergence results discussed in Part 2.)

Let us recall here (Chapter 3.2) that the estimation of the Lipschitz constant is probably the most crucial numerical issue in Lipschitzian optimization. In the present numerical study, two simple approximation methods are used. The first (trivial but often used) approach is to provide an overall estimate which might be a grossly exaggerated or a false value, while the second is an approximation on the basis of the sample points taken from the adaptively generated subintervals and the corresponding function values. The first of these approaches will be referred to as the 'overall' estimation method, while the second is called the 'adaptive' estimation technique.

In the first sequence of trial runs, 100 two-variate test problems were generated and solved with tolerances $\alpha = 1$, $\delta = 0.2$. In addition, the overall Lipschitz-constant estimate $L = 20$ was used: note that this might be an underestimate of the (random) problem-specific value L. In 93 cases, the system of equations was solved (up to seven digits of accuracy in the argument $x^\star$); in the remaining seven cases only 'good' local minima were found. Denoting the empirical mean and standard deviation of the number of global and local search phase steps respectively by $E(nglob)$, $D(nglob)$ and by $E(nloc)$, by $D(nloc)$, the following (approximate) run characteristics were obtained:

$$E(nglob) = 489, \quad D(nglob) = 290 \quad \text{and} \quad E(nloc) = 12, \quad D(nloc) = 5.4.$$

Recalling that in the global search phase steps only the function values are sampled, the above statistical characteristics seem to be reasonable. Note that, as reflected by the small number of local terminating steps, the global search delivered a truly close approximation of the (global or local) solution $x^\star$.

In the second set of runs, 50 two-variate test problems were generated and solved using the parameters $\alpha = 0.5$, $\delta = 0.0001$, and $L = 10$. The set value of δ implies that the second numerical stopping criterion is practically excluded. Moreover, the above value of L is often smaller than the minimal true value for the actually realized problem. The results clearly showed the negative consequences of the stated settings. Only 38 problems were solved exactly, and the run characteristics also became less favourable:

$$E(nglob) = 604, \quad D(nglob) = 664 \quad \text{and} \quad E(nloc) = 15, \quad D(nloc) = 9.7.$$

A closer look at the individual runs showed that although in some cases the solution was produced in a few (50–100) algorithm steps, a crude underestimate of L could cause some local deviations and resulted in long execution times.

On the basis of this observation, a third set of experiments was accomplished with the following parameters:

$$\alpha = 1, \quad \delta = 0.6, \quad L = 20\sqrt{2}, \quad maxfct = 200,\ 300,\ 500,\ 2000,$$

where *maxfct* denotes the maximal number of global search phase function evaluations introduced to model the effect of interrupting the 'long' runs.

Table 4.1.2
Random two-variable trigonometric equation systems: Summary of test results

maxfct	Number of successful solutions	*E(nglob)*	*D(nglob)*	*E(nloc)*	*D(nloc)*
200	32	189	35	17	7.1
300	36	263	71	17	7.2
500	44	335	177	15	6.5
2000	50	710	486	11	2.8

Solving 50 randomly generated problems in each case, the results are summarized in Table 4.1.2.

As Table 4.1.2 indicates, the increase of *maxfct* gradually leads to 'perfect' performance, but also to the increase of average global search effort and its variance (without much effect with respect to the local search characteristics). It is notable that relatively few (200) global search steps are already sufficient to solve some two-thirds of the problems.

Closing the illustrative study of this function-class, note that the successful solution of 50 three-variate problems (all solved using the parameters $\alpha = 1$, $\delta = 0.6$, $L = 50$) resulted in the following run characteristic:

$$E(nglob) = 1764, \quad D(nglob) = 384 \quad \text{and} \quad E(nloc) = 21, \quad D(nloc) = 13.$$

The simple run statistics presented highlight the expected difficulty of solving random instances of the problem class (4.1.13)–(4.1.14).

Shekel-Type Systems of Equations

The second problem class, discussed here, has its origin in the early global optimization literature, see, for instance, Dixon and Szegö (1978). The problem is to find the solution of the system of equations

$$f_i(x) = (x_i - x_i^\star)^2 - \sum_{j=3(i-1)+1}^{3i} \frac{1}{\|x - d_j\|^2 + c_j} + k_i = 0, \quad i = 1, \ldots, n, \tag{4.1.16}$$

($\|\cdot\|$ is the Euclidean norm) on the region

$$D = \{x = (x_1, \ldots, x_n) : 0 \leq x_j \leq 10,\ j = 1, \ldots, n\}. \tag{4.1.17}$$

The functions (4.1.16) are generated sequentially according to the following rules (applying again the simple pseudo-random number generator described earlier):

- the components of the solution vector $x^\star$ are i.i.d. uniform random variables (u.r.v.'s) on the interval $[0.5, 9.5]$;
- the term c_1 is a u.r.v. on $[0.1, 1]$; $c_2 = c_1 + r_2$; $c_j = c_2 + r_j$, $j = 3, \ldots, 3n$, where $r_2, \ldots, r_{3n}$ are i.i.d. u.r.v.'s on $[0, 1]$;
- the n-vector d_1 is identical to $x^\star$, the other vectors $d_2, \ldots, d_{3n}$ have i.i.d. components, all of them being u.r.v.'s on $[0, 10]$;
- the constants k_i are adjusted finally, in order to satisfy the equations (4.1.16) for $x = x^\star$.

Note that the above generating rules assure that $x \approx x^\star$ is the unique globally optimal solution; besides that solution, each equation in (4.1.16) has three 'locally best' (generalized) solutions. Consequently, the optimization problem derived, say, in the l_2-norm form

$$\min\{f(x) : x \in D\}, \qquad f(x) = \sum_{i=1}^{n} f_i^2(x), \tag{4.1.18}$$

has $m = 3n$ local generalized solutions. Note that the Shekel-type functions (namely, the sum term in the definition of f_i) are characterized by a rather 'flat plateau' which is modified by the 'valleys' corresponding to local optima (see Fig. 4.1.2). The additional terms $(x_i - x_i^\star)^2$ modify the somewhat 'flat' character of f, and thus have an 'orientating effect' towards the globally optimal solution.

For illustration, the parameters of the first univariate test problem generated by the above rules are as follows:

$$x_i^\star = 9.500000; \quad c_1 = 0.213798; \quad c_2 = 0.463416; \quad c_3 = 0.876831;$$
$$d_1^\star = 9.500000; \quad d_2 = 6.130334; \quad d_3 = 0.264490; \quad k_1 = 4.773540.$$

Similarly to the previously described test runs, the stopping criteria, parameterized by α (now the bound of the summarized squared errors) and δ (the bound of the l_1-norm of the subinterval selected for further partition), were used. In this example, however, interval-specific Lipschitz constants were estimated on the basis of those search points which are generated on the

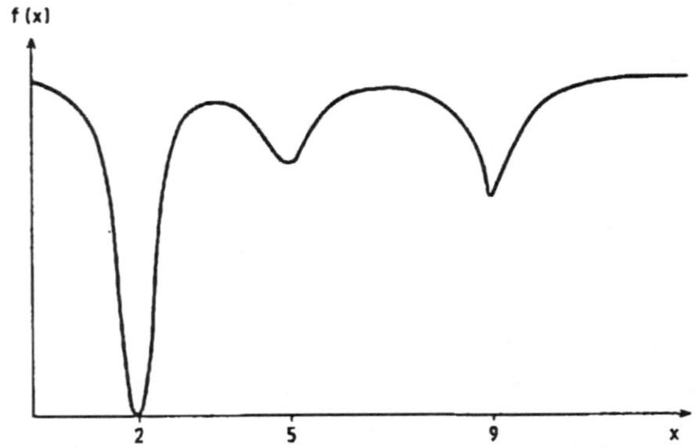

$$f(x) = -\frac{1}{(x-2)^2+0.1} - \frac{1}{(x-5)^2+0.4} - \frac{1}{(x-9)^2+0.3} + 10.126667;$$
$$0 \le x \le 10.$$

Figure 4.1.2
A Shekel-type univariate function

subinterval in question (namely, its central point, a quadratic interpolation point, another linear interpolation point, and the vertices of the main diagonal), before its bisection is done. Note that—in accordance with the related brief discussion in Chapter 3.1—this simple method was expected to have numerical advantages when compared to the (usually rather crude) 'overall' Lipschitz-constant estimation approach applied in solving the trigonometric test problems (4.1.13)–(4.1.14).

Five hundred test runs were accomplished, 100 randomly generated problems for $n = 1, \ldots, 5$. To decrease somewhat the computational burden of this experiment—and to see the 'success rate' of the global search phase—only the global search phase was carried out in each case. Therefore the solution was considered 'successful' in the global sense, if at least one of the following termination criteria was satisfied:

- the total squared error of the equations was less than $\alpha = 0.25n$, or
- when terminating a run by the stopping rule $\delta = 0.01n$, the l_1-distance between the global phase approximate solution and $x^\star$ was less than $0.25n$.

Table 4.1.3
Shekel-type systems of equations:
Summary of test results

Number of variables	Number of successful solutions	Expected number of global phase	Standard deviation function evaluations
1	100	11	13
2	98	137	203
3	100	522	407
4	100	2138	820
5	99	10471	6499

Let us note that, as several detailed test results indicated, in most cases these criteria yielded approximate solutions within the componentwise 0.01–0.05 neighbourhood of x^*. Even the few 'bad' results had only at most 0.3–0.4 componentwise errors. For comparison, note that the trivial uniform grid search applied on the region $[0, 10]^n$ would need about 20^n function calls for meeting the second criterion of 'successful search'. The results of the test runs are summarized in Table 4.1.3.

As noted earlier, even the (3 in 500) approximate solutions were situated in a close neighbourhood of x^*, although somewhat missing the α or δ threshold. The runtime printouts indicated that some test problems were quite difficult: the objective function changed in an extremely abrupt manner close to x^*. This circumstance explains the large number of steps in some cases. For example, one of the five-variate problems needed more than 22,000 function evaluations, though in most cases some 1,500 steps were sufficient and a number of problems were solved in less than 200 steps.

Without attempting a (probably not sufficiently sound) statistical analysis of these results, it is instructive to plot the expected number $E(nglob)$ of global phase function evaluations, on logarithmic scale, against n, the number of test problem variables (see Fig. 4.1.3). As this figure suggests, the number of function evaluations increases by a factor of about four (starting this tendency from $n = 2$: the univariate problems are 'too easy'). Plots similar to Fig. 4.1.3 make it possible to estimate the necessary effort to solve problems from a given class by applying, say, the normal distribution approximation induced by the central limit theorem (see, for example, Feller, 1971).

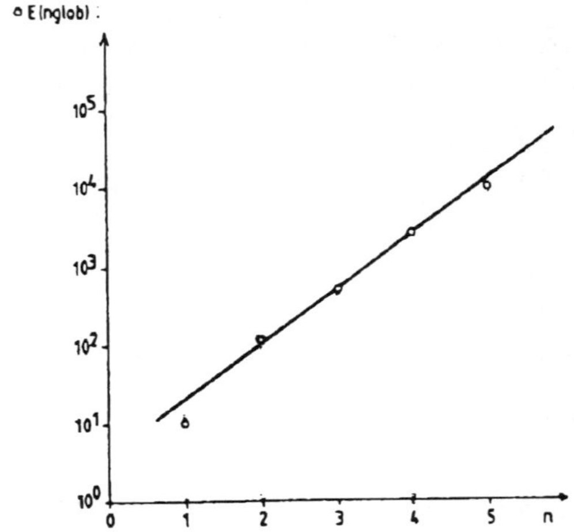

Figure 4.1.3
Solution of random equation systems: Empirical relationship between the number of variables (n) and the computational effort needed $(E(nglob)$, in log-scale)

4.1.5 Lipschitzian Equations and Inequalities

Let us remark that the solution of Lipschitzian equation/inequality systems can be accomplished, applying a simple extension of the global optimization approach presented here; see, for example, Horst and Thoai (1988), or Hendrix and Pintér (1991). Consider the problem

$$\begin{aligned} f_m(x) &= 0, & m &= 1,\ldots,M, \\ g_k(x) &\le 0, & k &= 1,\ldots,K, \\ x &\in D. \end{aligned} \tag{4.1.19}$$

Analogously to the methodology applied above, (4.1.19) can be transformed into the equivalent global optimization problem

$$\min\{f(x)+g(x)\}, \qquad x \in D, \tag{4.1.20}$$

in which

$$\begin{aligned} f(x) &= \|F(x)\|^M, & F(x)^T &= (f_1(x),\ldots,f_M(x)), \\ g(x) &= \|G(x)\|^K, & G(x)^T &= (\overline{g}_1(x),\ldots,\overline{g}_K(x)), \\ \overline{g}_k(x) &= \max\{g_k(x),0\}, & k &= 1,\ldots,K, \end{aligned} \tag{4.1.21}$$

and $||\cdot||^M$ and $||\cdot||^K$ denote arbitrary norms in $\mathbb{R}^M$ and $\mathbb{R}^K$, respectively. As can be immediately seen, the optimization problem (4.1.20) has solution(s) with zero objective function value exactly in the case where the system of equations and inequalities (4.1.19) has corresponding solution(s). Additionally, if the minimum of (4.1.20) is positive, then globally best approximate generalized solution(s) can be found in a well-defined, norm-dependent sense, following the approach of Section 4.1.3.

4.1.6 Concluding Remarks

In this chapter we discussed the relevance of global optimization in nonlinear approximation; as a special case, Lipschitzian global optimization methodology was applied for solving nonlinear systems of equations on relatively 'large' domains. Two classes of randomly generated test problems were investigated in detail. The extensive set of test runs indicates robustness of the approach proposed. Moreover, the methodology suggested makes possible fairly efficient implementations, and can be readily extended to the case of Lipschitzian systems of equations and inequalities.

Apparently, the solution of nonlinear equations—in the general sense studied here—is, and will remain, an inherently difficult numerical task. This point is illustrated by the exponentially increasing number of function evaluations needed in the global search phase to solve the problems (4.1.16)–(4.1.17). The computational demands, however, are significantly less than for the 'naive' grid search or other 'brute force' methods. Note finally that, for solving individual, practically important and complicated nonlinear systems of equations, a number of 'streamlining' refinements are applicable, for example,

- iterative reduction of the feasible region via nested subregions (on the basis of sequential runs with limited computational effort per run);
- multiple iterative local searches on 'promising' subdomains of the feasible set D;
- interactive or automatic problem decomposition by components of the vector x;
- inclusion of statistical decision principles and stopping criteria.

In conclusion, one can conjecture that the theoretical soundness of the global optimization approach helps to find new ways of numerically solving nonlinear approximation problems which were considered intractable earlier. The solution of nonlinear systems of equations can be regarded in this respect as a single—though particularly important—illustrative example; further examples will be discussed in the subsequent chapters.

Chapter 4.2

Data Classification (Clustering) and Related Problems

4.2.1 Introduction

Numerous decision problems related, for example, to environmental monitoring, regional solid waste management, manufacturing systems, transportation services, and so forth, depend essentially on the choice of a relatively small number of 'primary facilities' (such as, for example, water quality analysis laboratories, waste disposal sites, warehouses, bus terminals or other facilities serving customers, or multipurpose machines in a workshop environment). Once the strategic facility design/location decision is made, the long-term operational characteristics of the system depend on the assignment of a—usually (much) larger—number of given entities (data to be grouped, water quality measurement sites in a region, population affected by the waste disposal sites, customers, users of public transport, jobs in a workshop, etc.) to the 'primary facilities' selected at the first stage. In many such cases, the primary (strategic) locational decisions are to be optimized, while the overall system performance (for example, total investment/setup costs plus long-term discounted operational costs) can be assessed—for each primary decision—via 'expensive' evaluation methods: algorithmic solution of embedded optimization problems, stochastic simulation, and so forth. For related discussions and examples, the reader is referred, for example, to Cormack (1971), Hartigan (1975), Jacobsen and Madsen (1980), Dempster, Fisher, Jansen, Lageweg, Lenstra and Rinnooy Kan (1981), Gordon (1981), Muroga (1982), Bodin, Golden, Assad and Ball (1983), Florian (1984), Golden and

Assad (1986), Nitti and Speranza (1987), Pintér and Somlyódy (1987), Ermoliev (1988), Laporte (1988), Wallace (1988), Deininger and Lee (1989), Kaufman and Rousseeuw (1990), Galvão (1993), Ghost and Harche (1993). In a sense, the analysed system can be conceived as some 'black box' or 'oracle' that reacts in a complicated manner to decisions on the primary facility configuration setting. Naturally, there is no prior guarantee for unimodality of the overall performance indicator associated with such problems (since even its analytical form may be unknown). On the other hand, Lipschitz-continuity of the performance indicator—as depending on the basic design parameters—holds in many cases.

To describe a fairly general model that encompasses this problem-class, let us introduce the following notation:

$F_j \in \mathbb{R}^p$	'primary facilities' (basic design characteristics)
$F = \{F_1, \ldots, F_J\}$	'facility configuration'
$\mathcal{F}$	the set of all feasible configurations
$c_i \in \mathbb{R}^q$	'customers' ('secondary' elements in the evaluation of system performance)
$c = \{c_1, \ldots, c_I\}$	set of 'customers'
S_i	'service demand' of c_i; its satisfaction level affects system performance
$S = \{S_1, \ldots, S_I\}$	set of 'service demands'
R_{ij}	'service distance' between c_i and F_j (or, alternatively, 'effect' of F_j on c_i)
$R = [R_{ij}]$	'distance' ('effect') matrix
$SC = SC(F)$	basic construction/setup costs
$OC = OC(F, c, S, R)$	(long-term, average) operational costs, possibly including also other system characteristics.

Applying the above notation, a generic model for analysing a class of 'configuration design' problems can be formulated as

$$\min_{F \in \mathcal{F}} SC(F) + OC(F, c, S, R). \tag{4.2.1}$$

Put into words, best system performance is sought under constraints describing feasible basic configurations: location, size, type, and so forth.

Note that if the primary facility location can be continuously parameterized, then the constraint set in (4.2.1) often has a fairly simple structure (expressing, for example, lower/upper bounds on locations and sizes). On the other hand, as noted above, the overall performance indicator function often has a complicated form, and it may well be multiextremal.

As an instance of this generic model form, we shall consider below the set partition problem in cluster analysis in which 'most homogeneous' or 'most discriminative' grouping of a given set of entities is sought. Following Pintér and Pesti (1991), Lipschitz global optimization will be applied to solve this problem in a fairly general setting.

4.2.2 Data Classification: Problem Statement and Examples

The 'best' partition of a finite set of entities into groups of 'similar' objects (called clusters) is often required in operations research and statistics. Again, one may refer not only to facility location problems, mentioned above, but also to certain global optimization algorithms (see, for example, Törn, 1978; Boender, Rinnooy Kan, Stougie and Timmer, 1982; Rinnooy Kan and Timmer, 1987a,b; Törn and Viitanen, 1992, 1994).

Following the notation of Hansen and Jaumard (1987), let us define:

$O = \{O_k \mid k = 1, \ldots, N\}$	set of entities to be partitioned (grouped)
$D = \{d_{kl} \mid k, l = 1, \ldots, N\}$	set of dissimilarities $d_{kl} = d(O_k, O_l)$, between all pairs of entities
$P_M = \{C_1, \ldots, C_M\}$	a partition of set O into M clusters (assuming $C_j \neq \emptyset$, $C_i \cap C_j = \emptyset$, $\bigcup_j C_j = O$)
π_M	set of all possible partitions P_M.

Applying this notation, the prototype problem can be verbally stated as

Optimal set partition (OSP)

> Find a partition $P_M \in \pi_M$ which provides a 'most homogeneous' (or, alternatively, 'most discriminative') classification of the set of entities O as expressed in terms of the dissimilarities D.

As noted above, OSP is a problem of significance in numerous application areas. In statistical analysis, grouping of (medical, biological, environmental, etc.) data is often essential: clusters will simply indicate 'similarity' of entities belonging to the same subject. In facility location, clusters will be formed by customers who are to be served by the same facility (recall the examples from

Section 4.2.1). In both model types, the dissimilarity matrix D may be (but is not necessarily) defined by an underlying distance measure introduced in the space containing the set O: although the relations $d_{kk} = 0$, $d_{kl} \geq 0$, and $d_{kl} = d_{lk}$ are typically postulated, the 'triangle inequality' is not always required.

Clearly, the problem statement OSP can be quantitatively expressed in a number of non-equivalent ways. Defining, for example,

$$\begin{aligned} d(C_j) &= \max_{O_k, O_l \in C_j} d_{kl} && \text{diameter of cluster } C_j, \\ r(C_i, C_j) &= \min_{O_k \in C_i, O_l \in C_j} d_{kl} && \text{discrimination between } C_i \text{ and } C_j, \\ s(C_j) &= \min_{O_k \in C_j, O_l \notin C_j} d_{kl} && \text{split of cluster } C_j, \end{aligned}$$

one may want to find—typically different—partitions which solve the corresponding optimization problems

$$\min_{P_M \in \pi_M} \|\{d(C_1), \ldots, d(C_M)\}\|_p \tag{4.2.2}$$

$$\max_{P_M \in \pi_M} \|\{r(C_i, C_j, 1 \leq i < j \leq M\}\|_p \tag{4.2.3}$$

$$\max_{P_M \in \pi_M} \|\{s(C_1), \ldots, s(C_M)\}\|_p. \tag{4.2.4}$$

In the above formulas, $\|\cdot\|_p$ denotes an appropriate l_p-norm and $\{\cdot\}$ symbolizes the vector of diameters or discrimination measures, respectively. Well-known special cases of (4.2.2)–(4.2.4) include the following:

- minimum sum of diameters clustering (see, for example, Brucker, 1978; or Hansen and Jaumard, 1987):

$$\min_{P_M \in \pi_M} \sum_j d(C_j) \qquad (p = 1 \text{ in } (4.2.2)); \tag{4.2.5}$$

- minimum diameter partition (see, for example, Johnson, 1967; Rao, 1971; Brucker, 1978; Hansen and Delattre, 1978; or Welch, 1983):

$$\min_{P_M \in \pi_M} \max_j d(C_j) \qquad (p = \infty \text{ in } (4.2.2)); \tag{4.2.6}$$

- maximum sum of discriminations clustering:

$$\max_{P_M \in \pi_M} \sum_{i,j} r(C_i, C_j) \quad (p = 1 \text{ in } (4.2.3)); \tag{4.2.7}$$

- maximum discrimination clustering:

$$\max_{P_M \in \pi_M} \min_{i,j} r(C_i, C_j) \qquad (p = -\infty \text{ in } (4.2.3)); \qquad (4.2.8)$$

- maximum sum of splits clustering (see, for example, Hansen and Jaumard, 1988):

$$\max_{P_M \in \pi_M} \sum_j s(C_j) \qquad (p = 1 \text{ in } (4.2.4)); \qquad (4.2.9)$$

- maximum split partition (see, for example, Delattre and Hansen, 1980):

$$\max_{P_M \in \pi_M} \min_j s(C_j) \qquad (p = \qquad \text{in } (4.2.4)). \qquad (4.2.10)$$

The above (by far not exhaustive) list of problem formulations clearly indicates that 'optimal clustering' is to be defined and determined in a context-dependent manner.

Turning to numerical solution issues, let us point out first that all known (general and exact) set partitioning procedures are of combinatorial character, and thus may become computationally intractable for many realistic problem-sizes. To illustrate this point, note that most formulations of the clustering (Rao, 1971; Mulvey and Cowder, 1978) or facility location problem (see Vinod, 1969; or Kaufman and Rousseeuw, 1990) lead to an integer programming model, hence being NP-complete. For instance, Brucker (1978) and Welch (1983) showed that (4.2.5) and (4.2.6) are NP-hard for $M \geq 3$. For $M = 2$, the algorithm of Hansen and Jaumard (1987) requires $O(N^3 \log N)$ operations for solving (4.2.5). As another example, an $O(N^3)$ algorithm is provided by Hansen and Jaumard (1988) for the problem (4.2.9) (in which M is arbitrary). For this reason, only relatively small-scale problem instances can be solved exactly.

In contrast with the laborious exact algorithms, many feasible clustering procedures are heuristic: hence, they do not guarantee more than, at most, an approximate 'locally best' grouping. Clearly, these facts indicate the necessity of constructing algorithms which are not only sound in the theoretical sense, but are also numerically manageable for realistic problem-sizes: a possible such approach will be introduced below. We shall assume throughout in this chapter that the entities can be represented by real vectors of given (finite) dimensionality m: as indicated by the examples listed above, this case frequently occurs in modelling practical decision problems.

4.2.3 Classification Procedures Based on Cluster Seed Points

In the following discussion, we shall associate with each cluster C_j a seed point s_j: s_j is not necessarily an element of the set O, but it belongs to the m-dimensional real space S which contains O. For a given number of clusters, M, the introduction of the seed point set

$$\mathrm{SP}_M = \{s_1, \ldots, s_M\}$$

permits us to apply a general class of sequential grouping strategies. More precisely, let us assume that for all seed point induced subpartitions $\{C'_j\}$ for which there holds

$$s_j \in C'_j \qquad \text{and} \qquad \bigcup_j C'_j \subseteq O \cup \mathrm{SP}_M,$$

a context-dependent classification performance functional, $f(C'_1, \ldots, C'_M)$, can be evaluated. (Note that f will obviously depend on the concept of similarity/dissimilarity, but we will not express this fact in the notation.) Then—for any given set of seed points—one can formulate a general algorithm scheme:

Seed Point Clustering (SPC)

Step 0. (Initialization) Given O, D, and SP_M, set $k := 1$ (iteration counter); $O^{(1)} := \emptyset$ (initial set of points already clustered); $C_j^{(1)} := \{s_j\}$, $j = 1, \ldots, M$ (initial—seed point induced—clusters). Order (lexicographically or otherwise) the entities O_k.

Step 1. (Single linkage seed point clustering) Given $C_j^{(k)}$, $j = 1, \ldots, M$, and $O^{(k)}$, the next entity $O_k \notin O^{(k)}$ is classified into cluster $j = j(k)$, if

$$\begin{aligned} &f(C_1^{(k)}, \ldots, C_j^{(k)} \cup \{O_k\}, \ldots, C_M^{(k)}) \\ &= \operatorname*{ext}_i \; f(C_1^{(k)}, \ldots, C_i^{(k)} \cup \{O_k\}, \ldots, C_M^{(k)}) \end{aligned} \qquad (4.2.11)$$

holds, where 'ext' stands for the appropriate, context-dependent extremum of the criterion function f. In other words, each entity O_k is assigned to a cluster $C_j^{(k)}$ in such a way that the actual value of the performance indicator function f becomes 'best possible.' (In case of ties, the minimal index j satisfying relation (4.2.11) can be selected.)

Step 2. (Updating) Define

$$O^{(k+1)} = O^{(k)} \cup \{O_k\}, \quad C_j^{(k+1)} = C_j^{(k)} \cup \{O_k\}, \quad C_i^{(k+1)} = C_i^{(k)}, \quad i \neq j.$$

Setting $k := k + 1$, return to Step 1 while $k \leq N$; otherwise stop.

Several comments on the algorithm scheme (SPC) are in order. Note first that its philosophy is quite similar to those clustering algorithms which are based on the concept of 'medoids' (usually, but less appropriately, called also 'medians'), see, for example, Kaufman and Rousseeuw (1987, 1990) for discussions. The essential difference between the 'medoid approach' and the present one lies in the fact that we do not require that the seed points be actually observed entities, that is, they can be conceived as 'ideal medoids'. While in some actual problems this assumption is unacceptable (when special feasibility or observability requirements must hold for all medoids), it also has both conceptual and numerical advantages. First, it is not evident in general that the actually observed entities fully represent the data set to be partitioned (often the opposite case holds); thus, 'ideal medoids' may prove to be quantitatively better descriptors than observed ones. Second, if a search for continuously definable seed points can be justified, then the related numerical procedure—namely, the sequentially optimized choice of seed point sets $\{\mathrm{SP}_M\}$, each of them implying a corresponding embedded classification of the entities—will belong to the realm of continuous optimization. Hence, this approach avoids the complexity issues inherent in combinatorial methods. In this context, note that the algorithm scheme (SPC) needs at most $M - 1$ comparisons per iteration (see Step 1, in which each entry is classified into that cluster for which the actual performance indicator value is 'best'). In this way—for each fixed seed point configuration—$O(MN)$ operations are needed. The optimal choice of SP_M is, as a rule, still a computationally demanding procedure, but the remark above may indicate (problem-size and structure dependent) numerical advantages of (SPC) when compared to exact combinatorial approaches.

Another point which deserves mentioning is that if the criterion (4.2.11) depends only on the seed points chosen, then the clustering algorithm is independent of the actual sequencing of the entities: this is not necessarily the case for other types of approximate clustering rules. This feature of the algorithm scheme (SPC) reflects a certain 'objectivity'.

For the sake of illustration, in the following let us choose the nearest seed point clustering rule in Step 1 of (SPC): in other words, O_k is classified into C_j, $j = j(k)$, if

$$d(s_j, O_k) = \min_i d(s_i, O_k). \tag{4.2.12}$$

The rule (4.2.12) thus leads to a partition which exactly minimizes the total sum of in-cluster dissimilarities between the seed point and entities:

$$\sum_j \sum_{O_k \in C_j} d(s_j, O_k). \tag{4.2.13}$$

(The use of this objective function may be recommended, for example, in certain facility location problems.) Note further that having accomplished the algorithm (SPC)—applying (4.2.12) in Step 1—and defining the function $d(\cdot)$ by a norm-induced distance, one has the simple cluster diameter estimate:

$$\begin{aligned} d(C_j) &= \max_{O_k, O_l \in C_j} d_{kl} && (4.2.14) \\ &\leq \max_{O_k, O_l \in C_j} \{d(s_j, O_k) + d(s_j, O_l)\} \leq 2 \max_{O_k \in C_j} d(s_j, O_k). \end{aligned}$$

The estimate (4.2.14) shows that (4.2.12) leads also to well-quantifiable approximations of the diameter-based clustering rules (4.2.5) and (4.2.6). By contrast, note that 'nearest neighbour' clustering (that is, assigning the next point to that cluster which contains the nearest already classified entity) is also often applied with a similar objective of minimizing within-cluster dissimilarities: however, it does not imply any guaranteed estimate, similar to (4.2.14). Moreover, the latter clustering rule significantly depends on the order in which the entities are taken.

In the next section the issue of efficiently generating (approximate) globally optimal seed points will be investigated.

4.2.4 Selecting SP_M Applying Global Optimization

The widely known multiextremal character of clustering problems—namely, that a large number of locally best classifications may exist—evidently implies that 'standard' (convex) optimization techniques will, in general, fail to locate the globally best seed point configuration. For this reason, correct solution approaches should be based on global optimization methodology.

For the sake of simplicity, we shall assume that the seed points are to be selected on a rectangular domain. Such a region can easily be defined by knowing the entity coordinates $O_{k,i}$, $i = 1, \ldots, m$, as for the seed point coordinates $s_{l,i}$ one may typically postulate the relations

$$a_i = \min_k O_{k,i} \le s_{l,i} \le \max_k 0_{k,i} = b_i, \qquad i = 1, \ldots, m, \;\; l = 1, \ldots, M. \tag{4.2.15}$$

(Another reasonable choice for an initial feasible region may be the convex hull of the entities.) Note that if more exact—possibly cluster-specific—bounds can be provided, then this can lead to considerable numerical benefits.

Given the feasible seed point region $X = \{[a_i, b_i] \;\; i = 1, \ldots, m\}$, our classification model endowed with the clustering rule (4.2.12) can be reformulated as a global optimization problem of the general form

$$\min_{x \in X^M} g(x),$$

in which

$$g(x) = \sum_j \sum_{O_k \in C_j} d(s_j, O_k). \tag{4.2.16}$$

In (4.2.16), g denotes the objective function (overall performance measure, see (4.2.13)). Observe that, given the set O, g depends only on the seed point configuration

$$x = \{s_{1,1}, \ldots, s_{1,m}, s_{2,1}, \ldots, s_{2,m}, \ldots, s_{M,1}, \ldots, s_{M,m}\}, \tag{4.2.17}$$

which is defined by $M \cdot m$ components (the coordinates of the seed points being ordered in an arbitrary fixed sequence). Hence, the domain X^M is an interval region of the $n = M \cdot m$-dimensional space, on which a multiextremal function g is to be minimized. The function g is, by supposition, Lipschitz-continuous with respect to the seed point configuration x: this is essential for applying a Lipschitzian optimization strategy. It is not difficult to see that the l_1-norm Lipschitz property

$$|g(x) - g(y)| \leq L\|x - y\|_1, \qquad x, y \in X^M, \qquad L = L(X^M, g),$$

holds, for example, if the function $d(s_j, O_k)$ is defined by an l_p-distance, $1 \leq p \leq \infty$.

The solution approach, hence, can be based on an adaptive rectangular partition algorithm along the lines of Chapter 2.4. In the numerical examples below, the diagonally extended Piyavskii algorithm (Chapter 3.1) was applied.

4.2.5 Numerical Examples

To illustrate the viability of the suggested approach, test results from Pintér and Pesti (1991) are summarized below. The test problems are randomly generated, facilitating easy reproduction, or the construction of similar problems.

The problem construction consists of generating a finite number of pseudo-random entities on the two-dimensional unit interval, modelling, for example, sets of customers in a facility location problem. All entities follow properly truncated two-variate normal distributions with given mean values and standard deviations: they are produced sequentially, using a simple random number generator. The dissimilarity measure, leading to the matrix D and the values $d(s_j, O_k)$, is defined by the l_1-norm in $\mathbb{R}^2$. The clustering criterion (objective) function is defined by (4.2.16).

In the first example, the number of entities is 100, which are to be grouped into three clusters; in the second example 200 entities are to be classified into five clusters. As the entities are taken from three and five different two-variate normal probability distributions, respectively, the 'goodness' of the solution—the seed point configuration obtained—is easy to visualize. (Of course, this is not the case in general.) Note in passing that, although the selection of the proper number of clusters can be a nontrivial problem per se, it will not be investigated here: consult, for example, the graphical method described in Rousseeuw (1987) for validating decisions related to the number of clusters. In spite of the relative simplicity of the test problems, it would be a demanding task to produce exact (combinatorial) solutions (see, for example, Massart, Plastria and Kaufman, 1983; Klastorin, 1985; or Hansen and Jaumard, 1987, 1988).

Following the exposition in Sections 4.2.3 and 4.2.4, the classification problems above are solved by optimizing cluster seed points on the two-dimensional unit interval: consequently, these numerical examples lead to multiextremal optimization in 6 $(= 3 \cdot 2)$ and 10 $(= 5 \cdot 2)$ variables, respectively. Note that such relatively small dimensionalities may already lead to difficult numerical tasks in global optimization. Observe also the permutation-invariant structure of the seed point set: this implies that the induced global optimization problems would have a large number of formally different, but completely equivalent, solutions. To avoid this situation, one can apply a two-phase hierarchical optimization procedure: in the first phase, an approximate global solution is generated, which is then refined in the second phase by individually optimizing the seed point position for each cluster (associated with that point in the first phase). This simple idea will obviously help in circumventing the permutation problem referred to. Moreover, instead of solving an $n = M \cdot m$-dimensional optimization problem to great accuracy, the first-phase approximate solution is refined in a decomposed manner: namely, in the second search phase the n-dimensional problem is replaced by clusterwise subproblems, of dimensionality m (the dimension of the entities clustered).

To illustrate this approach, see Figs. 4.2.1 and 4.2.2, in which Phase 1 and Phase 2 approximations of the globally optimal seed points are shown, together with the entities to be clustered. This 100-entity, 3-cluster problem needed some 500 search steps in each $[0, 1]^2$ subspace. Recall that the search steps refer to gradient-free evaluations of the performance criterion (4.2.13): this way, given a configuration of seed points, the subsequent clustering procedure is rather fast. (The computation time of Phase 1 on an IBM PC/AT compatible machine was 1 minute 14 seconds, including the preparation of Fig. 4.2.1 and detailed run output files; the total Phase 2 runtime was some 23 seconds only, including again the preparation of Fig. 4.2.2 and of the result file.) As can be seen from Fig. 4.2.2, the algorithmically optimized seed points seem to be good approximate 'medoids', Let us remark at this point that many 'well-posed' classification problems typically will not be very sensitive with respect to slight misspecifications of the cluster seed points: thus an approximate globally optimal solution will most probably be acceptable by practical standards.

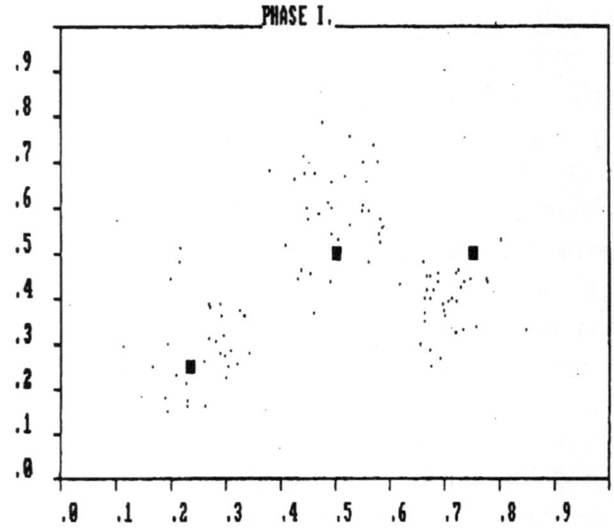

Figure 4.2.1
Globally optimized seed point estimates in Phase 1 (three clusters, 100 randomly generated entities)

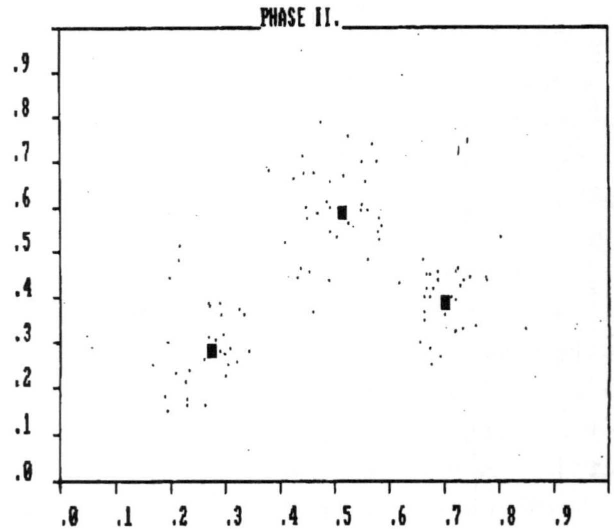

Figure 4.2.2
Globally optimized seed point estimates in Phase 2 (three clusters, 100 randomly generated entities)

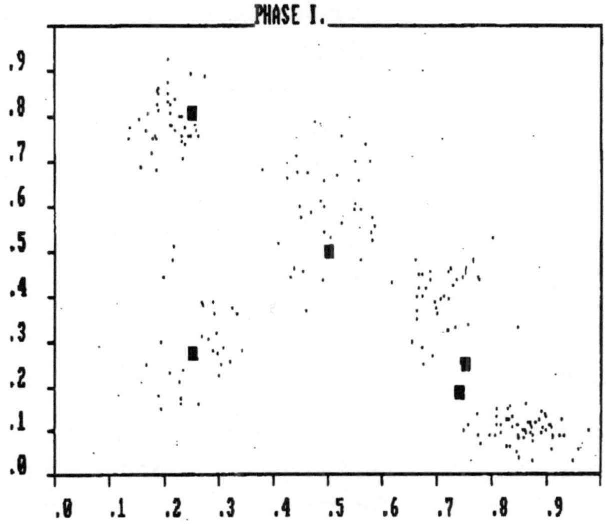

Figure 4.2.3
Globally optimized seed point estimates in Phase 1
(five clusters, 200 randomly generated entities)

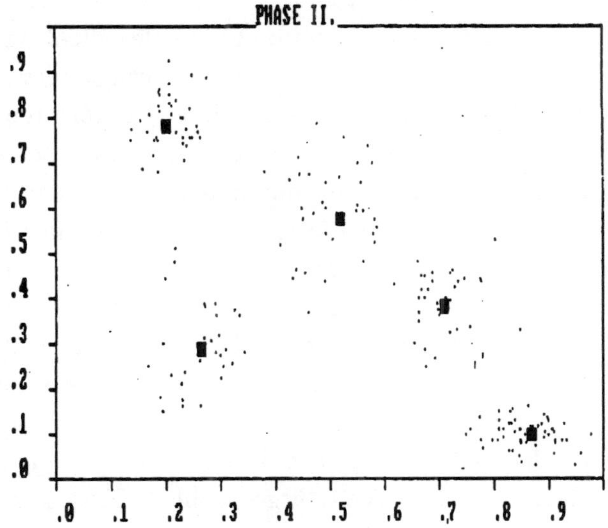

Figure 4.2.4
Globally optimized seed point estimates in Phase 2
(five clusters, 200 randomly generated entities)

Analogous comments can be made with respect to the second example, in which 200 points are automatically grouped into five optimized clusters. Here some 1,000 steps were accomplished in Phase 1, which were followed by some 300 steps clusterwise (on appropriately reduced subinterval regions) in Phase 2. The computational times were now 6 minutes 55 seconds in Phase 1, and 48 seconds (total) in Phase 2, including plots and result files. Again, the seed points obtained seem to be 'well-placed', both visually (see Figs. 4.2.3 and 4.2.4) and in quantitative terms.

4.2.6 Concluding Remarks

The 'best' classification of objects into groups is important in many fields of operations research and applied sciences. For large problem instances (great number of entities and/or clusters) this task is inherently difficult. Therefore exact (combinatorial) techniques can be applied, at present, only to problems of rather moderate size. Heuristic (combinatorial or continuous optimization) strategies guarantee, at best, the finding of some 'locally good' classification which might be quite far from the 'very best' one. Notable in this context is the strongly increasing computational complexity—primarily as a function of N and M—of all known exact algorithms.

The essence of the global optimization based classification procedure followed here is the algorithmic selection of cluster seed points which—given a seed point set related clustering criterion—directly determine the associated cluster formations. In contrast to the exact algorithms, the computational effort in the present approach does not depend as heavily on N, as the entities occur only in the relatively fast, automatic classification and performance evaluation steps of the procedure (SPC), see Section 4.2.3. However, it strongly depends on the number of clusters (M) and the dimensionality of objects (m), as the dimension n of the global optimization problem derived equals $M \cdot m$. As all known general (exact) global optimization methods seem to have exponentially increasing computational demand (in n), the present approach also has its numerical limitations. Still—as the test examples presented illustrate—relatively large problem instances can be solved easily, even on a (definitely 'ancient' by now) PC. It is probably worth emphasizing that the PC-related computational limitations are primarily of storage capacity character: that is, only a bounded number of search steps can be accomplished in a relatively high-dimensional feasible region. Hence, the resulting approximation of the global solution might be poor. Closing this discussion, one can conclude that—in spite of the numerical limitations of the global optimization approach—it is able to produce global estimates of the 'best possible' classification scheme, which can then be refined (for example, in the ways suggested above).

Finally, one can mention another advantage of this approach which lies in its flexibility. As illustrated in Section 4.2.2, numerous non-equivalent mathematical model forms can express the objective of 'optimal' classification. All these problem forms would then need a corresponding combinatorial method for their exact solution. By contrast, the continuous global optimization methodology suggested here will not vary, only the form of the objective function (the clustering criterion and associated performance indicator) is to be provided in a seed point-dependent way. To illustrate this point, let us mention that, given a global optimization algorithm for 'best' facility location as expressed by the objective function (4.2.13), the very same method can be applied for the respective weighted facility location problem (in which 'customer weights' representing, for example, demands are simply included as multipliers of the summands in (4.2.13)), using an arbitrary definition of the distance measure. This remark indicates the potentials of the present approach, also in connection with combining 'expert opinions' in an 'optimally negotiated' way: the latter problem can be mathematically modelled as finding the seed point of a set (or those of several sets) of entities which represent weighted expert judgements. This problem will be investigated in Chapter 4.3. By similar reasoning, many other 'configuration design' problems (mentioned in Section 4.2.1) which depend on a number of primary design objects and some—possibly a great many—entities in an arbitrarily complicated, but algorithmically definable manner, can be treated in the framework presented.

Chapter 4.3

Aggregation of Negotiated Expert Opinions

4.3.1 Introduction

There are many reasons to combine expert assessments and opinions. To list a diversity of areas, where such combinations are relevant, one may think of the weather forecast at a particular location and time, the estimated yield of a natural resources exploration project, the reliability/risk of a new device or technology, the location of a new shopping centre or a waste disposal site, the real-time response of an expert system in a given situation, and so forth. As reflected by these examples, the words 'expert', 'assessment', and 'opinion' will be used here in the broadest possible sense. The aim of this chapter is to propose a unifying framework to analyse and solve problems of aggregating opinions in an objective, quantitatively established manner. The discussion draws on Pintér and Cooke (1987), providing model extensions, much improved solution methodology, and nontrivial numerical examples.

As experience shows, combining (reasonable) expert opinions generally produces 'better' results: more accurate, reliable and stable estimates, reduced random forecasting errors, more realistic forecast ranges, and so on. In other cases, a decision maker typically asks or considers the opinion of more than one individual (relying, for example, on the work of a committee, a board, or several independent advisors), and then—rather than deciding which single opinion is the 'best'—combines them to arrive at some 'generally acceptable' compromise. Lack of data (for example, rare occurrences or unknown outcomes in environmental sciences or aerospace engineering, etc.) may also create a situation in which one simply has to rely on expert estimates and opinions.

The examples listed above indicate that there are at least two distinct areas in which optimization concepts may be brought to bear on the problem of combining expert opinions. One area concerns forecasting in the broadest sense of the word—a decision maker wants to improve his or her forecast by combining the assessments of several experts: what is the 'best' combination? A second area has to do with group decision making (GDM): a group has agreed upon some allocation of relative influence among its members, and now it has to find a decision which properly aggregates the stated individual opinions. What is the 'best' way to determine the group decision?

The problem of forecasting the values of a stochastic process and combining forecasts has received a great deal of attention in the operations research and statistical literature. See, for example, Box and Jenkins (1970), Savage (1971), Morris (1977), McConway (1981), Abraham and Ledolter (1983), DeGroot and Fienberg (1983, 1986), Martz and Bryson (1983), Winkler and Makridakis (1983), Apostolakis (1985, 1988), Ashton and Ashton (1985), Berger (1985), Bopp (1985), Bras and Rodriguez-Iturbe (1985), Ferrell (1985), French (1985), Hipel (1985), Genest and Zidek (1986), Lindley and Singpurwalla (1986), Mosleh and Apostolakis (1986), Hipel and McLeod (1987), Lock (1987), Clarotti and Lindley (1988), Cooke, Mendel and Thys (1988), Singpurwalla (1988), Cooke (1990, 1992), Chibber and Apostolakis (1992), with numerous further references therein. If various properties of the underlying stochastic process are (can be) postulated, then well-defined statistically optimal decisions can be found: see, for example, Genest and Zidek (1986) for a survey.

In more general forecasting problems, however, there may be no stochastic process with known properties which would generate the values of the variables and factors being assessed. Experts estimating failure rates in risk analysis (Apostolakis, 1985), evaluating military intelligence reports (Morris and D'Amore, 1980), bookmakers giving odds on horses (Snyder, 1978)—these are all examples of expert opinion in which the underlying process cannot be properly modelled. (In a sense, if we have a valid stochastic model for the underlying process, then the forecasting problem becomes almost trivial.) When we are deprived of an appropriate underlying stochastic model, experts are generally asked to state their opinion, and then these are to be combined, ideally in an 'optimal' manner (Saari, 1993).

The objective of this chapter is, hence, to show that

- there exist conceptually simple and consequent ways to combine expert opinions in a quantitatively 'best' sense
- this concept has a great variety of specifications tailored to the issue at hand
- robust numerical methods can be applied to find (approximate) the globally best solution of the optimization problems derived.

Obviously, this approach reflects our conviction that there is no single 'universal' model or method which would be optimal for studying all aggregation problems arising in practice.

4.3.2 A General Model for Combining Negotiated Expert Opinions

To formulate a prototype model of the GDM problem, the following notation is introduced:

$\{e_i : i = 1, \ldots, n\}$	set of 'experts' (individuals, groups, etc.) expressing their quantitative opinion
A	set of possible ('feasible') stated opinions
$a = \{a_1, \ldots, a_n\}$	vector of given expert opinions ($a_i \in A$ for $i=1, \ldots, n$)
$\{w_i \geq 0,\ i = 1, \ldots, n\}$	fixed 'negotiating weights' of experts (reflecting their influence, knowledge, experience, and so forth)
$w = \{w_1, \ldots, w_n\}$	vector of weights
T	fixed, parameterized 'pooling formula' which defines the method of combining the opinions
$\{x_i,\ i = 1, \ldots, n\}$	parameters of pooling operator T, subject to optimization (x_i can also be called the pooling weights, when this terminology is appropriate)
$x = \{x_1, \ldots, x_n\}$	vector of opinion pool parameters
D	set of feasible parameterizations
$a_c = T\{a, x\} \in A, \quad x \in D$	opinion resulting from combining the individual opinions and selecting a parameterization of the pooling formula

$d = d(u, v)$	difference measure, expressing the discrepancy between two feasible opinions u and v
$\|\| \cdot \|\|$	given norm functional in $\mathbb{R}^n$.

Applying this notation, one can verbally state the

Prototype Group Decision Making Problem (PGDMP)

Given the opinions a, the expert weights w, the pooling operator T, and the difference measure d, find the 'best possible' parameter vector x in D which solves the following optimization problem:

$$\min_{x \in D} \|(w_1 d(a_1, a_c), \ldots, w_n d(a_n, a_c))\|, \qquad a_c = T\{a, x\} \in A. \qquad (4.3.1)$$

In other words, for the optimal parameterization $x^\star \in D$ of the pooling operator T, $a_c = T\{a, x^\star\} \in A$ minimizes in norm the weighted difference between the individual experts' opinions and the combined opinion.

Note that it is tacitly assumed that the experts e_i are in 'dialectical equilibrium' (DeGroot, 1974), that is, their individual opinions have already been fixed and will not change by renegotiation.

It is worth emphasizing again at this point that there is no unique rational way of aggregating opinions. For example, the surveys of French (1985), and Genest and Zidek (1986) clearly show that distinctly different 'optimal' (suitable) pooling function forms can be deduced from diverse 'reasonable' sets of postulated requirements. This fact obviously implies that such a pooling operator cannot be found which would simultaneously meet all 'expectable' statistical desiderata. Moreover, even arriving at an 'acceptable' pooling formula, its parameterization remains largely 'user-dependent'. In light of these comments, we shall present model variants which then can be 'optimally' parameterized, in some well-defined quantitative sense.

As seen from the general problem statement (4.3.1), its concretizations depend primarily on the mathematical form of the following problem components:

- expert opinions $a_i \in A$
- pooling operator T
- discrepancy measure d

- norm $\|\cdot\|$
- structure of the feasible parameterization set D.

Observe that these problem components are not necessarily independent: a brief discussion of several possible choices and the resulting model variants follows.

4.3.3 Model Specifications

Expert Opinion Forms, Norms, and Related Discrepancy Measures

Deterministic ('Point') Opinions, Expressed by Vectors of Linear Normed Spaces

Assume first that the set A belongs to a linear normed space S: specifically, this includes the case $S = I\!R^K$, in which the opinions are represented by real K-vectors. Then it is most natural to define the discrepancy measure by the norm of S, that is, $d(u,v) = \|u-v\|$, $u,v \in S$. In particular, if $A \subset I\!R^K$, then d is often defined by some fixed l_p- norm

$$\|u-v\| = \|u-v\|_p = \left\{\sum_{k=1}^{K} |u_k - v_k|^p\right\}^{1/p}, \qquad 1 \le p \le \infty. \tag{4.3.2}$$

The most frequently considered variants are: $q = 1$ (Manhattan distance), $q = 2$ (Euclidean distance), and $q = \infty$ (maximum norm). Another notable example is the binary l_1-distance, defined for $\{0,1\}$ ('no/yes') binary vectors u, v by

$$d(u,v) = j, \quad \text{if} \ \sum_{k=1}^{K} |u_k - v_k| = j. \tag{4.3.3}$$

Statistical Opinions, Expressed by Probability Distributions

Suppose now that the individual opinions are stated in the form of probability distribution functions (p.d.f.'s). This is typical, for example, in the context of uncertainty analysis, risk assessment, and management (see Martz and Bryson, 1983; Lindley and Singpurwalla, 1986; Mosleh and Apostolakis, 1986; Cooke and Pintér, 1989; Van Noortwijk, Dekker, Cooke and Mazzuçchi, 1992; Cooke, 1994). We shall assume that all opinion p.d.f.'s F_i are defined with respect to a common underlying measurable space $\{\Omega, \mathcal{A}\}$. For simplicity, the present discussion will be restricted to scalar (one-dimensional) p.d.f.'s, but it can certainly be extended to multidimensional p.d.f. vectors, applying, for example, the concepts and tools of multiobjective analysis, or the approach of DeGroot and Fienberg (1986). Observe that even in this simplest case the set of possible opinions A will not be closed, per se, under the addition operator. However, if we restrict our attention to convex combinations of p.d.f. opinions, then all these will be again p.d.f.'s.

Let us turn now to the question of finding appropriate discrepancy measures. Note first of all that, in several cases of known statistical relevance, the functional forms introduced below will not have all properties of the distance notion in analysis: more precisely, while $d(u,v) \geq 0$ and $d(u,v) = 0 \Leftrightarrow u = v$ will hold in all cases, $d(u,v) = d(v,u)$ and $d(u,v) \leq d(u,w) + d(w,v)$ will not necessarily be valid. We shall assume the existence and finiteness of all integrals below, and shall also suppose the existence of the density function F' (of the p.d.f. F) when it appears.

Note first that the continuous extensions of l_p-norms can be directly applied: the difference of p.d.f.'s F and G is defined then by

$$d(F,G)_p = \left\{ \int_{-\infty}^{\infty} |F(z) - G(z)|^p dz \right\}^{1/p}, \qquad 1 \leq p \leq \infty. \tag{4.3.4}$$

A well-known special case of (4.3.4), corresponding to $p = \infty$, is

$$d(F,G) = \sup_{-\infty<z<\infty} |F(z) - G(z)| \qquad \text{(Kolmogorov statistics)} \tag{4.3.5}$$

or its one-sided variant, $d(F,G) = \sup_{-\infty<z<\infty} \{F(z) - G(z)\}$.

Other discrepancy measures can be defined by somewhat less common statistics. One may consider, for instance, the following forms of d:

$$d(F,G) = \int_{-\infty}^{\infty} |F(z) - G(z)| F'(z) dz \qquad \text{(Fisher-von Mises statistics);} \tag{4.3.6}$$

$$d(F,G) = \int_{-\infty}^{\infty} |F(z) - G(z)|^2 F'(z) dz \qquad \text{(Cramer-von Mises statistics);} \tag{4.3.7}$$

$$d(F,G) = \frac{1}{2} \int_{-\infty}^{\infty} |F(z) - G(z)| dz \qquad \text{(Lindefors statistics);} \tag{4.3.8}$$

$$d(F,G) = 1 - \int_{-\infty}^{\infty} [F'(z) \cdot G'(z)]^{\frac{1}{2}} dz \qquad \text{(Hellinger distance);} \qquad (4.3.9)$$

$$d(F,G) = \int_{-\infty}^{\infty} (\ln F'(z) - \ln G'(z))F'(z)dz \quad \text{(Relative entropy).} \quad (4.3.10)$$

For a theoretical discussion of these statistics, the so-called 'divergence geometry' of p.d.f.'s and induced minimization problems, see, for example, Sherman (1967), Csiszár (1975), and Ermakov (1975). Klafszky, Mayer and Terlaky (1989) describe an application in which an optimized mixture of discrete p.d.f.'s is compared to a target p.d.f., applying several discrepancy criteria: linear and convex programming techniques are then used to solve the problems derived. It should be stressed at this point that although the evaluation of formulas like (4.3.6) to (4.3.10) can be computationally more demanding, they also provide more 'sensitive' discrepancy measures than some of the most usual statistical criteria. Observe additionally that all nonsymmetric discrepancy formulas cited above lead naturally to a symmetric counterpart d_s, defining, for example,

$$d_s(F,G) = \frac{1}{2}\{d(F,G) + d(G,F)\}. \qquad (4.3.11)$$

Pooling Operators

In many situations there is an explicit statistical rationale to select one of several standard functional forms of the pooling operator T. Among these, probably most well known are the linear pool

$$a = T\{a_1, \ldots, a_n\} = \sum_{j=1}^{n} x_j a_j, \qquad (4.3.12)$$

$$\text{(usually, } 0 \le x_j \le x_{j\max} \text{ or } 0 \le x_j, \ \sum_{j=1}^{n} x_j = 1\text{);}$$

and the logarithmic pool

$$a = T\{a_1, \ldots, a_n\} = \prod_{j=1}^{n} a_j^{x_j}, \qquad (4.3.13)$$

$$\text{(again, } 0 \le x_j \le x_{j\max} \text{ or } 0 \le x_j, \ \sum_{j=1}^{n} x_j = 1\text{).}$$

For details, see French (1986) or Genest and Zidek (1986). In the context of the present discussion, the main point is that—identifying (selecting) a pool structure—the optimization problem is reduced to the optimal parameterization of the fixed pooling operator.

Expert Weights

As mentioned earlier, the experts may have different weights reflecting, for example, their experience or negotiating power. The issue of eliciting expert weights applying probabilistic scoring rules is discussed in detail by Cooke (1990, 1992), with numerous further references therein. Therefore we shall assume here that the weights are given, possibly resulting from a relatively straightforward assessment or a detailed prior study.

For providing a brief motivation concerning the objective function form in (4.3.1), namely,

$$\|w_1 d(a_1, a_c), \ldots, w_n d(a_n, a_c)\|, \qquad a_c = T\{a, x\} \in A, \tag{4.3.14}$$

one can observe that the most common weighted linear aggregation formula

$$a_c = \sum_{j=1}^{n} w_j a_j \tag{4.3.15}$$

has some apparent deficiencies. First of all, it may well happen that the opinion a_c determined by (4.3.15) is simply not acceptable, because it does not belong to the feasible set A. (An example is the aggregation of integer vector opinions which—by requirement—should result in an integer vector.) Second, accepting (4.3.15), the pairwise discrepancies between the aggregated and individual opinions are subsequently expressed by

$$d_i = d(a_c, a_i) = d\Big(\sum_{j=1}^{n} w_j a_j, a_i\Big), \qquad i = 1, \ldots, n, \tag{4.3.16}$$

and one may question whether these discrepancies are indeed 'optimal' or at least 'acceptable'. Evidently, it is not automatically so, and one may think of other aggregation concepts which explicitly consider the 'differences' d_i and the expert weights w_i. This argument naturally leads to the introduction of the objective function (4.3.14), which—besides (4.3.16)—enables a great many diverse specifications.

Additional Constraints and Requirements

Concluding the discussion of various model forms, let us note that besides the explicit constraints, expressed by $x \in D$ and the pooling formula $a_c = T\{a, x\} \in A$, further requirements can also be considered. For instance, one may require that the combined opinion should be sufficiently 'close' to the individual ones,

$$d(a_c, a_i) \leq \delta_i, \tag{4.3.17}$$

or that it should be sufficiently 'far' from all a_i's,

$$d(a_c, a_i) \geq \varepsilon_i \tag{4.3.18}$$

($\delta_i > 0$ and $\varepsilon_i > 0$ are given tolerance levels for $i = 1, \ldots, n$). As for possible examples, one may think about allowable maximal or required minimal distances between the location of some 'desirable' (shopping mall) or 'undesirable' (waste disposal site, nuclear power plant, etc.) facility and several neighbouring customers, or municipalities. (For a detailed analysis of the second problem type, see, for example, the study of Lober (1993) in which public attitudes towards waste facility siting are discussed, providing also an extensive list of further references.)

The 'menu' of model specifications presented is, of course, far from being exhaustive: it only serves to illustrate the flexibility of the unifying problem formulation (4.3.1). The general modelling framework naturally implies the possibility of specifying and solving a sequence of diverse related problem statements. Below a global optimization strategy will be proposed which can be applied to solve a broad range of the resulting model variants.

4.3.4 Solution Approach Based on Lipschitz Global Optimization

In order to encompass a wide range of continuous specifications of the problem statement (4.3.1), we shall consider the mathematical programming problem

$$\begin{aligned} &\min f_0(x) \\ &f_m(x) \leq 0, \qquad m = 1, \ldots, M, \\ &x \in D_0 \end{aligned} \tag{4.3.19}$$

under the following specifications (recall the analytical assumptions discussed in detail earlier, in Part 2):

1. The explicit constraint set D_0 is a given box (interval) or simplex in the corresponding Euclidean real space; for example,

$$D_0 = \{x \in I\!R^n : 0 \le x_j \le 1,\ j = 1, \dots, n\}, \text{ or} \tag{4.3.20}$$

$$D_0 = \{x \in I\!R^n : 0 \le x_j,\ j = 1, \dots, n,\ \sum_{j=1}^{n} x_j = 1\}. \tag{4.3.21}$$

   (The simplex D_0 in (4.3.21) lies in an $(n-1)$-dimensional subspace of $I\!R^n$.)

2. The feasible set

$$D = \{x \in D_0 : f_m(x) \le 0,\ m = 1, \dots, M\} \tag{4.3.22}$$

   is either empty, or it is the closure of a nonempty open set.

3. The functions f_m, $m = 0, \dots, M$, are Lipschitz-continuous on D_0, that is, the inequalities

$$|f_m(x) - f_m(y)| \le L_m \|x - y\|, \quad x, y \in D_0, \quad m = 0, \dots, M, \tag{4.3.23}$$

   hold: the constants $L_m = L_m(f_m, D_0)$ being valid, known (over)estimates of the (minimal) Lipschitz constant of the corresponding function f_m on the embedding set D_0.

4. The function $f(x) = f_0(x)$ is (possibly) multiextremal on D, but the set X^* of globally optimal solutions to (4.3.19) is at most countable.

One can observe that the above problem statement reflects model (4.3.1) in a fairly general manner. For instance, if the pooling operator T, the distance d in (4.3.1), and the functions f_m in (4.3.23) are all defined by continuously differentiable functions, then—possibly after simple scaling transformations—such problems can be expressed in the form (4.3.19).

As discussed earlier in detail (see especially Chapters 2.4 and 2.5), for solving the LGO problem (4.3.19) a broad class of adaptive partition methods can be introduced. These methods generate on D_0 a sequence of partition subsets (in the present case, respectively, intervals or simplices) such that the sample points taken from these subsets converge to the set of globally optimal solutions. The results presented below were attained using a recent (1994) version of the PC implementation of this adaptive partition strategy.

4.3.5 Numerical Examples

To illustrate the approach suggested, in the following a few—simple, but nontrivial—test problems will be stated and solved, applying Lipschitzian global optimization. The expert opinions will be modelled by discrete probability distribution functions, and 'best' parameterized p.d.f. approximations to this set of p.d.f.'s will be sought. Hence, we shall analyse specific instances of the general problem statement (PGDMP) formulated in Section 4.3.2. All these problems are expressed in the following form:

Given a set of p.d.f.'s, their weights (to be considered in the aggregation), an appropriate p.d.f. discrepancy measure, and an approximating function form, find the optimally parameterized approximation which results in the best match with the weighted set of p.d.f.'s. Explicit (range) constraints are also given, related to the parameters of the functional p.d.f. approximation.

Although the numerical examples below are illustrative, the actuality of such p.d.f. aggregations is obvious, for example, in the context of diverse engineering systems reliability and maintenance studies, in resource exploration projects, and in environmental risk assessment (reference is made to Beck and van Straten, 1983; Martz and Bryson, 1983; Berger, 1985; Flåm and Pintér, 1985; Lindley and Singpurwalla, 1986; Apostolakis, 1988; Singpurwalla, 1988; Cooke and Pintér, 1989; Chibber and Apostolakis, 1992; Cooke, 1992, 1994; Van Noortwijk, Dekker, Cooke and Mazzucchi, 1992).

In all subsequent illustrative examples, seven input p.d.f. 'opinions' will be given: four different generalized (shifted) Poisson p.d.f.'s and three different binomial p.d.f.'s. The shifted Poisson p.d.f.'s are defined, for given $\lambda > 0$ and nonnegative integer shift parameter s, by

$$p_k = \text{Prob}\{\text{Poisson}(\lambda, s) = k\} = \exp(-\lambda)\lambda^{k-s}/(k-s)! \qquad (4.3.24)$$
$$k = s, s+1, s+2, \ldots, \qquad (p_k = 0 \text{ for } k < s).$$

The standard binomial p.d.f.'s are defined, for given $0 \le p \le 1$ and positive integer n, by

Table 4.3.1
Input data of Examples 1 and 2

			Poisson p.d.f.'s		
p.d.f.	λ	s	mean value	variance	weight
1	2	3	5	2	0.07
2	3	22	25	3	0.16
3	10	8	18	10	0.23
4	5	10	15	5	0.12
			Binomial p.d.f.'s		
p.d.f.	n	p	mean value	variance	weight
5	5	0.2	1.0	0.80	0.22
6	12	0.6	7.2	2.84	0.12
7	25	0.5	12.5	6.25	0.08

$$b_k = \text{Prob}\{\text{Binomial}(n,p) = k\} = n!p^k(1-p)^{n-k}/[k!(n-k)!] \quad (4.3.25)$$
$$0 \le k \le n, \qquad (b_k = 0 \text{ otherwise}).$$

The input p.d.f. parameters applied in the first two numerical examples are given in Table 4.3.1.

Obviously, the p.d.f.'s defined above can be interpreted as discrete approximations of corresponding three-parameter gamma p.d.f.'s and normal p.d.f.'s: respectively, these are among the most important models of random variables having a skewed or a symmetric unimodal distribution. One can also observe that the input ('opinion') p.d.f.'s are all noticeably different, as reflected by the corresponding mean/variance data.

Example 1

In the first example, the weighted mixture of the above p.d.f.'s will be approximated by the parameterized linear combination of one generalized Poisson and one binomial p.d.f.:

$$F = F(\lambda, s, n, p, \alpha) = (1-\alpha)\text{Poisson}(\lambda, s) + \alpha\text{Binomial}(n,p). \quad (4.3.26)$$

The corresponding discrete p.d.f. $\{f_k\}$

$$f_k = \text{Prob}\{F(\lambda, s, n, p, \alpha) = k\} \qquad (k \ge 0 \text{ integer})$$

is given by the following expressions. If $s \le n$, then

$$f_k = \alpha n! p^k (1-p)^{n-k} / [k!(n-k)!] \qquad \text{if } 0 \le k < s; \qquad (4.3.27\text{a})$$

$$f_k = (1-\alpha)\exp(-\lambda)\lambda^{k-s}/(k-s)! + \alpha n! p^k (1-p)^{n-k}/[k!(n-k)!] \qquad \text{if } s \le k \le n;$$

$$f_k = (1-\alpha)\exp(-\lambda)\lambda^{k-s}/(k-s)! \qquad \text{if } k > n.$$

For $s > n$, the composite p.d.f. is

$$f_k = \alpha n! p^k (1-p)^{n-k}/[k!(n-k)!] \qquad \text{if } 0 \le k \le n; \qquad (4.3.27\text{b})$$

$$f_k = 0 \qquad \text{if } n < k < s;$$

$$f_k = (1-\alpha)\exp(-\lambda)\lambda^{k-s}/(k-s)! \qquad \text{if } k \ge s.$$

Absolute value (l_1) norm will be applied to aggregate the component discrepancies. In other words, if the discrepancy of F and some other F_i is expressed by the function $d(F, F_i)$ and w_i is the 'weight' of F_i, then the discrepancy between F and the weighted collection of F_i's is defined by $d(F, \{F_i\}) = \sum_i w_i d(F, F_i)$.

In Example 1, the discrepancy measure d is defined as follows. For brevity, we shall introduce the notation

$$p_k^{(i)} = \text{Prob}\{\text{Poisson}(\lambda^{(i)}, s^{(i)}) = k\}, \qquad i = 1, 2, 3, 4,$$

and

$$b_k^{(j)} = \text{Prob}\{\text{Binomial}(n^{(j)}, p^{(j)}) = k\}, \qquad j = 1, 2, 3,$$

for the p.d.f.'s described in Table 4.3.1.

Then the following aggregated discrepancy (objective) function will be used (which symmetrizes (4.3.6)):

$$d(F, \{F_i\}) = \sum_{k=0}^{\infty} \Big(\sum_{i=1}^{4} w_i |f_k - p_k^{(i)}| (f_k + p_k^{(i)}) + \sum_{j=1}^{3} w_j |f_k - b_k^{(j)}| (f_k + b_k^{(j)}) \Big). \qquad (4.3.28)$$

The feasible region, on which $d(F, \{F_i\})$ will be minimized, is defined by the—intentionally broad, 'safely chosen'—range constraints:

$$0.05 \le \lambda \le 25\,;\ 0 \le s \le 30\ (s \text{ integer})\,;\ 1 \le n \le 40\ (n \text{ integer})\,;$$
$$0 \le p \le 1\,;\ 0 \le \alpha \le 1. \qquad (4.3.29)$$

Table 4.3.2
A few search points generated by the LGO solver procedure (Example 1)

argument					objfct	argument					objfct
λ	s	n	p	α		λ	s	n	p	α	
0.050	0	1	0.000	0.000	0.4971	25.000	30	40	1.000	1.000	0.5643
12.525	15	20	0.500	0.500	0.0776	12.067	14	19	0.482	0.482	0.0776
15.268	24	14	0.131	0.829	0.1091	07.701	22	30	0.035	0.329	0.9814
17.182	8	1	0.693	0.298	0.9287	10.972	5	8	0.325	0.162	0.0804
3.545	8	32	0.108	0.254	0.1065	15.461	15	7	0.507	0.804	0.0820
8.852	26	27	0.446	0.198	0.0989	14.277	18	14	0.005	0.032	0.1024
11.403	4	33	0.900	0.945	0.0691	15.575	2	21	0.647	0.514	0.0732
11.403	30	40	1.000	1.000	0.5461	11.403	0	10	0.000	0.000	0.0988
5.726	15	20	0.500	0.500	0.0778	5.518	14	19	0.482	0.482	0.0775
3.755	4	23	0.877	0.644	0.0745	7.876	19	12	0.525	0.101	0.0999

(For comparison, recall the input parameter data displayed in Table 4.3.1.)

To illustrate the variability of the objective function (4.3.28), the first 20 sample points, generated by the global search phase of the optimization procedure, and the corresponding objective function values (*objfct*) are listed in Table 4.3.2.

The first few sampling results listed seem to suggest that, apart from strongly suboptimal parameterizations, function (4.3.28) may be quite 'flat' on the 'good' parameter regions. This tentative observation was confirmed by the additional calculations (not quoted here in detail), indicating a certain 'insensitivity' of the discrepancy measure (4.3.28). The 'insensitivity' referred to is induced primarily by the probability multipliers $f_k + p_j^{(k)}$ and $f_k + b_j^{(k)}$ which are quite small, for most values k. Therefore only minor differences can be observed among the 'better' solutions, in terms of the objective function (4.3.28).

On the basis of an automatic mode computation, consisting of 25,000 global search phase function evaluations and some 500 local refinement steps, one of the suggested 'best' composite p.d.f.'s is parameterized by the following values:

$$\lambda^\star = 22.525, \quad s^\star = 7, \quad n^\star = 17, \quad p^\star = 0.776, \quad \alpha^\star = 0.864. \qquad (4.3.30)$$

The corresponding value of the objective function (4.3.28) equals 0.0692.

As indicated above, similarly good fit has been attained by a number of other (noticeably different) parameter combinations. For this reason, in the following illustrative examples the discrepancy indicator (4.3.28) has been replaced by a more sensitive function.

Example 2

All problem data are the same as in Example 1, but an—apparently more sensitive—absolute discrepancy statistic will be used in the definition of the objective function. That is—on the basis of (4.3.8)—function d from (4.3.28) is replaced by

$$d(F, \{F_i\}) = \frac{1}{2}\sum_{k=0}^{\infty}\Big(\sum_{i=1}^{4} w_i|f_k - p_k^{(i)}| + \sum_{j=1}^{3} w_j|f_k - b_k^{(j)}|\Big). \qquad (4.3.31)$$

The weighted aggregate of the seven input p.d.f.'s is approximated again by the parameterized p.d.f. form (4.3.26):

$$F = F(\lambda, s, n, p, \alpha) = (1-\alpha)\text{Poisson}(\lambda, s) + \alpha\text{Binomial}(n, p).$$

To allow for a more thorough global search in this experiment, in comparison with (4.3.29), the feasible region is increased now to

$$0.01 \le \lambda \le 40\,;\ 0 \le s \le 40\ (s\ \text{integer});\ 1 \le n \le 50\ (n\ \text{integer});$$
$$0 \le p \le 1;\ \ 0 \le \alpha \le 1. \qquad (4.3.32)$$

After 25,610 search steps, the estimated globally optimal solution is the following:

$$\lambda^\star = 4.627, \quad s^\star = 21, \quad n^\star = 44, \quad p^\star = 0.409, \quad \alpha^\star = 0.976. \qquad (4.3.33)$$

The corresponding 'optimal' discrepancy value equals 0.652.

A comparison of (4.3.30) and (4.3.33) immediately shows the significant influence of the objective function form on the 'optimal' solution derived. This point will be discussed later, within the more general context of model calibration under various problem specifications.

Table 4.3.3
Input Data of Example 3

p.d.f.	**Poisson** λ	s	**p.d.f.'s** mean value	variance	weight
1	2	4	6	2	0.07
2	3	5	8	3	0.16
3	1	1	6	5	0.12
4	5	10	15	5	0.12
p.d.f.	**Binomial** n	p	**p.d.f.'s** mean value	variance	weight
5	7	0.8	5.6	1.12	0.22
6	12	0.4	4.8	2.88	0.12
7	10	0.6	6.0	2.40	0.08

Example 3

In the third illustrative problem, the set of given input p.d.f.'s is changed: this is reflected by the modified parameterization shown in Table 4.3.3.

Comparing these data to Table 4.3.1, one can immediately see that the input p.d.f.'s are much more 'similar' to each other, reflected by the mean/variance data, than in the previous two examples. (The p.d.f. 'weights', as well as all other components of the problem description are left unchanged.)

Optimizing again the parameterized approximation (4.3.26),

$$F = F(\lambda, s, n, p, \alpha) = (1-\alpha)\text{Poisson}(\lambda, s) + \alpha\text{Binomial}(n, p)$$

with respect to the objective function (4.3.31), the feasible region is defined now as

$$0.01 \le \lambda \le 15. ;\ 0 \le s \le 10 \ (s \text{ integer});\ \ 1 \le n \le 15 \ (n \text{ integer});$$
$$0 \le p \le 1;\ \ 0 \le \alpha \le 1. \tag{4.3.34}$$

In 25,248 optimization steps, the estimated best approximation is parameterized by

$$\lambda^\star = 3.496, \qquad s^\star = 3, \qquad n^\star = 6, \qquad p^\star = 0.757, \qquad \alpha^\star = 0.422.$$

The corresponding discrepancy measure now equals 0.358, reflecting a significantly better fit than was found in the second numerical example.

All computations reported above were done on an Intel 386&387 DX/33 MHz processor based, 8 Mbyte RAM PC, in a Lahey Fortran 77 environment. Due to the computationally intensive form of the discrepancy measures used, the runtimes were between 20 and 25 minutes in each numerical example quoted here.

4.3.6 Conclusions

The objective of this chapter has been to present a general optimization framework for combining negotiated expert opinions. The generic problem statement introduced here allows for a broad variety of specifications, depending on the main problem components: mathematical form of opinions, norms and distance measures, pooling formulas, expert weights and, possibly, additional constraints. On the basis of the postulated Lipschitz structure of the underlying general optimization problem, adaptive partition strategies can be applied to solve a broad class of such problems. Relatively simple, but nontrivial numerical experiments illustrate the practical viability of the approach.

The mathematical problem statement presented is sufficiently general to cover many standard models in combining expert opinions. More generally, it permits also the analysis of a variety of seemingly unrelated applications such as, for example, certain product/mixture design, data classification, or facility location problems, discussed in other chapters.

Chapter 4.4

Product (Mixture) Design

4.4.1 Introduction

In this chapter we shall discuss a rather general global optimization problem originating from product design in mixing and processing industries. (The subsequent exposition draws on Hendrix and Pintér, 1991; Pintér, 1991c; and Chapter 2.5.)

Assume that we are given n base materials, $b_1, \ldots, b_n$, and we want to identify mixture products which shall meet certain stated usability, quality, etc. requirements. For instance, one may think of mixture products of coal processing plants, chemical factories, or pharmaceutical companies.

The set of a priori possible mixtures is mathematically defined by

$$S = \left\{ x \in I\!R^n : x = (x_1, \ldots, x_n),\ x_j \geq 0,\ \sum_{j=1}^{n} x_j = 1 \right\}.$$

The variables x_j express the rate of the components in product $x = \sum_{j=1}^{n} x_j b_j$. Note that S lies in the $n-1$ dimensional hyperplane given by $\sum_{j=1}^{n} x_j = 1$: in other words, for each feasible setting of $x_1, \ldots, x_{n-1}$ there is a uniquely determined $x_n = 1 - x_1 - \ldots - x_{n-1}$.

By assumption, all stated requirements (target quality levels, demands, and properties) will be expressed by the inequalities $g_i(x) \leq 0$, $i = 1, \ldots, m$.

The functions g_i could be arbitrary nonlinear smooth functions, but, for simplicity, we shall assume that they are quadratic. (Note that these quadratic relationships can be derived on the basis of analytical or empirical second order approximations of smooth functions.) Accordingly, the functions $g_i(x)$ will be written explicitly as

$$g_i(x) = x^T Q_i x + d_i^T x + c_i, \qquad i = 1, \ldots, m, \tag{4.4.1}$$

in which Q_i is a possibly indefinite symmetric $n \times n$ matrix, d_i is an n-vector, c_i is a scalar, and (as earlier) ^{T} denotes the operator of transposition. In this way

$$D = \{x \in I\!R^n : \; g_i(x) \leq 0, \, i = 1, \ldots, m\}$$

characterizes the set of 'satisfactory' (feasible) products.

Applying the above notation, the problem to be solved can be verbally stated as

Find an element (or, possibly, a list of elements) of the set $D \cap S$. (4.4.2)

In many practical situations, lower and upper bounds or other linear restrictions on x are also present. In addition, process variables x_j, that are not themselves determining components of the mixture, can be added to the problem formulation (by increasing the dimensionality of x).

An important aspect of the problem statement (4.4.2) is that industry is typically interested in stable (robust) solutions, as irregularities, or fluctuations, may appear during the production process. Mathematically, this means that a subset of $D \cap S$ with a given volume ε should be looked for, or alternatively, an interior point of $D \cap S$ located at a given distance from the boundary of D is sought. This aspect of robustness will be discussed briefly later; for additional expositions, see, for example, Thach (1988), Csendes (1989, 1990), Zabinsky, Graesser, Tuttle and Kim (1992), Csendes and Pintér (1993a,b), Hendrix, Mecking and Hendriks (1993), or Kristinsdottir, Zabinsky, Csendes and Tuttle (1993), with further references. The dimensions of the underlying practical problems are often characterized by several tens of components and mixture properties.

For illustration, consider the following simple example.

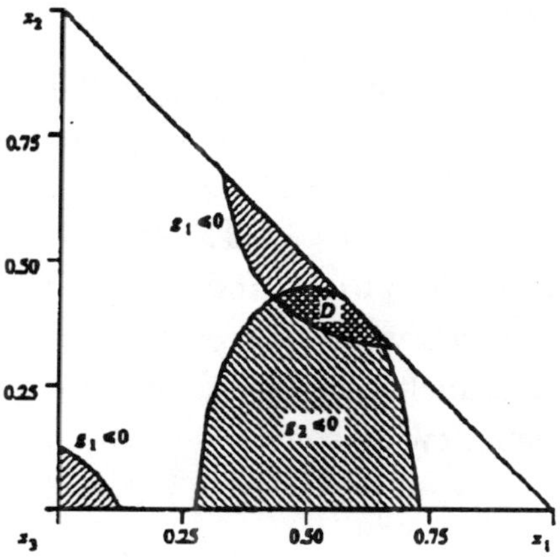

Figure 4.4.1
The feasible set D in Example 4.4.1

Example 4.4.1. A mixture product consisting of three components ($n = 3$) has to be found which meets the following two requirements (see Fig. 4.4.1):

$$g_1(x) = 2 + 8x_1 + 8x_2 - 32x_1x_2 + 1 \leq 0,$$

$$g_2(x) = 4 - 12x_1 - 4x_3 + 4x_1x_3 + 10x_1^2 + 2x_3^2 - 0.4 \leq 0.$$

In Fig. 4.4.1, vertex x_p corresponds to a product x consisting of 100 percent of component p, $p = 1, 2, 3$. The area which corresponds to the feasible products is given by the set D.

Following the approach highlighted in Section 4.1.5, the problem of finding an element of $D \cap S$ can be solved, for example, by minimizing the sum of infeasibilities

$$\min_{x \in S} g(x), \quad g(x) := \sum_{i=1}^{m} \max\{g_i(x), 0\}. \tag{4.4.3}$$

Problem (4.4.3) has a local optimum, for example, at $x_{\text{loc}} = (0.125,\ 0,\ 0.875)$, $g(x_{\text{loc}}) = 0.725$; of course, the global optimum is 0, for all elements of $D \cap S$. This simple illustrative example shows the potential multiextremality of the infeasibility indicator function g in (4.4.3). This is a direct consequence of the possible indefiniteness of the quadratic functions $g_i(x) \leq 0$, which implies the potential nonconvexity of the feasible region.

4.4.2 Solution Approach

In accordance with the exposition in Chapter 2.5, the applied solution algorithm is based on a partition of S via adaptively generated subsimplices $\{S_k\}$. The information obtained from the n vertices of S_k is used to calculate lower bounds for the functions g_i. The lower bounds can also be used to possibly eliminate infeasible subsets, and to decide on which subset is to be split further. The algorithm will be denoted by SPAS/B&B (simplicial partition algorithm scheme including branch-and-bound operations), and can be summarized by the following steps.

Algorithm SPAS/B&B

Step 0. Set $S_1 := S$ (initial simplex partition), $r := 1$ (number of currently active subsimplices).

At every iteration there exists a list of active subsimplices of $S : S_1, \ldots, S_r$.

Step 1. For every (constraint index) $i = 1, \ldots, m$ and $k = 1, \ldots, r$, calculate lower bounds φ_{ik} of $g_i(x)$ on $S_k : \varphi_{ik} \leq g_i(x)$, $x \in S_k$.

Step 2. If $\varphi_{ik} > 0$ for any i, delete S_k (as it is infeasible), and set $r := r-1$. If the list of subsets becomes empty ($r = 0$), stop: there is no feasible product composition; in the opposite case, proceed to Step 3.

Step 3. Split subset S_k with the lowest value of $\sum_{i=1}^{m} \varphi_{ik}$ into two parts of equal volume, halving it over the longest edge. Evaluate x_{new}, the midpoint of the longest edge of S_k. If $g_i(x_{\text{new}}) \leq 0$ for $i = 1, \ldots, m$, stop: a solution is found. Otherwise, delete S_k and add the two new simplicial subsets to the list, $r := r + 1$; return to Step 1 (of the next iteration cycle).

Obviously, algorithm SPAS/B&B can be seen as a special case of the approach discussed in Chapters 2.5 and 2.7. Observe that if one is interested only in robust solutions, then by deleting at Step 2 also subsets S_k which are 'too small' (for example, vol$(S_k) < \varepsilon$, $\varepsilon > 0$, is a given threshold parameter), it is clear that SPAS/B&B converges to either a robust solution of (4.4.2) or produces the answer that there is no such solution. Unfortunately, considering exclusively this termination criterion, an exponential number of iterations in n may (theoretically) be necessary to verify that all generated partition subsets are smaller than ε.

Let us also remark that algorithm SPAS/B&B can be modified in such a way that it is not always necessary to calculate all lower bounds at every iteration. Checking inequality $\varphi_{ik} > 0$ in Step 2 can be interpreted as the question: is it possible that S_k contains a vector x for which $g_i(x) \leq 0$? The answer to this question is certainly 'yes', if for one of the vertices x_{pk} of C_k, $g_i(x_{pk}) \leq 0$ holds for $i = 1, \ldots, m$. Algorithm SPAS/B&B can easily be modified so that in such situations the lower bound is not calculated. Note that—as a consequence—in this case the selection criterion in Step 3 should be changed accordingly.

4.4.3 Calculation of Lower Bounds

Following Chapter 2.5, the lower bounds in Step 1 of an implemented SPAS/B&B algorithm can be derived, making use of the (sole) fact that g_i is a Lipschitz-continuous function. Let x_{pk}, $p = 1, \ldots, n$ denote the vertices of S_k. Then the value of φ_{ik} can be based on the following relations:

$$g_i(x) \geq g_i(x_{pk}) - L_{ik}\|x - x_{pk}\|, \quad x \in S_k, \quad p = 1, \ldots, n. \tag{4.4.4}$$

In (4.4.4), L_{ik} is the Lipschitz-constant of g_i on S_k; it can be (over)estimated by solving

$$L_{ik} = \max_{x \in S_k} \| \nabla g_i(x)\|. \tag{4.4.5}$$

Observe now that as the function g_i is quadratic, its gradient is linear, and thus also convex. Since the norm function is monotonically increasing and also convex, therefore the composite function $\| \nabla g_i(x)\|$ is convex. In this way, problem (4.4.5) leads to the maximization of a convex function over a polyhedron, which can be solved now by simply evaluating $\| \nabla g_i(x)\|$ at every vertex x_{pk}. Note also that the estimate of L_{ik} can be made sharper, by projecting first the gradient $\nabla g_i(x)$ on the hyperplane of S. In the practical implementation of algorithm SPAS/B&B, the lower bound φ_{ik} applied is

$$\varphi_{ik} = \max_{x_{pk}}\{g_i(x_{pk}) - L_{ik} \max_{x_{vk}} \|x_{vk} - x_{pk}\|\} \tag{4.4.6}$$

(x_{vk} being the vertices of S_k, different from x_{pk}).

Alternative ways to determine lower bounds—based either on the (sole) Lipschitzian or on the quadratic structure of the problem—are discussed briefly by Hendrix and Pintér (1991), with further references therein.

4.4.4 Numerical Examples and Remarks

As referred to earlier in Part 3, one of the main computational problems of implementing a branch-and-bound type algorithm is that information concerning the partition sets has to be kept in the computer memory. This can be done now quite simply, by maintaining a list of active subsets and a list of points generated, but more special data structures are also possible. The important point is that the program should not run out of memory before a feasible solution has been found. In branch-and-bound methods, efficient use can be made of the fact that memory can be 'recycled' if certain partition sets are deleted from further consideration.

Another computationally relevant aspect is that, by the structure of SPAS/B&B, a generated point may be used as a splitting point several times. For saving computer memory and reducing the runtime, it is important that such a point is evaluated only once and will not be added repeatedly to the list of points. Applying this idea, notice that only a single new point which is not already part of the search information has to be evaluated at every iteration cycle.

According to the numerical experience described in Hendrix and Pintér (1991), the algorithm suggested can solve problems having a few variables (say, up to $n = 5$) in several hundred iteration steps (hence, it certainly can be applied also to larger problems, even in a PC environment). This will be illustrated by the following simple example.

Example 4.4.2. We shall consider a test problem originating from a practical application with $n = 3$ components and $m = 5$ postulated properties $y_1, \ldots, y_5$. Assume that the following requirements are given for the properties:

$$y_1 \leq 1.496, \quad y_2 \geq 0.92, \quad y_3 \leq 10, \quad y_4 \geq 179, \quad y_5 \leq 85.$$

The calculation of the functions y_i in Example 4.4.2 is based on the following relations. Property y_i is calculated, as a function of $x = (x_1, x_2, x_3)$, by

$$\begin{aligned} y_i(x) = & c_i + d_{i1}\hat{x}_1 + d_{i2}\hat{x}_2 + d_{i3}\hat{x}_3 + q_{i12}\hat{x}_1\hat{x}_2 + q_{i13}\hat{x}_1\hat{x}_3 + q_{i23}\hat{x}_2\hat{x}_3 \\ & + q_{i11}\hat{x}_1^2 + q_{i22}\hat{x}_2^2 + q_{i33}\hat{x}_3^2, \end{aligned}$$

in which $\hat{x}_i = x_i - 0.5$. The coefficients of $y_i(x)$ are given in Table 4.4.1.

Table 4.4.1
Coefficients in numerical examples

property		$y_i(x)$ coefficients								
i	c_i	d_{i1}	d_{i2}	d_{i3}	q_{i12}	q_{i13}	q_{i23}	q_{i11}	q_{i22}	q_{i33}
1	1.495	-.006	-0.024	0.050	-0.002	0.017	-0.021	0.001	0.008	-0.021
2	0.951	-.014	-0.048	0.108	-0.001	0.004	-0.006	0.004	0.001	-0.014
3	15.986	4.729	-16.657	-8.974	-10.174	-21.977	-86.952	20.605	32.003	-81.278
4	178.708	-0.687	12.800	-7.347	0.241	-4.947	-3.967	-0.766	-0.528	7.822
5	52.002	-5.217	-201.326	192.989	-6.180	337.106	1030.228	116.750	-67.424	-845.215

Applying SPAS/B&B to solve this problem, a feasible solution is found after 80 points have been generated. This solution is given by

$$x = (0.375, 0.4375, 0.1875) \text{ and } y(x) = (1.48, 0.921, 9.114, 180.77, -42.81).$$

Obviously, it is possible to generate additional feasible points by simply continuing the algorithm. For the problem given, 100 additional points lead to 10 more feasible points. Note that it can be instructive to carry out an 'exhaustive' search, say, in a ball around the first solution found to estimate how large one of the connected components of the set $D \cap S$ is.

The fact that the number of iterations becomes larger if the minimum of $g(x)$ (see (4.4.3)) is a small positive number can be illustrated by modifying the above example as follows:

Example 4.4.3. Let $y_1 \leq 0.85$, $y_2 \geq 0.963$; all other requirements of Example 4.4.2 are left unchanged.

The algorithm now needs 56 iterations to conclude that the problem does not have a solution. If the problem is 'near to feasibility', for example, $y_2 \geq 0.94$, then concluding that there is no solution needs 157 iterations; if $y_2 \geq 0.93$, the 'no solution' conclusion requires 239 iterations. Keeping these simple examples in mind, it is intuitively clear that such 'bad' cases can be very hard to solve, especially in higher dimensions. Let us also remark that in practically relevant cases of infeasibility, the approach suggested aims at a 'best compromise' infeasible solution, in the sense of the objective form $g(x)$ suggested (see, for example, (4.4.3)). In the terminology of the product design problem, one can obtain the 'best nearly feasible' mixture, even if the postulated quality requirements cannot be exactly met (or were misspecified). This comment obviously points towards the possibility of formulating and solving a sequence of parametric product design model formulations, in order to assess a range of potential solutions.

Chapter 4.5

Globally Optimized Calibration of Complex System Models

4.5.1 Introduction

The predominantly incomplete or poor understanding of environmental (as well as many other complex) systems calls for model development as an essential tool of the related scientific and applied research. For discussions—in the context of environmental management—see, for instance, Kneese and Bower (1968), Kneese, Ayres and d'Arge (1970), Brubaker (1973), Rich (1973), Park et al. (1974), United States Environmental Protection Agency (1974, 1980), Munn (1975), Parvin and Grammas (1976), Bower (1977), Russell and Spofford (1977), Holling (1978), Nijkamp (1980), Loucks, Stedinger and Haith (1981), Haith (1982), Novotny and Chesters (1982), Jørgensen (1983), Orlob (1983), Beck (1985a, 1987a,b), Beck and van Straten (1986), Somlyódy and van Straten (1986), Kleindorfer and Kunreuther (1987), Pintér (1987a, 1990c,e, 1991a), Richardson (1988), Stokoe, Lane, Côté and Wright (1989).

Generally speaking, the following main phases of quantitative systems modelling can be distinguished:

- identification: formulation of principal modelling objectives, setup of model structure
- calibration: model 'fitting' to available data information
- model validation and application (analysis, forecasting, control, management).

Consequently, calibration—the 'suitable' or 'best' parameterization of (identified) descriptive models—is an important stage in the process of understanding and managing the evolution of environmental—or other complex —systems. The significance of this issue is illustrated by the fact that among the most frequently cited 100 papers in natural sciences there is a single mathematical one: Marquardt's (1963) classical work on (locally) optimized parameterization of nonlinear system models.

Confronting model results with expertise, background information, and available observations/measurements, difficulties inherent to the model-fitting procedure arise. These difficulties stem primarily from

- lack of 'perfect' scientific knowledge (and, hence, of 'perfect' models)
- complicated model structure
- necessary decomposition of reality, followed by (re)aggregation in the frame of a model
- inadequacy/errors in monitoring and data processing
- subjectivity in the interpretation of results.

Frequently, because of the difficulties mentioned, model calibration takes place solely via consulting experts (for example, of the particular field of environmental sciences), and/or simply 'checking by the eye' the results of manual 'trial and error' parameterization procedures. Naturally, the results obtained in this way are questionable, and there is room for more objectively established model calibration methodology.

Before turning to a mathematical formulation of the calibration problem, see Fig. 4.5.1, in which the basic structure of an iterative model calibration procedure is shown.

To illustrate the procedure summarized in Fig. 4.5.1, we shall consider a (simplified) water quality model calibration problem. Let the ambient water quality in a lake at time t be characterized by a vector $s(t)$: its components may be, for example, biological oxygen demand, chemical oxygen demand, dissolved oxygen concentration, concentrations of micropollutants and heavy metals, etc. Naturally, the water quality is influenced by a number of external factors, for example by the stochastically varying pollution load transported to the lake (by rivers and discharges), by regional hydrometeorological conditions (streamflow rates, temperature, etc.), and by the resulting physical and biochemical reactions. Some of these factors can be

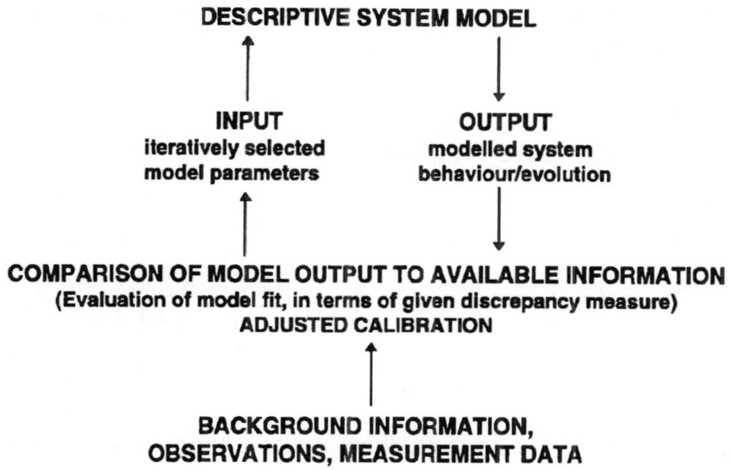

Figure 4.5.1
Iterative model calibration scheme

observed, while some others may not be completely known. Suppose further that we have a theoretical model form for the description of water quality evolution in time (such as, for example, the classical Streeter and Phelps (1925) differential equations for characterizing the longitudinal evolution of biochemical oxygen demand and dissolved oxygen concentrations in a river). This model, symbolically denoted by F, usually has a number of non-universal, unknown parameters x (for example, for quantifying the location- and time-dependent 'speed' of relevant physical and biochemical reactions). As is known, for different (scientific, technical, financial, etc.) reasons it may be difficult or even impossible to measure all relevant water quality indicators. In this way, the model of the system has to provide observable output quantities which are then compared to actual measurements. Naturally, both model outcome and measurements may depend also on different kinds of uncertainties and errors. Under the circumstances outlined, it seems quite reasonable to attempt selecting the model parameters in 'good' or 'best' agreement with the set of available observations.

On the basis of this verbal problem statement, in the following chapters a variety of calibration problem formulations and their influence on the results will be investigated. The exposition draws on the work of Pintér (1990c,e) with co-authors (noted when appropriate). In particular, a detailed discussion will be devoted to the following issues:

- generic model calibration problem statement
- global optimization based solution approach
- diverse (deterministic, scalar) discrepancy measures, in order to express the 'difference' between model output and corresponding observation data
- multiple criteria calibration: selection of vector-valued discrepancy function forms, in order to reflect several model-fitting objectives simultaneously
- uncertainty analysis: choice of stochastic discrepancy function forms that reflect the inherent system uncertainties and/or model errors
- application examples.

4.5.2 Model Calibration: A General Problem Statement

Throughout the discussion of generic model calibration issues, the following notation will be used:

t	time
$s(t)$	l-dimensional vector of state variables: $s(t)$ characterizes the modelled dynamic behaviour of the system under study
$d(t)$	M-dimensional vector of measured external factors and input disturbances, influencing the state of the system
$\xi(t)$	I-dimensional vector of unknown/unmeasured input disturbances: $\xi(t)$ is modelled as a vector of component stochastic processes
x	n-dimensional vector of model parameters: x is to be chosen 'optimally', as compared to available observations
X	the n-dimensional set of theoretically admissible (a priori feasible) parameterizations
F	the model structure functional which describes the temporal evolution of the system, as a function of its modelled state, the disturbances and the parameters
G	the model output functional which determines the output, as a function of parameters, system state, and disturbances $(x, s(t), d(t), \xi(t))$

$\widehat{y}(t)$ J-dimensional vector of model output variables, depends on $(x, s(t), d(t), \xi(t))$

$y(t)$ J-dimensional vector of measured output variables, corresponding componentwise to $\widehat{y}(t)$

$\eta(t)$ K-dimensional vector of random output measurement errors, modelled as a vector stochastic process

$\overline{s}(t)$ l-dimensional vector of 'true' state variables corresponding to the real environmental state: this way, generally speaking, $s(t)$ only approximates $\overline{s}(t)$

H the measurement functional which determines the observable output as a function of real system state, disturbances, and measurement errors $(\overline{s}(t), d(t), \xi(t), \eta(t))$

D discrepancy measure for evaluating the 'overall deviation/distance' between parameterized model output and actual observations.

Applying this notation, the state variable dynamics of the model system, its output, and the related observation process can be formally described by the following relations:

$$\frac{ds}{dt} = F\{x, s(t), d(t), \xi(t)\} \tag{4.5.1}$$

$$\widehat{y}(t) = G\{x, s(t), d(t), \xi(t)\} \tag{4.5.2}$$

$$y(t) = H\{\overline{s}(t), d(t), \xi(t), \eta(t)\}. \tag{4.5.3}$$

Given this symbolic description, it is quite natural to seek for such calibrations (parameterizations) which, at least, lead to 'plausible' (modelled) system behaviour. Ideally, one can attempt to find the 'best' theoretically admissible parameterization which results in model output that is 'as close as possible' to the set of available observations. In mathematical terms, this objective can be expressed by the following optimization problem statement:

$$\min_{x \in X} D\{\widehat{y}(t), y(t)\}, \qquad \widehat{y}(t) := \widehat{y}(t, x). \tag{4.5.4}$$

It is understood that in (4.5.4) the model output $\widehat{y}(t)$ implicitly depends on the actual model parameterization x and, in general, also on $s(t)$, $d(t)$ and $\xi(t)$ (although the latter are suppressed by the notation in (4.5.4)).

Before proceeding further, several comments concerning the above general problem formulation are in order: these will be provided under the subheadings below. Throughout the discussion which follows, the illustrative calibration problem outlined in the introduction may be kept in mind for reference.

4.5.3 Basic Underlying Assumptions

The mathematical problem formulation (4.5.1)–(4.5.4) tacitly presupposes that the system description (F), the model output (G), the measurement mapping (H), and the discrepancy measure (D) are all unambiguously known (and hence they can be prefixed). In reality, F, G, and H will often significantly depend on the available scientific knowledge, the information concerning system behaviour, and the measurement methodology applied. Moreover, the choice of D largely depends on our particular objectives of expressing the deviation between model output and observations. The identification, calibration, and validation of environmental—in particular, ecological—models, and the complex relationships between theoretical research, modelling, and experimental studies are illustrated, for example, by the discussions in Beck (1985b), Cochrane, Ashton, Jarvis, Twinch and Zohary (1987), Bartell, Cale, O'Neill and Gardner (1988), Caswell (1988), Hall (1988), Onstad (1988), Ulanowicz (1988), Klepper (1989), Chang and Delleur (1992), Klepper and Hendrix (1994), with many additional references therein. Consequently, in a system identification and calibration problem, the 'most appropriate' selection of F, G, H, and D may (separately or jointly) be subject to investigation. According to this larger picture, the calibration of a selected model/measurement structure is merely a 'technical'—though essential and, typically, quite complicated—stage of the analysis.

It is also worth mentioning at this point that the concept of selecting model parameters optimally may not always be a reasonable objective. Frequently, because of improper scientific knowledge (or simply looking only for 'acceptable' parameterizations) there is a tendency towards modelling environmental systems without having in mind any sort of 'optimal' description: see, for example, the related discussions in Beck (1985a, 1987a), with many further references therein. Without debating the context-dependent validity of one or the other concept, one may say that in cases where the 'acceptable' model parameterizations are not 'equally good', and there are quantitative ways to evaluate their performance, the optimization framework proposed can be useful in a search for 'the very best' calibration (given all previously detailed assumptions).

Let us note finally that, for applying the continuous or Lipschitzian global optimization approach, some fairly mild mathematical requirements concerning the problem structure (4.5.1)–(4.5.4) are to be satisfied. These requirements—discussed earlier in detail—are, in essence, related to the closed and bounded structure of the feasible parameter set X and the continuity of the 'goodness-of-fit' measure D, with respect to x. The analytical properties above hold for many descriptive model structures, but certainly there are exceptions: for example, such system models that show discontinuous or 'chaotic' behaviour as a function of their initial state conditions and/or admissible parameterizations. For illuminating expositions on the chaotic behaviour of systems and processes, see Mandelbrot (1983), Casti (1990), Hall (1991), or Briggs (1992).

4.5.4 Discretization

The relations (4.5.1)–(4.5.4) are based on a continuous system description, in terms of state evolution in time. Frequently, however, only discrete observations are available, say, at (not necessarily equidistant) time moments $t_1, \ldots, t_P$. In such cases—in order to compare model outcome and available observations—the relations (4.5.1)–(4.5.4) are to be replaced by appropriate discretizations. This leads to the following model form:

$$s(t_{p+1}) = F_d\{x, s(t_p), d(t_p), \xi(t_p)\}, \qquad p = 1, \ldots, P-1, \qquad (4.5.1')$$

$$\widehat{y}(t_{p+1}) = G_d\{x, s(t_p), d(t_p), \xi(t_p)\}, \qquad p = 1, \ldots, P-1, \qquad (4.5.2')$$

$$y(t_{p+1}) = H_d\{\overline{s}(t_p), d(t_p), \xi(t_p), \eta(t_p)\}, \qquad p = 1, \ldots, P-1, \qquad (4.5.3')$$

$$\min_{x \in X} \sum_{p=1}^{P} D_d\{\widehat{y}(t_p), y(t_p)\}, \qquad (4.5.4')$$

where F_d, G_d, H_d, and D_d are derived—if necessary—by suitable numerical approximation (discretization) of F, G, H, and D respectively.

4.5.5 Statistical Uncertainties and Robust Calibration

The proper consideration and handling of model uncertainties (when present) is essential. Incomplete knowledge, inadequate model structure, process decomposition and aggregation, natural variability, observation errors, and differences in the (partly) subjective interpretation of results all contribute to uncertainty in modelling: see, for example, Hornberger and Spear (1981), Gardner, Cale and O'Neill (1982), Somlyódy (1982a,b), Beck and van Straten (1983), Hornberger and Cosby (1985), Beck (1985a, 1987a,b), Cosby, Wright, Hornberger and Galloway (1985), Cosby, Hornberger, Galloway and Wright (1985), Fields and Miller (1988), Loehle (1988), Warwick and Cale (1988), Cooke (1994), Klepper and Hendrix (1994), and Masliev and Somlyódy (1994) on these issues, primarily in the context of environmental systems.

Here we shall only deal briefly with the problem of appropriately handling stochasticities (that is, those uncertainties that can be described using probabilistic concepts and methods) of the system modelled. This endeavour is already reflected by the problem formulation (4.5.1)–(4.5.4), in which (scalar or vector) random variables or stochastic processes $\xi(t)$ and $\eta(t)$ are to represent all significant random input disturbances and measurement/interpretation errors. Under these conditions, it is natural to combine the mentioned probabilistic concepts and tools with appropriate optimization techniques. This leads to problem formulations in terms of statistical decision theory or stochastic programming, highlighted earlier in Chapter 3.5.

In the present context, the main sources of uncertainty are described as observable or unobservable system disturbances. Assuming that these can be modelled (approximated) in a suitable probabilistic form—having, for example, an appropriate joint probability distribution to describe all relevant fluctuations—it is desirable to search for such model calibrations that are robust, that is, are sufficiently stable, considering, to the extent possible, all realizations (occurrences) of the random factors in question.

Evidently, the above verbal definition of 'robust' parameterizations may be expressed in many different mathematical forms. To highlight this concept, we shall introduce a simple general optimization model form which can be specified to include also the calibration problem (4.5.1)–(4.5.4), or its discretized variant (4.5.1′)–(4.5.4′).

As above, let X denote the set of admissible parameterizations, and, for $x \in X$, let $f(x)$ be the objective/performance functional, expressing the quality of the parameterization x. Typically—as also in the present context—x is an n-component real vector; furthermore, the set X is defined by a finite number of range constraints and also by implicit relations with respect to the components of x. In this way—in the case of a scalar objective function—the model calibration problem can be concisely written as

$$\min_{x \in X} f(x), \tag{4.5.5}$$

or, in a more detailed form, as

$$\begin{aligned} &\min f(x) \qquad x \in X, \\ &X = \{x \in I\!R^n : x_{\min} \leq x \leq x_{\max}, \; g_m(x) \leq 0, \; m = 1, \ldots, M\}. \end{aligned} \tag{4.5.6}$$

In the context of the calibration problem $f(x) := D\{\widehat{y}(x,t), y(t)\}$, the functions g_m symbolize the possibly present implicit constraints with respect to the parameter vector x, which has to be in the n-dimensional 'box' $[x_{\min}, x_{\max}]$.

Assuming now that, in general, both the objective function and the constraints in (4.5.6) may nonnegligibly depend on random (uncertain, unforeseen) events, we shall model these occurrences by the random realizations of the vector ζ. In this way, for obtaining stable parameterizations, diverse extended model forms can be suggested. Replacing formally $f(x)$ by $f(x,\zeta)$, $g_m(x)$ by $g_m(x,\zeta)$, $m = 1, \ldots, M$, one may consider, for example, the stochastic model variants (4.5.7) to (4.5.9):

$$\begin{aligned} &\min E\{f(x,\zeta)\} \\ &E\{g_m(x,\zeta)\} \leq 0, \qquad m = 1, \ldots, M, \\ &x_{\min} \leq x \leq x_{\max}. \end{aligned} \tag{4.5.7}$$

In (4.5.7) E denotes the expected (mean) value operator, that is, the constraints are to be met 'on average', and the mean value of the objective function is to be minimized.

$$\begin{aligned} &\min E\{f(x,\zeta)\} \\ &P\{g_m(x,\zeta) \leq 0, \; m = 1, \ldots, M\} \geq PB, \\ &x_{\min} \leq x \leq x_{\max}. \end{aligned} \tag{4.5.8}$$

In (4.5.8), $P\{\cdot\}$ denotes the probability of joint occurrence of the embedded random event. In this way, the implicit constraints are to be met simultaneously, with a probability which is bounded from below by the prefixed value $0 < PB < 1$ (PB is usually chosen near to one). Again, the expected value of the objective function is to be minimized.

$$\begin{aligned} &\min \alpha E\{f(x,\zeta)\} + (1-\alpha) SD\{f(x,\zeta)\} \\ &P\{g_m(x,\zeta) \le 0\} \ge PB_m, \qquad m = 1,\dots,M, \\ &x_{\min} \le x \le x_{\max}. \end{aligned} \tag{4.5.9}$$

In (4.5.9), $0 < \alpha < 1$ is an appropriately chosen model parameter and SD symbolizes the standard deviation of the random variable $f(x,\zeta)$. That is, the implicit constraints are to be met separately, at least with corresponding probabilities $0 < PB_m < 1$. Under these conditions, a suitable linear combination of the mean and standard deviation of the random objective function value is minimized.

The above problem formulations merely serve for illustration. For additional details, recall the discussion of stochastic model forms and corresponding solution approaches from Chapters 3.5 to 3.7.

4.5.6 Multiple Calibration Objectives

So far it has been tacitly assumed that the system state observations and corresponding model output are one-dimensional quantities: this supposition justifies the formulation of the scalar optimization problems presented. Typically, the evolution of an environmental system is better characterized by modelling and measuring a number of descriptors: for instance, water quality observations may be related, inter alia, to organic micropollutants and heavy metals.

Formally, if $j = 1,\dots,J$ denote the indices of (finitely many) system descriptors, then some suitable vector-valued discrepancy measure

$$D\{\widehat{y}(t), y(t)\} = \left(D_j\{\widehat{y}_j(t), y_j(t)\},\ j = 1,\dots,J\right) \tag{4.5.10}$$

is to be considered, in order to evaluate and compare different model parameterizations. This leads to multiobjective (vector) generalizations of the problem (4.5.1)–(4.5.4), or its modifications treated above. Generally, in such problems it is not guaranteed that an 'ideal' solution can be found which leads to the 'very best' model parameterization, with respect to all state descriptors considered. Therefore it is only rational to search for nondominated (Pareto-optimal) decisions, for which there is no 'universally superior' parameterization (in terms of all state descriptors).

There exists a rich variety of multiple objective decision models and solution approaches, from rather simple methods to sophisticated techniques: for discussions, see, for example, Cohon (1978), Hwang and Masud (1980), Goicoechea, Hansen and Duckstein (1982), Romero (1986), Vincke (1986), or Lewandowski and Wierzbicki (1989). It suffices to say here that—among others—there are two correct and well-interpretable approaches to rewriting vector optimization problems into parametric scalar forms in such a way that the solutions found will be Paretian (at least under certain conditions concerning the analytical structure of the multiobjective optimization problem). Although these two model forms are closely related in a mathematical sense, their context-dependent use can be highly recommended. We shall briefly analyse these approaches below.

Assume first that the vector discrepancy criterion is made up of commensurable components. This implies that the modeller/user is able to provide weights $\{w_j\}$ that express the relative importance of the individual descriptors. In such cases, it is natural to form a weighted scalar objective function

$$\sum_{j=1}^{J} w_j D_j\{\widehat{y}_j(t), y_j(t)\} \tag{4.5.11}$$

$$\left(w_j \geq 0, \ \sum_{j=1}^{J} w_j = 1 \ \text{ are given user-dependent weights}\right)$$

with respect to $x \in X$; then this function is minimized to yield a 'good' (nondominated) model calibration option. Of course, the 'best' option will depend on the actual choice of the weight factors $\{w_j\}$.

Another mathematically correct and tractable approach can be followed in the case where the components of model output and measures cannot be (are not to be) directly compared, in a coherent manner. Under such circumstances, the most important system descriptor—or one of them—is to be chosen for defining the (primary) objective function, while all others can be considered in the form of explicit accuracy constraints. Formally, if, for example, $D_1\{\widehat{y}_1(t), y_1(t)\}$ is selected as the principal calibration quality indicator, then one may state the optimization problem

$$\begin{aligned}
&\min D_1\{\widehat{y}_1(t), y_1(t)\} && (4.5.12)\\
&D_j\{\widehat{y}_j(t), y_j(t)\} \le \delta_j, \qquad j = 2,\ldots,J,\\
&(\delta_j > 0 \text{ are user-dependent tolerance bounds})\\
&x \in X.
\end{aligned}$$

As mentioned above, under simple analytical conditions, the solution of the scalarized problem formulations (4.5.11) and (4.5.12) leads to nondominated model calibration results. It should be emphasized, however, that none of the above presented multiobjective optimization approaches yields unambiguous results (as the models (4.5.11) and (4.5.12) are parameterized by $\{w_j\}$ and $\{\delta_j\}$ respectively). In this way—systematically changing $\{w_j\}$ and/or $\{\delta_j\}$—a corresponding 'menu' of solutions can be generated, among which one should choose ideally the 'best' or, more realistically, the most acceptable ones.

4.5.7 Multiextremality

It is well known that many physical, biochemical, ecological, etc. processes—and thus their realistic descriptive models—tend to be highly nonlinear. As a direct consequence of this fact, the discrepancy measure D typically will be multiextremal, with respect to the parameter vector x. In other words, initiating the search from different points of the feasible parameter domain X may lead to largely differing results, both in terms of the parameterization found and its 'performance' (as expressed by D). For this reason, if the problem of optimal calibration is meaningful and well-posed, then there is a strong rationale to search for the 'globally best' x in X. For example, Csendes, Daróczy and Hantos (1986, 1992), Yeh (1986), Csendes (1988), Hendrickson, Sorooshian and Brazil (1988), Loehle (1988), Hantos, Daróczy, Csendes, Suki and Nagy (1990), Suki, Csendes and Daróczy (1990), Chang and Delleur (1992), Hemker and Kok (1993), Porto and Lopes (1993), Everaars, Hemker and Stortelder (1995), Stortelder (1995) discuss various models in which globally optimal parameterization is sought, in order to improve system description.

To illustrate the potential multiextremality of the model calibration problem, we shall consider a simple (but probably the most often studied) specification of the nonlinear parameter estimation problem. In this model version the scalar, deterministic, discrete model output values $\widehat{y}(t_p, x)$ should match 'as well as possible' the corresponding time-series of observations $y(t_p)$, $p = 1, \ldots, P$. This requirement can be expressed, for example, in the form of the optimization problem

$$\min_{x \in X} f(x), \qquad f(x) = \sum_{p=1}^{P} [\widehat{y}(t_p, x) - y(t_p)]^2. \tag{4.5.13}$$

Let us assume that $\widehat{y}(\cdot)$ is a twice continuously differentiable nonlinear function of the model parameter vector x. Then we have the following expressions for the gradient vector $\nabla f(x)$ and the Hessian matrix $\nabla^2 f(x)$:

$$\nabla f(x) = 2 \sum_{p=1}^{P} [\widehat{y}(t_p, x) - y(t_p)] \nabla \widehat{y}(t_p, x)$$

$$\nabla^2 f(x) = 2 \sum_{p=1}^{P} [\widehat{y}(t_p, x) - y(t_p)] \nabla^2 \widehat{y}(t_p, x) + 2 \sum_{p=1}^{P} \nabla \widehat{y}(t_p, x) \, [\nabla \widehat{y}(t_p, x)]^T.$$

(Again, the symbol ^{T} denotes the transpose of the column vector $\nabla \widehat{y}(t_p, x)$.)

Now, in $\nabla^2 f(x)$ the second sum consists of positive semi-definite diagonal matrices, but the first sum is typically made up by indefinite matrices, as x varies on the feasible region X. Consequently, the Hessian matrix $\nabla^2 f(x)$ frequently becomes indefinite, indicating local convexity or concavity of f: but this implies the multiextremality of the corresponding nonlinear parameter estimation problem. Csendes (1988, 1993) also discusses this point, showing in particular that an arbitrary nonnegative function $g(x)$ can be identical to the objective function $D(x)$ of an associated model calibration problem (the form of $D(x)$ depends on the discrepancy measure used, and the set of observation points and observed values). Given the fact that the most widely applied 'classical' techniques in numerical optimization guarantee, at best, only the approximation of a locally optimal calibration, the use of suitable global optimization algorithms is justified. In the following chapter the calibration problem will be considered in a number of different versions; this is followed by model calibration case studies that have been solved by global optimization.

Chapter 4.6

Calibration Model Versions, Illustrated by Examples

4.6.1 Introduction

We shall consider now important—from a practical point of view—specifications of the general calibration model stated by (4.5.1)–(4.5.4). The discretized model form will be studied, noting that the continuous case could be discussed analogously. (The exposition draws on Pintér and Van der Molen, 1991.)

To introduce a somewhat simpler notation than the one used in (4.5.1′)–(4.5.4′), let us assume that the discrepancy measure (objective function) x is given in a discretized form:

$$\min_{x \in X} D\big(\{\widehat{y}_t(x)\}, \{y_t\}\big), \tag{4.6.1}$$

in which both

$$\{\widehat{y}_t(x)\} = (\widehat{y}_1(x), \ldots, \widehat{y}_T(x)) \quad \text{and} \quad \{y_t\} = (y_1, \ldots, y_T) \tag{4.6.2}$$

are T-dimensional vectors. Additionally, we shall suppose first that both the observations and model output values are scalar and deterministic quantities, that is, a single type of model result and corresponding observations are considered, and these values are 'exactly' known. Further generalizations, including those outlined in the previous chapter, will be introduced later.

We shall analyse diverse—optionally applicable—functional forms of $D(x)$. Under the given specifications, D will measure the 'difference' of two T-variate real vectors: several options of defining D will be discussed briefly below.

4.6.2 Average l_p-Distance and Variants

Basic Case

Let us consider first the discrepancy measure defined by the family of functions

$$D(\{\widehat{y}_t(x)\}, \{y_t\}) = \frac{1}{T}\Big(\sum_{t=1}^{T} |\widehat{y}_t(x) - y_t|^p\Big)^{\frac{1}{p}}, \qquad 1 \leq p \leq \infty. \tag{4.6.3}$$

The discrepancy measure D (4.6.3) is defined, in essence, by an appropriate l_p-norm, for some given p, $1 \leq p \leq \infty$. Note that the factor $1/T$ allows for comparing calibration results for varying sample sizes. Again, the best known special cases of (4.6.3) include the minimization of the (averaged) absolute deviations

$$D(\{\widehat{y}_t(x)\}, \{y_t\}) = \frac{1}{T}\sum_{t=1}^{T} |\widehat{y}_t(x) - y_t| \qquad (p = 1), \tag{4.6.4}$$

the (averaged) least squares estimation problem

$$D(\{\widehat{y}_t(x)\}, \{y_t\}) = \frac{1}{T}\Big(\sum_{t=1}^{T} |\widehat{y}_t(x) - y_t|^2\Big)^{\frac{1}{2}} \qquad (p = 2), \tag{4.6.5}$$

and the minimization of the maximal discrepancy between the set of model output values and the observations (minimax or Chebyshev objective function form):

$$D(\{\widehat{y}_t(x)\}, \{y_t\}) = \frac{1}{T}\max_t |\widehat{y}_t(x) - y_t| \qquad (p = \infty). \tag{4.6.6}$$

Note that in (4.6.6) one could omit the factor $1/T$.

Before proceeding further, some remarks are in order. The mathematical form of the function (4.6.3) implies that for increasing values of p the 'outliers' (that is, the more significant discrepancies between model output and measurements) are more heavily 'penalized'. This way, the choices $p = 1$ and $p = \infty$ represent two extreme points of view: the function form (4.6.4) includes all pairs of output and observation values with equal weight, while (4.6.6) considers only the maximal deviation. This comment may help to choose the 'right' value of the parameter p in applications, or rather, to consider a suitable variety of them. Note also that, in order to avoid typical numerical instabilities, the exact minimax objective function form (4.6.6) can be recommended, instead of using 'large' values of p, in many practical cases.

Let us also observe that, evidently, the class of discrepancy measures (4.6.3) is invariant with respect to translations. That is, if all model output values and observations are replaced by adding a constant to them (the value of y is replaced by $y + c$), then none of the above objective function values will change.

Relative Error Based Discrepancy

Let us consider now the general class of objective functions

$$D(\{\widehat{y}_t(x)\},\{y_t\}) = \frac{1}{T}\Big(\sum_{t=1}^{T} |[\widehat{y}_t(x) - y_t]/y_t|^p\Big)^{1/p} \quad (y_t \neq 0), \quad 1 \leq p \leq \infty. \tag{4.6.7}$$

Comparing this function form to (4.6.3), one can see that the absolute deviation between $\widehat{y}_t$ and y_t is replaced by its relative value (normalized by the nonzero measurements y_t). This objective function can be recommended, when relative function values are preferable to their 'real' values, as characterizing the goodness-of-fit. The underlying idea leading to this function form is that larger absolute valued model results and measurements may also have larger inherent errors. Evidently, (4.6.7) is invariant with respect to all multiplicative transformations that replace the value y by $c \cdot y$, c being an arbitrary real multiplier. Note at the same time that (4.6.7) is not invariant with respect to translations: this way, the scale dependencies eliminated by (4.6.3) or (4.6.7) are different.

Again, the l_p-norm based general objective function form (4.6.7) can be specified in a number of ways; recall, for example, (4.6.4)–(4.6.6). Regarding the choice of the parameter p, analogous comments to those made previously are valid.

Standardized Error Based Discrepancy

Frequently, the natural fluctuations of the investigated system and/or the inherent error of the measurements are also important to consider when evaluating the 'goodness' of diverse model parameterizations. For instance, one can estimate the ratio of model discrepancy to the 'natural' (inherent) measurement data fluctuations. In such cases, for example, the following statistically established discrepancy measure can be recommended. First, the expected value and the variance of the measurements is estimated by the relations

$$E(y_t) \approx \frac{1}{T}\sum_{t=1}^{T} y_t, \quad V(y_t) \approx \frac{1}{T-1}\sum_{t=1}^{T}[y_t - E(y_t)]^2.$$

Then, applying these estimates, one can define the standardized model output and measurement values by

$$\widehat{y}_t^s = \frac{\widehat{y}_t(x) - E(y_t)}{V(y_t)^{1/2}} \quad \text{and} \quad y^s = \frac{y_t - E(y_t)}{V(y_t)^{1/2}}.$$

Now, similarly to the discrepancy functions presented above, one can consider the mean l_p-distance between the vectors consisting of the standardized model output and measurement values

$$D\big(\{\widehat{y}_t^s(x)\}, \{y_t^s\}\big) = \frac{1}{T}\Big(\sum_{t=1}^{T} |\widehat{y}_t^s(x) - y_t^s|^p\Big)^{1/p}, \qquad 1 \le p \le \infty. \qquad (4.6.8)$$

With respect to the choice of p, analogous comments to those provided earlier can be made. Note also that (4.6.8) is invariant, if linear transformations are accomplished (that replace y by $ay + b$), for given real numbers a and b. Consequently, the discrepancy measure (4.6.8) will not vary, if both the scale unit and location parameters are changed (or just any of them).

To provide an example of a possible application of the function form (4.6.8), one can consider the case when the model output and measurements at two different monitoring stations on the same river are to be considered in a single objective function. The standardizing transformation (4.6.8) 'filters out' the site-specific variances, and the objective function can be defined as the sum of two functions of the type (4.6.8).

Weighted Discrepancy Measures

For various reasons (for example, timing, place, reliability, etc., of the measurements taken), different pairs of model output and system observation values—although related to the same system characteristics (state variable)—may be considered as of different importance. This fact can be explicitly taken into consideration, for example, by defining a coherent system of relative weights (multiplicators) which are then incorporated into the objective function. For instance, a possible direct generalization of (4.6.3) has the form

$$D\big(\{\widehat{y}_t(x)\},\{y_t\}\big) = \Big(\sum_{t=1}^{T} |w_t[\widehat{y}_t(x) - y_t]|^p\Big)^{1/p}, \qquad 1 \le p \le \infty, \tag{4.6.9}$$

in which $w_t \ge 0$, $\sum_{t=1}^{T} w_t = 1$ are suitably chosen weights. (Observe that (4.6.9) reduces to (4.6.3), if $w_t = 1/T$, $t = 1,\dots,T$.)

Evidently, specifications and extensions of (4.6.9)—similarly to those expressed by (4.6.4)–(4.6.8)—are again possible, depending on the importance of 'outliers' and the invariance structure required. Note that, for increasing p, measurements with greater weight factors w_t tend to have an even more pronounced role than in the basic (uniformly weighted) cases considered above.

4.6.3 Discrepancy Measures for Calibrating 'Soft' Systems

Before proceeding to important generalizations of the discrepancy measure types presented, some remarks with respect to the optimization concept applied are in order. In a number of cases—because of our limited knowledge and/or the 'soft' character of the system studied, the scarcity of measurements and 'exact' information, etc.—a strict optimization paradigm may be more of theoretical than of practical interest. To mention a trivially simple example: it rarely makes sense to calibrate n model parameters, on the basis of T measurements, if T is merely of the order of n (such as, for example, $n = 17$ and $T = 38$). Moreover, some implicit ('hidden') preferences of the modeller are often difficult to quantify, and they are only gradually elicited during an iterative model identification-calibration-verification procedure. Under such circumstances, less 'ambitious' objectives than formally 'best' model calibration may be more appropriate.

Some Nonoptimizing Approaches

The issue of 'soft modelling' versus 'formal optimization' is extensively discussed in the literature, with a number of alternative nonoptimizing (predominantly statistical or empirical) approaches: see, for example, the works of Beck and van Straten (1983), Beck (1985a,b, 1987a), Hornberger and Cosby (1985), Somlyódy and van Straten (1986), Klepper (1989), Banks and Fitzpatrick (1990), Cooke and Vogt (1990), Kleijnen, Van Ham and Rotmans (1990), Scheffer (1990), Scholten, De Hoop and Herman (1990), Van Raaphorst, Kloosterhuis, Cramer and Bakker (1990), Van der Molen (1991), Klepper and Hendrix (1994), with many additional references therein. As argued earlier, if there exists a possibility of quantitatively evaluating the goodness-of-fit of an arbitrarily selected model parameterization, and there is a sufficient background of data/information, then the optimization concept is justified, at least in a formal (mathematical) sense. For the sake of completeness—and also for their apparent relevance—some basic (single objective, deterministic) model forms, aiming at just 'acceptable' or at 'sufficiently good' model parameterizations will be discussed briefly below.

In the first case, one may require that the model output trajectory is realized in the range of the available (often just a few) measurements: for illustration, see Fig. 4.6.1. This requirement can be expressed, for example, by the constraints

$$\max_t \ \widehat{y}_t(x) \le \max_t y_t, \qquad \min_t \ \widehat{y}_t(x) \ge \min_t y_t, \tag{4.6.10}$$

where it is understood that maximization (or minimization) on the left- and right-hand sides of the inequalities may be taken over sets of different size (cardinality). This way, we are seeking for feasible model parameterizations (all such parameter-combinations, or just any of them).

Another—in general, more strict—acceptability criterion can be to require that the distance between the modelled system trajectory and the observation data be uniformly bounded. This requirement can be expressed, for example, by the relation

$$\max_t |\widehat{y}_t(x) - y_t| \le \delta \qquad (\delta > 0 \text{ is a fixed tolerance parameter}). \tag{4.6.11}$$

Note that in (4.6.11) a one-to-one correspondence between model output and corresponding data is assumed.

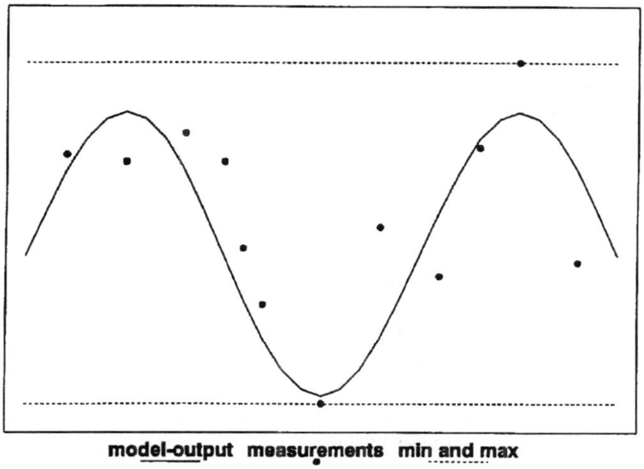

Figure 4.6.1
A feasible model result, with respect to the given measurement data and criterion (4.6.10)

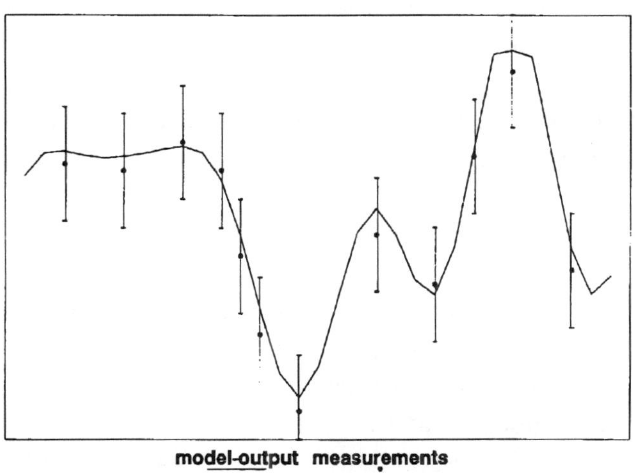

Figure 4.6.2
A feasible model result, with respect to the given measurement data and criterion (4.6.11)

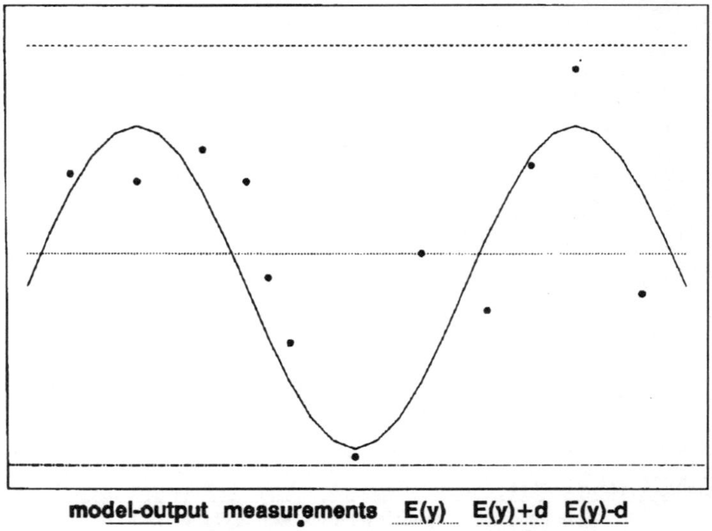

Figure 4.6.3
A feasible model result, with respect to the given measurement data and criterion (4.6.12)

To provide an example, assume that a certain water quality component is modelled, but it is measured rather infrequently. Due to spatial (horizontal and vertical) heterogeneity and observation errors, realistically, a range d around the measurements can only be taken, instead of just considering the 'exact' data. This idea can be expressed by relation (4.6.11), see Fig. 4.6.2.

Let us remark that in similar cases another simple criterion can be based on the feasibility constraints

$$\max_t \widehat{y}_t(x) \leq E(y_t) + \delta, \qquad \min_t \widehat{y}_t(x) \geq E(y_t) - \delta \qquad (\delta > 0), \quad (4.6.12)$$

in which $E(y_t)$ is the (estimated) average of the measurement data and δ is again a fixed tolerance parameter.

An example of applying the feasibility requirement (4.6.12) can be the following. Let us suppose that the variations of certain measurements over a given time period are (expectably) caused more by random errors and fluctuations than by real changes: for example, this is often the case for sediment characteristics in a period less than a year. To illustrate (4.6.12), see Fig. 4.6.3.

Partial Optimization Approaches

Let us consider now a somewhat more general type of acceptance criterion that can be specified to yield not only feasibility constraints, but also certain partially optimized calibration objectives. Suppose first that the model results can be broadly classified as either being (equally) 'acceptable' or 'nonacceptable', without further differentiation between them. Symbolically, let 'A' and 'U' denote the sets of acceptable and unacceptable model output, respectively. Then a feasible parameterization x is determined by the relations

$$\widehat{y}_t(x) \in A, \qquad t = 1, \ldots, T. \tag{4.6.13}$$

Taking one of the most often studied special cases of criterion (4.6.13), 'sufficiently good' model parameterizations can, for example, be characterized by defining the set of acceptability as

$$A = \{x \in X : D(\{\widehat{y}_t(x)\}, \{y_t\}) \leq \delta\}. \tag{4.6.14}$$

In (4.6.14), $\delta > 0$ is a given tolerance level; thus, the specification of the discrepancy measure D defines the concrete form of (4.6.14). For instance, (4.6.11) can be interpreted as a special case of (4.6.14), if D is defined by the l_∞-norm.

Taking another example, let $T_1 \subset T$ denote the subset of measurements that are 'the most important', that is, they are the ones to be followed most closely. A suitable criterion function form can then be determined by using

$$A = \{x \in X : D(\{\widehat{y}_t(x)\}, \{y_t\})_{t \in T_1} \leq \delta\}, \tag{4.6.15}$$

additionally defining the set T_1 (for example, by simply listing the indices t and corresponding measurements of primary interest).

Consider now two practically important examples of defining the set T_1. In a number of calibration studies, it is of high significance that the more extreme measurement values are reproduced well by the model. A suitable criterion can then be based on choosing the index set of relevance T_1 by defining

$$T_1 = \{t \in T : |y_t| \geq d\} \qquad (d > 0 \text{ is an appropriate threshold value}). \tag{4.6.16}$$

Similarly, if the measurements taken in a (sub)period of time $[t_{\min}, t_{\max}]$ are of particular interest, then we can define

$$T_1 = \{t \in T : t_{\min} \leq t \leq t_{\max}\}. \tag{4.6.17}$$

Note that, as a matter of course, these or similar definitions of the subset T_1 may naturally lead, not only to feasibility criteria, but also to appropriate (restricted, partial) optimization objectives. To obtain such problem statements, it suffices to replace the index set T by T_1 in the discrepancy measures discussed in the earlier optimization model variants.

Let us turn now to another aspect of calibration. The essential underlying objective of model-fitting, namely,

'find parameterizations that follow expected tendency'

or

'the model output (curve) should be close to measurements'

cannot always be appropriately expressed by the criterion functions presented so far, as they are tacitly based on the coherent timing of model output and observations. At the same time, for example, in environmental models certain data (nutrient loads, meteorological forcing functions, etc.) are often given on a weekly or ten-day time-scale. These larger time-steps often have a certain 'equalizing' effect (possibly 'hiding away' some extremes that, on the other hand, may be followed by the model output).

For another illustration of the time-scale dependence aspect, suppose that the model output exactly matches the set of observations, except for a (small) constant time translation. That is to say, for a given x,

$$\widehat{y}_t(x) = y_{t+\Delta} \qquad (\Delta > 0 \text{ is an acceptable shift in time}).$$

Then in practice the fit could be regarded as 'really good', while in terms of, say, (4.6.3) the discrepancy can be significant.

These or similar examples raise the issue of defining alternative measures of 'closeness' to those introduced earlier. For instance, if the (possibly discretized) model output is $\widehat{y}_\tau(x)$, $\tau = 1, \ldots, T_0$, then the objective function can be defined by the distance between the data and the linearly interpolated function passing through the model output points. For instance, the

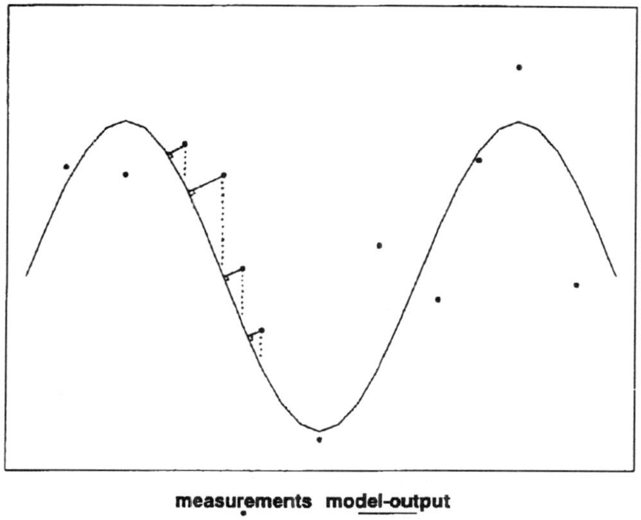

Figure 4.6.4
Linearly interpolated model output and its points associated with the corresponding measurements

objective function

$$D\big(\{\widehat{y}_t(x)\}, \{y_t\}\big) := \frac{1}{T}\Big(\sum_{t=1}^{T} |\widehat{y}_t^l(x) - y_t|^p\Big)^{1/p} \tag{4.6.18}$$

can be introduced: here $\widehat{y}_t^l(x)$, $t = 1, \ldots, T$, are those points of the linear interpolation curve that are 'closest' to the corresponding measurement y_t in the usual geometric sense. To illustrate (4.6.18), see Fig. 4.6.4.

A final remark is concerned with the typical variability of model results as a function of the time-step chosen. For instance, in evaluating dynamic environmental models, the appropriate time-step in the calculations—depending on the case studied—can be an hour, a day, a week, a month, a year, or even more (wind conditions can be observed hourly, while sediment characteristics are often described on a yearly time scale). Therefore, when used simultaneously, some of the available data have to be aggregated, and evidently the length of the time-step limits the precision of describing the temporal evolution. This comment implies that—apart from variations and discrepancies of the model output and measurements—the role of the time-step chosen should also be carefully considered. A straightforward way of doing that is to compare the 'optimal' parameterizations obtained when applying a sequence of appropriate time-steps in the model.

Closing the discussion in this section, let us emphasize that the acceptability criterion functions presented above do not necessarily require the application of any formal optimization approach per se. Nevertheless, to find admissible parameterizations can frequently be a nontrivial issue and—global, or more conventional—optimization approaches can be helpful in finding feasible solutions (recall in this context the corresponding expositions from Chapters 4.1 and 4.4).

The discrepancy function forms analysed in Sections 4.6.2 and 4.6.3 illustrate a fairly broad choice of possible basic calibration model types. Additional model variants arise by including an appropriate consideration of uncertainty and/or multiobjectivity aspects of the calibration problem at hand.

In the following sections and the two subsequent chapters, model calibration case studies will be discussed in more detail. The discussion will also illustrate further calibration problem statements, together with modelling and solution approaches.

4.6.4 A Sediment-Water Interaction Model for Shallow Lakes

Wind induced sediment-water interaction has a considerable impact on artificial eutrophication of shallow lakes by two phenomena:

- the change in the concentration of particulate and dissolved phosphorus (P) fractions, due to particle resuspension and possible sorption
- the variation of light transparency properties influencing algal growth, due to the change of suspended solids (SS) concentration.

From the viewpoint of both influences, the determination of the SS dynamics is of primary importance. As shown in Somlyódy (1980), the temporal evolution of the volume-averaged SS concentration (C) of a lake can be approximately described by the simple differential equation (d.e.)

$$\frac{dC}{dt} = -B_1 C + B_2 W^N, \tag{4.6.19}$$

in which W is the absolute value of the wind speed (at 10 metres high above lake surface); C and W are expressed in g/m^3 and m/s, respectively. In order to provide a brief explanation of (4.6.19), note that—after division by water depth—the first term on the right-hand side of (4.6.19) represents the flux of deposition (classically assumed to be proportional to the concentration), while the second term symbolizes that of the resuspension. Resuspension is described on the basis of energy transformation principles and depends solely upon the wind speed. In the d.e. (4.6.19), B_1, B_2, and N are lumped parameters which are to be estimated on the basis of time series observations for C and W. Lower and upper bounds can be easily provided for all three parameters which are known to be positive real numbers: hence, the estimation problem can be regarded as a special case of the model calibration paradigm, discussed in Chapter 4.5.

The simple model (4.6.19) was first applied for the Szemes Basin of Lake Balaton, a large shallow lake in Hungary, where daily SS observations were performed for nearly six months, including a month-long period of absence. Hourly wind data were available for the entire period. Note that in a related study of a shallow lake in the Netherlands, four sediment-water interaction models were tested and compared (Aalderink, Lijklema, Breukelman, van Raaphorst and Brinkman, 1984): among them the model outlined above resulted in the best performance. Below we illustrate the optimization methodology on the original Lake Balaton example, in which both variables W and C are characterized by very fast temporal changes.

During the calibration procedure (Pintér, Szabó and Somlyódy, 1986), the numerical solution of (4.6.19) was derived for subsequent time-steps, which are described by approximately constant wind speeds. In other words, the parameters of the approximation

$$C_t = C_{t-\triangle t} e^{-B_1 \triangle t} + \frac{B_2}{B_1} W_{t-\triangle t}^N (1 - e^{-B_1 \triangle t}) \tag{4.6.20}$$

($\triangle t$: time-step of the approximation) were calibrated, applying the objective function $\sum_{t=1}^T [\widehat{y}_t(\boldsymbol{x}) - y_t]^2$. Global optimization yielded the following parameter values:

$$B_1 = 1.03 \cdot 10^{-5} s^{-1}; \quad B_2 = 6.21 \cdot 10^{-5} \text{ gm}^{-4}; \quad N = 0.926.$$

Comparison of observed and computed concentrations, applying the 'optimal' parameters, can be seen in Fig. 4.6.5: the horizontal axis is scaled in days, and the vertical axis shows C in g/m^3.

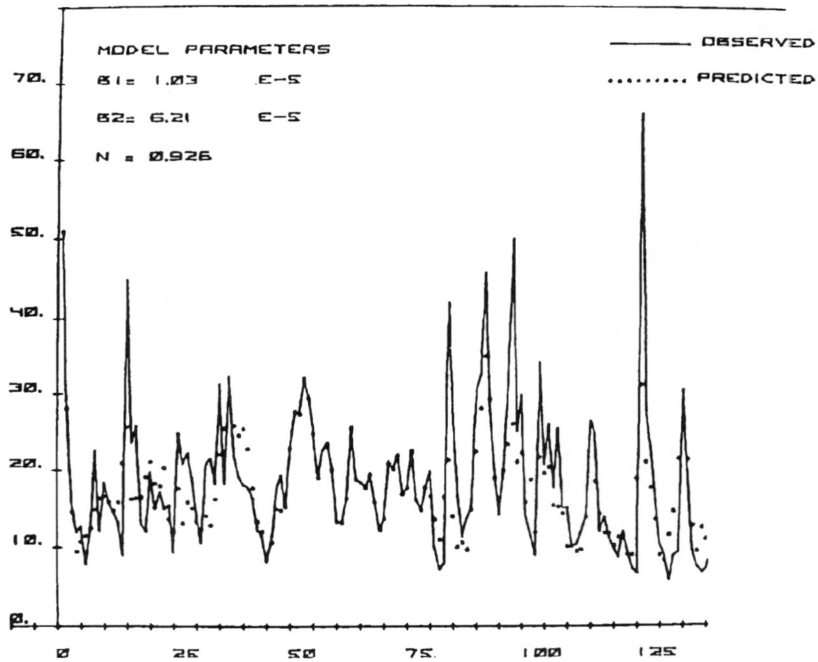

Figure 4.6.5
Calibration of a sediment resuspension model:
Results of globally optimized parameterization

From this figure it can be concluded that

- the model captures quite well the observed behaviour of the system for the first 120 days (including also the period when sampling started again right after absence)
- peaks are generally underestimated; the largest discrepancy occurs around day 125
- model results deviate more considerably from observations during the last 30 days of the monitoring program (not all shown in figure).

Note also that previous calibration exercises for the same process resulted in basically similar parameter values (Somlyódy, 1982a,b; Beck, 1985b). The Extended Kalman Filter (EKF) yielded an exponent slightly larger than 1. Subsequently, N was fixed as equal to one, since this value has a direct physical interpretation (Somlyódy, 1982a). In this case the values $B_1 = 1.5 \cdot 10^{-5} s^{-1}$ and $B_2 = 9 \cdot 10^{-5}$ gm^{-4} were obtained; approximately the same values were determined by the Marquardt algorithm (MA) (Somlyódy and Szilágyi, 1984). Note that, of course, EKF and MA will not, in general, find globally optimal parameter values. Moreover, the parameterization found above by global optimization led to some 15 per cent reduction in the objective function value when compared to the best solution found by repeated applications of MA.

In order to verify model stability, the globally optimized model parameters were carefully checked. The diagnostic validation procedure included the following steps:

- calibration problem solution under various initial state conditions
- calibration using data records of different length and initial time period
- calibration on the basis of a partial data set, followed by a subsequent check based on additional data.

The model validation procedure outlined indicated the stability of the parameter values found, concluding this calibration exercise. For additional details, consult Pintér, Szabó and Somlyódy (1986).

4.6.5 A Chemical Fate Model

The DELWAQ-IMPAQT model system was developed at Manhattan College (New York), and Delft Hydraulics (Netherlands), in cooperation with the Dutch National Institute for Inland Water Management and Waste Water Treatment (DBW/RIZA), Lelystad, in the period 1985 to 1990. It is a chemical fate model which describes the transport, distribution, and degradation of chemicals in the aquatic environment. In this section only a very brief description of the model is given; more information can be found in De Vries (1987), Delft Hydraulics (1990), and ten Hulscher, Bak and Pintér (1991).

Model Description

For modelling purposes, an aquatic system can be divided into several parts. In the model system DELWAQ-IMPAQT, the transport of water, particulates, and the modelled compounds is calculated by DELWAQ; IMPAQT then calculates the distribution and degradation of chemicals during a time-step in each of the compartments. The relationship between DELWAQ and IMPAQT is shown in Fig. 4.6.6; the major processes described by the IMPAQT model are presented in Fig. 4.6.7. As seen, micropollutants can enter the system by inflow of water, or by direct load to the system. The micropollutants are distributed between sorbed forms and the dissolved form, according to their distribution coefficients. Micropollutants that are sorbed to particles exchange with the bottom sediment as a result of settling and resuspension. Both in the water column and in the bottom sediment, degradation (biodegradation or photochemical degradation) of organic micropollutants may occur.

The DELWAQ-IMPAQT model system was programmed in Fortran. It consists of several thousand source code lines, and its single execution—that is, the evaluation of any given parameterization—can take seconds to minutes (on a PC, depending on the solution accuracy of the underlying system of partial differential equations). The model has several tens of such parameters which, in principle, could all be subject to calibration. Given the computational demand implied, 'exact' model fitting (especially at the time of the study) was prohibitively expensive. Therefore—on the basis of a preliminary sensitivity analysis—only four important model parameters were optimized. Following the original notation, these are

- K_{oc} a parameter that describes sorption of organic micropollutants to sediments (K_{oc} is expressed in volumetric units, for instance in ml)
- X_{doc} a parameter that describes the efficiency of sorption of a compound to dissolved organic carbon (X_{doc} is dimensionless)
- H_e Henry's constant that characterizes the tendency of a chemical to volatilize from water (H_e is dimensionless)
- V_s settling velocity of particles (expressed in m/day).

For more details on these parameters and their role in the model, see ten Hulscher, Bak and Pintér (1991).

The calibration of these four major model parameters was based on the error function type

$$f(x) = \frac{1}{I}\sum_{i=1}^{I} \mid \widehat{y}(t_i, x) - y(t_i) \mid / \mid y(t_i) \mid \tag{4.6.21}$$

(assuming that $y(t_i) \neq 0,\ i = 1, \ldots, I$); $\widehat{y}(t_i, x)$ is the model outcome at

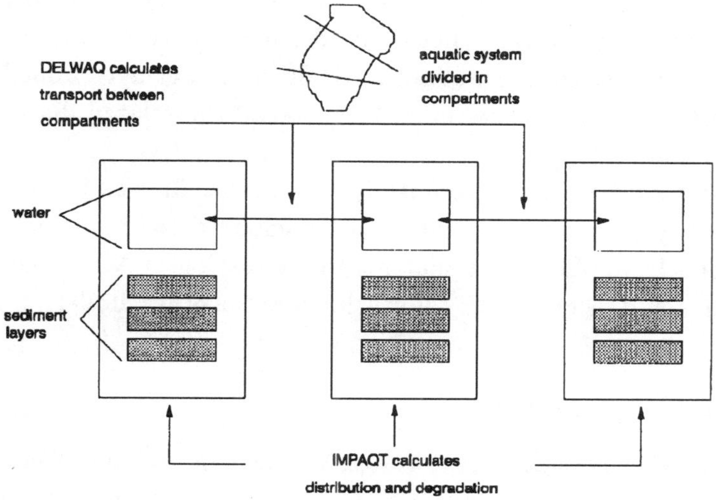

Figure 4.6.6
Relationship between DELWAQ and IMPAQT
(adapted from Delft Hydraulics, 1990)

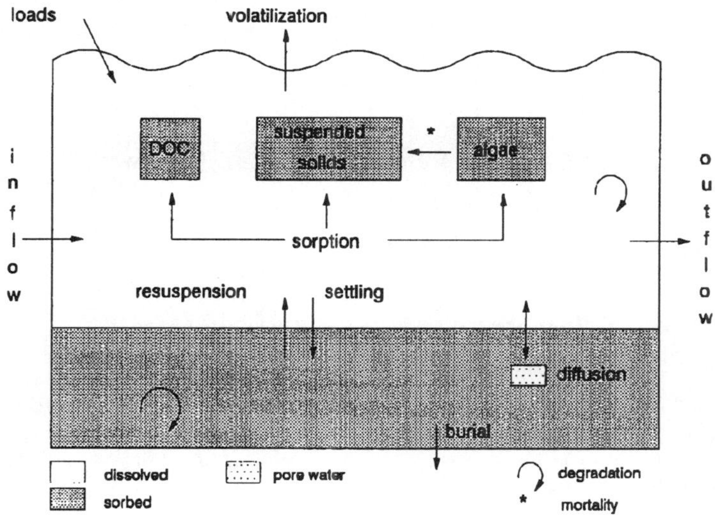

Figure 4.6.7
Major processes in IMPAQT
(adapted from Delft Hydraulics, 1990)

time moment t_i, under model parameterization x; $y(t_i)$ is the corresponding observation. As several types of model output were confronted with observations, a weighted, normalized error function was determined, on the basis of components analogous to (4.6.21) (recall the related discussions in Sections 4.5.2 and 4.6.2).

As a result of our calibration study, the model fit was improved very significantly (by some 40 per cent) when compared to the error resulting from—previously applied—nominal parameterizations. Note that further improvement could be expected, calibrating also the other model parameters.

4.6.6 A River Flow Model

ZWENDL (Rijkswaterstaat, 1991; Voogt, Van Zetten, Bak and Pintér, 1991) is a longitudinal (one-dimensional) flow model that computes water levels (H) and discharges (Q) in a river network. The principal governing model relations are the long-wave equations for describing the evolution of the momentum and continuity of water flow (Cunge, Holly and Verwey, 1980):

$$\frac{\partial Q}{\partial t} + \frac{\partial (Q^2/A)}{\partial x} + \frac{A\partial P}{\rho \partial x} + \frac{gQ|Q|}{C^2 AR} = B_s \gamma W^2 \cos(\theta - \phi) \quad (4.6.22)$$

$$\frac{\partial Q}{\partial x} + B\frac{\partial H}{\partial t} = 0. \quad (4.6.23)$$

In (4.6.22) and (4.6.23) the following notation (followed by their units, L = length, T = time, M = mass) is used:

x	location (in one-dimensional flow)	$[L]$
t	time	$[T]$
A	cross-sectional flow area	$[L^2]$
B	cross-sectional storage width	$[L]$
B_s	cross-sectional flow width	$[L]$
C	Chezy friction coefficient	$[L^{1/2}T^{-1}]$
g	constant of gravity	$[LT^{-2}]$
H	water level	$[L]$
P	pressure	$[ML^{-2}]$
Q	flow rate (discharge)	$[L^3T^{-1}]$
R	hydraulic radius	$[L]$
W	wind velocity	$[LT^{-1}]$

γ	wind conversion coefficient	$[-]$
ϕ	wind direction	$[-]$
ρ	density of water	$[ML^{-3}]$
θ	direction of channel axis	$[-]$.

The partial differential equations (4.6.22) and (4.6.23) are solved numerically, following discretization and applying an implicit method based on Verwey's variant of the Preissman scheme (see Cunge, Holly and Verwey (1980) for details).

Apart from the straightforward numerical solution of the coupled p.d.e.'s (4.6.22) and (4.6.23), the program system ZWENDL has many special facilities, including, for example, the simulation of diverse engineering structures like weirs, sluices, pumping stations, and shiplocks. Moreover, several types of boundary conditions, such as water levels, discharges, head-discharge relations, and so on, can be prescribed. Since these features are not of direct relevance here, they will not be treated in any detail.

The parameter that is chosen to 'tune' the flow model is the Chezy friction coefficient, C (see the denominator of the fourth term of the momentum equation (4.6.22)). C is a lumped parameter of the model, practically serving also to incorporate ('collect') a significant part of the model errors.

The friction coefficient can be described by the weakly water level dependent function

$$C = (RH^{1/6})/Ma, \tag{4.6.24}$$

in which

R	hydraulic radius	$[L]$
H	water level	$[L]$
Ma	Manning friction coefficient	$[L^{2/3}T]$.

Alternatively, it can be given directly as a water level dependent Chezy value, according to a continuous, piecewise linear function form

$$C = C(H). \tag{4.6.25}$$

In general, the function (4.6.25) is derived numerically. Let us remark further that both the Chezy and Manning friction coefficients may depend on the flow direction and are usually a function of location, but—under steady-state conditions—not of time.

An important application area of the program system ZWENDL is the model of the so-called Noordelijk Delta Bekken (NDB) region located in

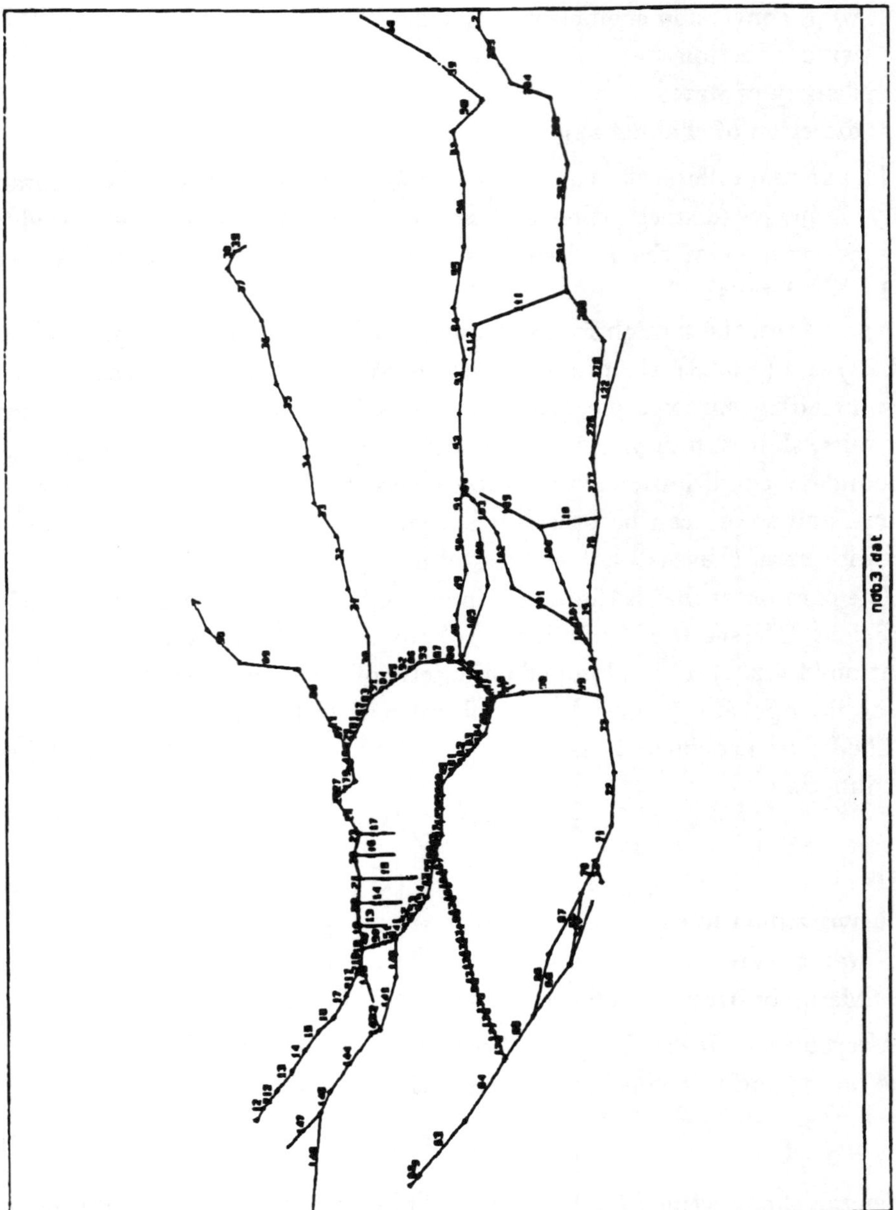

Figure 4.6.8
The river network of the Noordelijk Delta Bekken

the Netherlands. Fig. 4.6.8 shows the schematic layout of the NDB river network model. It is instructive to summarize at this point the most relevant computational characteristics of the NDB model. It has 178 sections; to simulate a single day takes about 80 seconds on a mainframe (UNIVAC 1100) computer, or about 520 seconds on a PC (at the time of this study, on a TULIP AT with 80287 mathematical coprocessor). A normal run of the model includes the simulation of several days, somewhat depending on the specific hydraulic conditions studied.

In the past, the model was calibrated by hydraulic experts of Rijkswaterstaat, using data sets of measured water levels collected over a wide range of hydraulic conditions. As an example of the model performance, computed and measured water levels are displayed in Fig. 4.6.9. As can be seen, the model performance is quite satisfactory. Now, if we mean to improve the quality of description by optimizing (some of) the model parameters, then the following facts should be kept in mind. From the above-mentioned computational characteristics it is clear that—whatever type of 'exact' optimization method is chosen—the 'precise' calibration of such a large model as ZWENDL would take a considerable (most likely prohibitive) amount of computer time. Therefore a simplified model has been introduced as a test case for the optimization methodology proposed, in order to keep the computational demands within acceptable limits.

Since the Chezy coefficient is a lumped parameter, the information concerning its a priori admissible range can typically be rather vague. In the model runs—on the basis of related practical experience—the initial ranges of Chezy coefficients were chosen, in most cases, as $[40, 60]$, independently for each river section. No implicit relations among these Chezy values were postulated: this way, the a priori feasible parameterizations belong to a multidimensional 'box' region of corresponding dimensionality.

Next, a standardized objective function form (recall (4.6.8)) was introduced. To describe this function, let $m = 1, \ldots, M$ denote the measurement stations that are considered during the model calibration procedure. In addition, let $t = 1, \ldots, t_m$ be the time moments in which measurements $\{y_{m,t}\}$ were taken at station m. Finally, let us apply the notation

$$\overline{y}_m = \frac{1}{t_m} \sum_{t=1}^{t_m} y_{m,t}$$

and

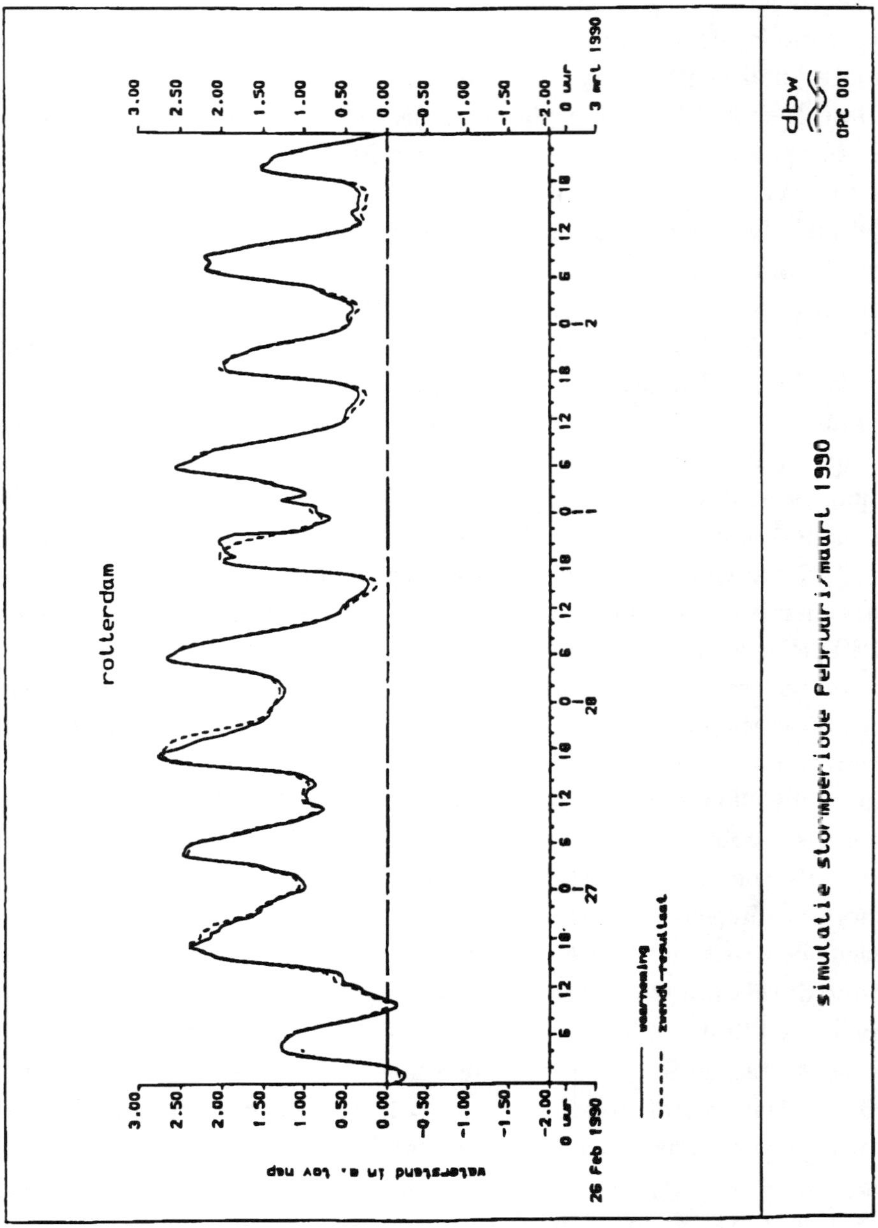

Figure 4.6.9
ZWENDL model output versus measurements:
Simulation of the storm period February–March 1990

$$V^{\frac{1}{2}}(y_m) = \Big(\frac{1}{t_m - 1}\sum_{t=1}^{t_m}(y_{m,t} - \overline{y}_m)^2\Big)^{\frac{1}{2}}, \qquad m = 1, \ldots, M,$$

for the empirical mean value and standard deviation of the measurements recorded at station m. Then an appropriate discrepancy function can be defined by

$$D(x) = \frac{1}{M}\sum_{m=1}^{M}\frac{1}{t_m}\sum_{t=1}^{t_m}\frac{|\overline{y}_{m,t}(x) - y_{m,t}|}{V^{\frac{1}{2}}(y_m)}, \tag{4.6.26}$$

which expresses the average (unit) standardized difference between measurements $y_{m,t}$ and corresponding model output values $\widehat{y}_{m,t}(x)$.

Results

Nominal calibration. For the sake of comparison with the results of the optimized calibration model runs, first the outcome (the corresponding discrepancy) obtained by using nominal Chezy values was generated. The value of the objective function $D(x)$ (see (4.6.26)) equals 0.307 in this case, corresponding to about 31 per cent average standardized discrepancy between model output and observations (under the given spatial and temporal specifications). As we shall see below, a significant improvement can be attained.

Single (uniform) parameter calibration. In the next cycle of computational runs, an attempt was made to find a single (that is, constant) Chezy coefficient, for all sections considered. The 'optimal fit' (obtained at Chezy coefficient 51.66) yields a discrepancy function value 0.935. This result is significantly worse than the objective function value attained applying river section-specific nominal parameter values. (The physical reason for this 'bad performance' is that we attempt to fit a Chezy value to some average environmental state which, in fact, does not occur.) Consequently, the approach of splitting the complete ZWENDL model into river parts and then fitting a 'uniform' Chezy coefficient for all of them is certainly not a good calibration strategy.

River section-specific, multiparameter calibration. Next, for every section a corresponding Chezy value was searched for. The best performance function value found was 0.534. In this way, the objective function value improved by some 40 per cent, compared to the results of the single-parameter run.

Although further improvement can be expected, this result already shows the merits of assigning site-specific Chezy values.

Water-level-dependent calibration. Finally, for every river section five water-level-dependent Chezy coefficients were selected via global optimization. The calibration was done, section by section, in four consecutive steps. Naturally, in this case, for every section the objective function value was defined by taking into account only the data of that particular section. The 'optimal' Chezy coefficients of the sections that are calculated in earlier steps are sequentially used during the following computations. (This can be done because the water level is mainly influenced by the state of the downstream section, while the influence of a section on the upstream section can be neglected.) The final evaluation of the discrepancy, however, includes all sections and the objective function value is also calculated as previously. The best objective function value equals 0.190. As this result indicates, among the model fitting exercises investigated above, this attempt is the most successful. It provides some 37 per cent decrease of the discrepancy, as compared to the previously applied ('expert judgement' based) model parameterization. Given the fact that the objective function is defined in terms of average standardized error, this improvement is quite significant. One should remark here, however, that the 'expert' model calibration is based on several high discharge flow periods (instead of the single time period used here for calibration). Therefore, an improvement can naturally be expected when calibrating only with respect to a single discharge period. Ideally, more calculations—and, possibly, model refinements—are needed, in order to fully evaluate the potentials of advanced model calibration in ZWENDL.

Chapter 4.7

Dynamic Modelling of Phosphorus Release from Sediments

4.7.1 Introduction

The analysis of nutrient release from sediments in shallow eutrophic lakes is an important stage in properly describing such ecological systems. (On the subject of lake ecosystem modelling, consult, for instance, Park et al., 1974; Scavia, 1980; Orlob, 1983; Somlyódy and van Straten, 1986; Riley and Stefan, 1988; and also Chapters 4.6 and 4.11.) SED is a dynamic model that was developed (and tested on Lake Veluwe, Netherlands data) for describing the release of phosphorus from lake sediments (Van der Molen, 1991). The model includes three environmental state variables: dissolved phosphorus, organic particulate sediment phosphorus, and active sediment layer. It considers only the basic relationships between these variables. Specifically, processes of the diffusive transport of dissolved phosphorus to the overlying water body, mineralization of organic phosphorus in the sediment, and adsorption/desorption and fixation of inorganic phosphorus are modelled by SED. To describe this model, the following notation is introduced:

t	time [d]
z	depth of the active sediment layer [m]
$C_{s,op}$	reactive organic phosphorus in the sediment $[\mathrm{g}\cdot\mathrm{m}^{-3}]$
$S_{s,ip}$	dissolved inorganic phosphorus in the sediment $[\mathrm{g}\cdot\mathrm{m}^{-3}]$
$S_{w,ip}$	dissolved inorganic phosphorus in the overlying water body $[\mathrm{g}\cdot\mathrm{m}^{-3}]$

P_{sed} phosphorus load to the sediment $[\text{g} \cdot \text{m}^{-2} \cdot \text{d}^{-1}]$
K_m mineralization rate $[\text{d}^{-1}]$
K_d rate constant for diffusive transport $[\text{m} \cdot \text{d}^{-1}]$
K_f fixation rate $[\text{d}^{-1}]$
K_a partition rate for absorbed phosphorus $[\text{d}^{-1}]$.

In SED two differential equations are solved numerically: one for organic phosphorus, and one for dissolved inorganic phosphorus in the sediments, see equations (4.7.1) and (4.7.2) below. The thickness of the active sediment layer is derived from the amount of reactive organic matter present in the sediment. The differential equations have the following form:

$$\frac{\partial C_{s,op}}{\partial t} = \frac{P_{sed}}{z} - K_m \cdot C_{s,op} \tag{4.7.1}$$

$$\frac{\partial S_{s,ip}}{\partial t} = K_m \cdot C_{s,op} - \frac{K_d}{z}(S_{s,ip} - S_{w,ip}) - K_f \cdot S_{s,ip} - K_a \cdot S_{s,ip}\frac{\partial z}{z}. \tag{4.7.2}$$

For further details of SED that are not discussed here, see Van der Molen (1991), Pintér and Van der Molen (1991), Van der Molen and Pintér (1993).

Six model parameters are to be calibrated: they are used in the numerical approximation of the d.e.'s (4.7.1)–(4.7.2), applying Euler's method. The description is highly nonlinear, so that multiextremality of the resulting calibration model variants can reasonably be expected (recall the related discussion in Chapter 4.5).

Model calibration was accomplished, considering data—monthly mass balances between 1978 and 1983—for Lake Veluwe (Netherlands). Before presenting a summary of the results, a few general comments are in order: these are listed below.

1. The case study (Pintér and Van der Molen, 1991) was aimed at exploring potentially useful calibration problem statements. A broad class of problem formulations has been analysed and compared, instead of finding the 'precise' solution of a preselected single model form. Therefore no local optimization (refinement) of the globally optimized calibration was carried out, reducing the computational burden of the study. The results obtained are, hence, to be interpreted more in a qualitative and comparative manner, than accepting (only) the results of some single 'best' model variant.

2. In the course of evaluating the calibration results, the 'optimal' argument values—that is, the parameter combinations selected by the optimization procedure—are of primary interest. The objective function values cannot be directly compared, as they are run-specific, that is, depend on the objective function form actually used.
3. Because of the underlying model of global optimization on 'box' regions, in general one would prefer interior point solutions to such solutions whose several components fall to the boundary of the search 'box'. Therefore, besides the global optimum estimate, several 'promising' solutions were also listed and checked. Additionally, a number of 'bad' parameterizations were also recorded in a number of cases, in order to provide a picture concerning the role of the parameters in the 'black-box' model, and also to exhibit some clearly inferior parameter combinations.
4. The detailed analysis of generated search points and the corresponding discrepancy values indicated the multiextremality of the diverse objective functions. Apart from this fact, in some cases the objective functions considered were quite 'flat' with relatively small function value ranges. Note, however, in this context that—as in many cases average error function values are considered—even 5 per cent difference between the performance of two model parameterizations can be considered as quite significant.

4.7.2 Numerical Results and Discussion

For comparison with the calibration results obtained subsequently, Fig. 4.7.1 shows the SED model output (dotted line) and the observations (large dots), when the model is parameterized on the basis of expert judgement, that is, without applying a formal optimization procedure. (All subsequent figures in this chapter show the evolution of phosphorus retention—the difference between sedimentation and release—over time.)

Mean l_p-Norm Objective Function

In the first sequence of optimization runs, the mean l_p-norm based objective function form (4.6.3)

$$D(\{\hat{y}_t(x)\},\{y_t\}) := \frac{1}{T}\Big(\sum_{t=1}^{T}|\hat{y}_t(x)-y_t|^p\Big)^{\frac{1}{p}}, \qquad 1 \le p \le \infty,$$

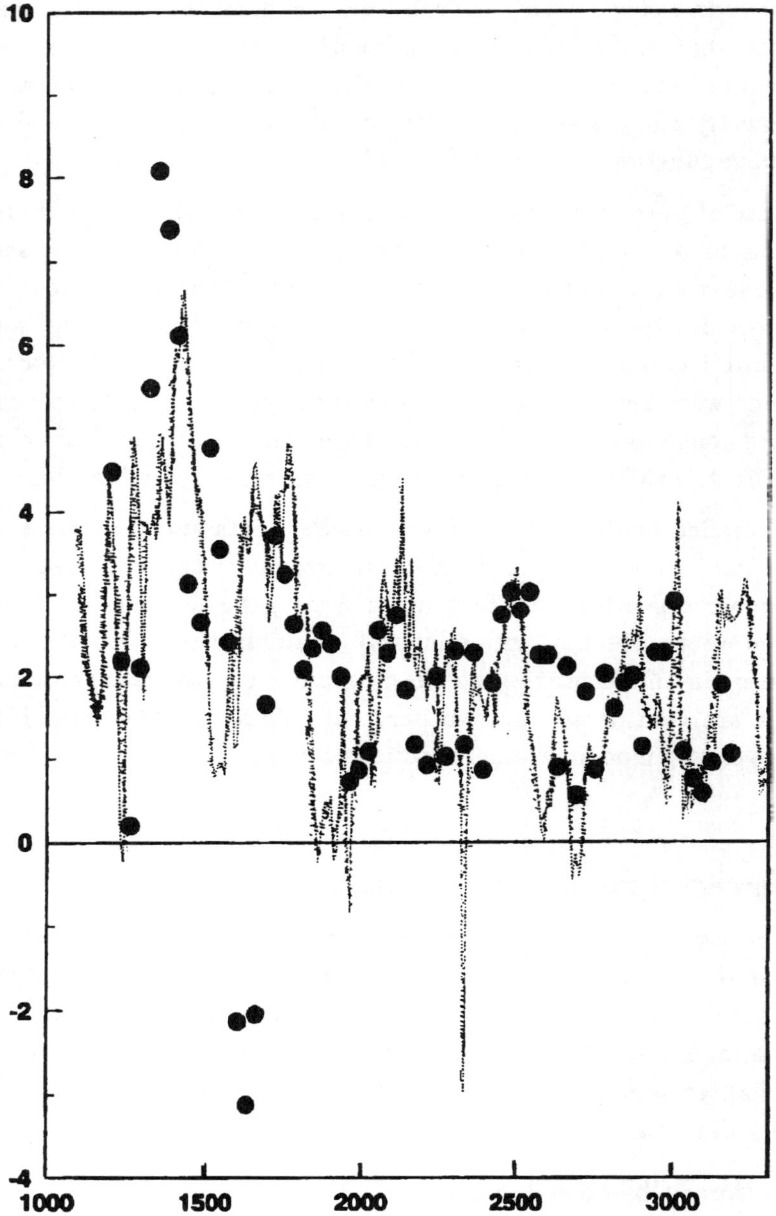

Figure 4.7.1
Comparison of SED model output and observation data:
Nominal (expert judgement based) model parameterization

was applied, choosing $p = 1, 2, 3, 5$ and 10. In the second set of runs, the original scaling of the calibrated parameters was replaced by a suitable linear transformation into the unit 'box' $[0, 1]^6$.

The results obtained in this case have shown a remarkable stability around the parameter vector $x \approx [0.12, 0.001, 1.05, 1.05, 3.0, 0.10]$. Unfortunately, although in terms of the l_p-norm based discrepancy form chosen the solutions referred to perform quite satisfactorily, they do not capture properly some essential characteristics of the observed environmental system behaviour. To be more specific, applying the parameterization found, the high and low extremal measurements ('peaks') are not followed as closely as would be desirable. This ex post negative finding indicates that—in spite of the formal correctness of the objective function class (4.6.3)—in the studied case it is not capable of 'forcing' the 'best' model parameterizations to imitate properly some of the important process characteristics. To show an example, the 'optimal' parameterization found (for $p = 3$) is

$$x = [0.119,\ 0.005,\ 1.020,\ 1.057,\ 3.036,\ 0.005].$$

As Fig. 4.7.2 shows, these model parameters result in a reasonably good match between the model output and the set of measurements, but some other state variables—which are not represented explicitly by the objective function form applied here—take values that are not realistic. This fact calls for some sort of multiple objective model formulation, discussed briefly later.

To give another example, the best solution found for $p = \infty$ (minimax objective function form) equals

$$x = [0.150,\ 0.006,\ 1.111,\ 1.048,\ 2.814,\ 0.061],$$

leading to a generally good SED model behaviour.

As seen from the previous discussion, the variety of parameterizations tested gives possibilities to the modeller to check these (automatically produced) model settings. In this way, non-quantified or semi-quantitative judgement can also be included into the model-fitting procedure.

Relative Error Based Objective Function

In the following set of computations, the discrepancy function form (4.6.7)

$$D(\{\hat{y}_t(x)\}, \{y_t\}) := \frac{1}{T}\Big(\sum_{t=1}^{T} |[\hat{y}_t(x) - y_t]y_t^{-1}|^p\Big)^{\frac{1}{p}} \quad (y_t \neq 0), \quad 1 \leq p \leq \infty,$$

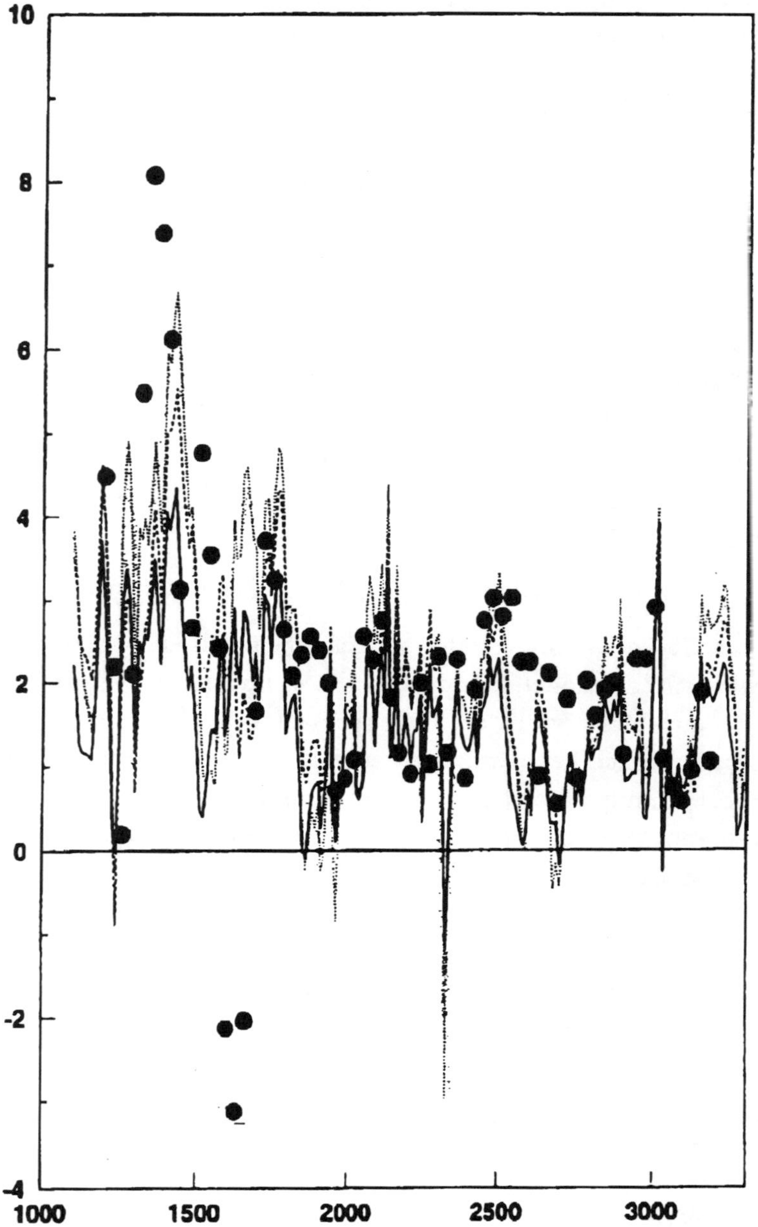

Figure 4.7.2
Comparison of several six-parameter model calibration results: Nominal (dotted line), $p = 1$ (small lines), $p = 3$ (continuous line), observation data (big dots)

was used: recall that this objective expresses the relative error of the actual model parameterization. We shall not go into extensive detail concerning these runs, because they did not lead to improved model calibration when compared to the previous runs. The best parameterization found was

$$s = [0.118,\ 0.007,\ 1.042,\ 1.030,\ 3.250,\ 0.029],$$

with a corresponding ($p = 1$) value of (4.6.7) that equals 0.733: Fig. 4.7.3 shows the resulting model output. Again, it is instructive to note that although in a mathematical sense somewhat better parameterizations were also found (with objective function values around 0.72), the corresponding parameter combinations were practically unacceptable. This comment indicates the merits of a globally optimizing, 'exhaustive' search approach, from yet another angle.

Closing this discussion, let us remark that—because of the above negative findings with respect to the relative error objective function forms—we did not accomplish calculations using the standardized error as a discrepancy measure.

Four-Parameter Calibration of the SED Model

In the following cycle of calculations, a four-parameter version of SED has been calibrated. Two of the model parameters—related to the description of temperature dependence—were fixed, because their ranges were quite well known, compared to the others. Besides, it was expected that this action might help to 'capture' better the 'outlier' observations.

A detailed evaluation of the results has shown that in this case a number of almost equally good parameterizations were obtained in a few (ca. 100) global search steps. This finding indicates again the significance of selecting appropriate model forms and their most important parameters. The SED model output, corresponding to one of these 'good' parameterizations, is displayed in Fig. 4.7.4.

Calibration With Respect to a Restricted Set of Observations

Having in mind the experience that in most cases the 'optimal' parameterization did not lead to a proper reproduction of the observed peaks (that were considered as being realistic and important), in the next stage calculations

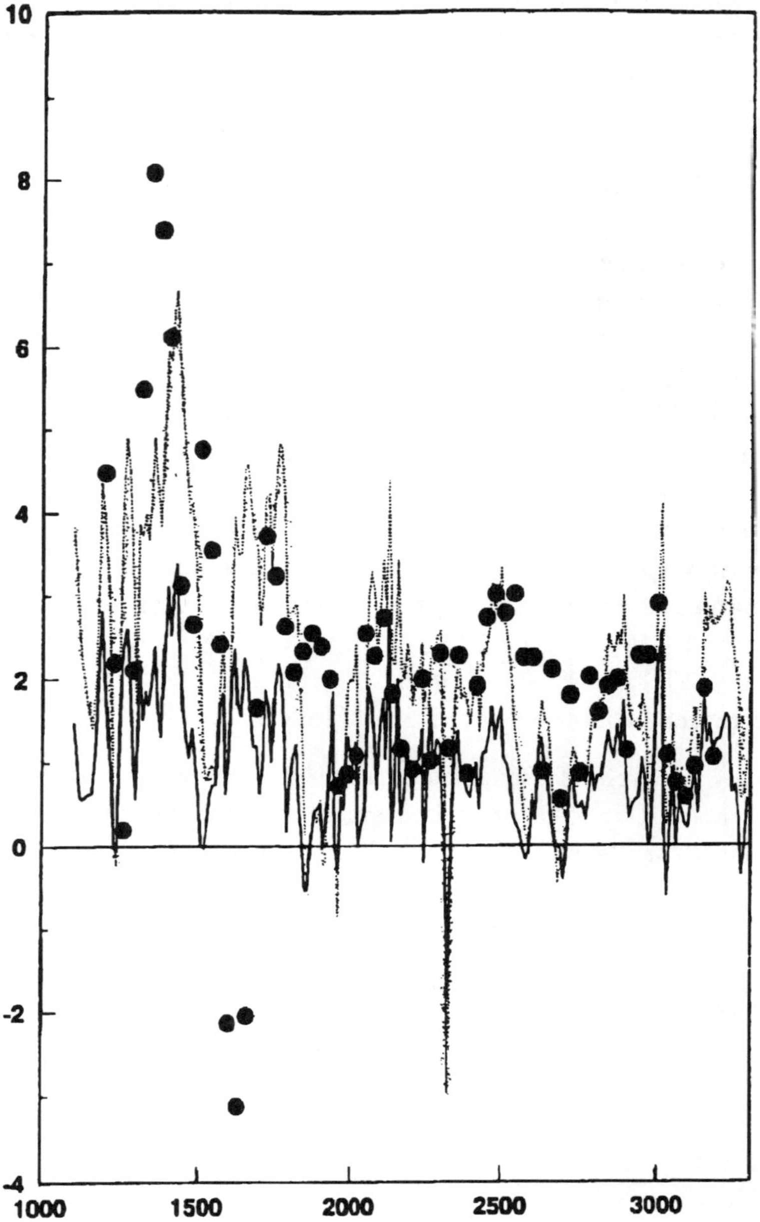

Figure 4.7.3
Comparison of two six-parameter model calibration results: Nominal (dotted line), relative error type discrepancy measure, with $p = 1$ (continuous line), observation data (big dots)

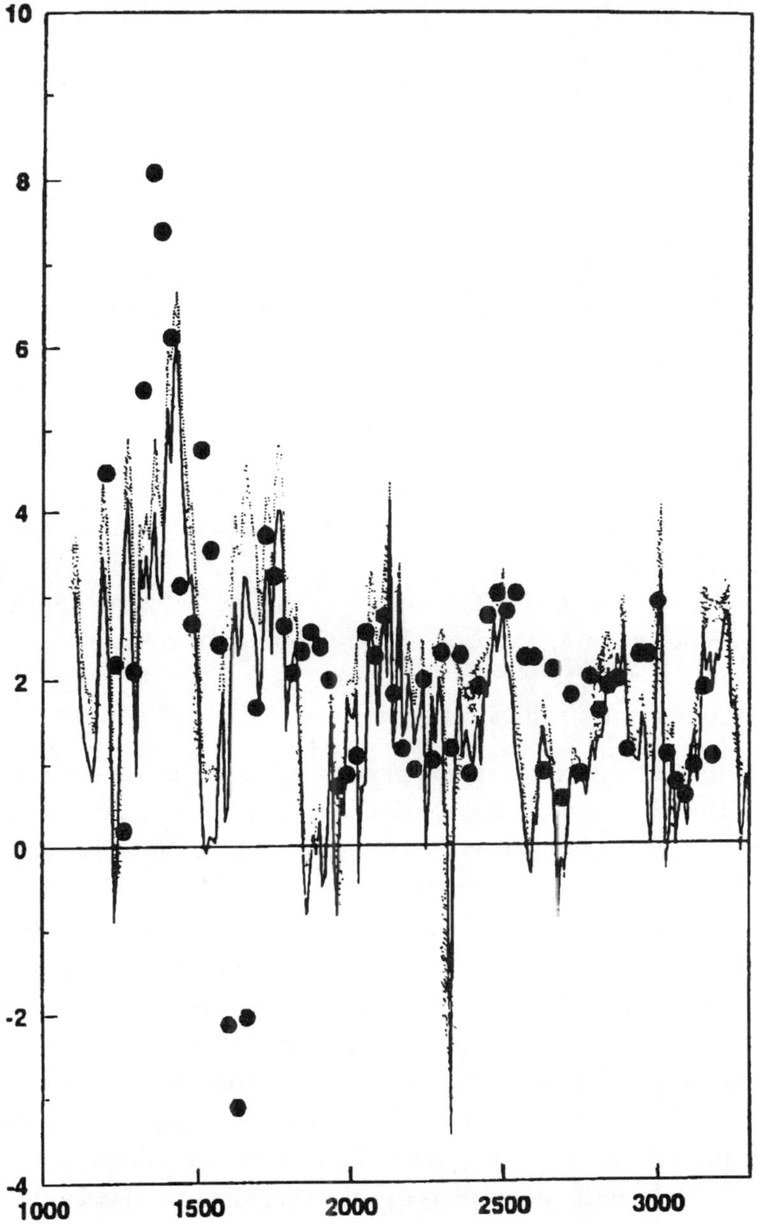

Figure 4.7.4
Comparison of two four-parameter model calibration results: Nominal (dotted line), optimized (continuous line), observation data (big dots)

were carried out which place a primary emphasis on these measurements. Specifically—recalling the related discussion in Section 4.6.3—a subset T_1 of the observation time-moments T was considered that is defined by the relation

$$T_1 = \{t \in T : |y_t| \geq d\}.$$

The y_t's are the measured phosphorus retention values; $d = 2$ is chosen as the threshold value. The objective function considered is the average l_1-norm error with respect to T_1, that is, it is expressed by

$$\frac{1}{T_1}\Big(\sum_{t \in T_1} |\hat{y}_t(x) - y_t|\Big).$$

In some 450 global search steps, several almost equally good SED parameterizations were found. As expected, the model fit with respect to the 'peaks' became appreciably better, see Fig. 4.7.5. Note, however, that none of the model variants investigated was capable of 'catching' quite well the lowest observation data (a reason also could be some measurement errors).

Multiobjective Analysis: Illustrative Results

During some of the previous model parameterization and ex post simulation runs, questionable values were obtained with respect to several state variables not considered explicitly. Therefore—recalling the related discussion (Section 4.6.2) and the specific features of the model SED—the following simple way of including several non-commensurable objectives was considered. Besides directly optimizing a selected discrepancy function (the primary objective being the 'good fit' with respect to the corresponding state variable), explicit constraints were also placed on several other state variables, and the solution was sought among those parameterizations that satisfy also the added constraints. On the basis of the previous experience with diverse scalar calibration model forms, again the l_1-norm specification of the average error (with respect to phosphorus retention) was applied as the primary objective.

The subsequent calculations resulted in several good model parameterizations with respect to the discrepancy function chosen, and provided realistic values also for the state variables in question. This finding supports the appropriateness of the description and the multicriteria formulation.

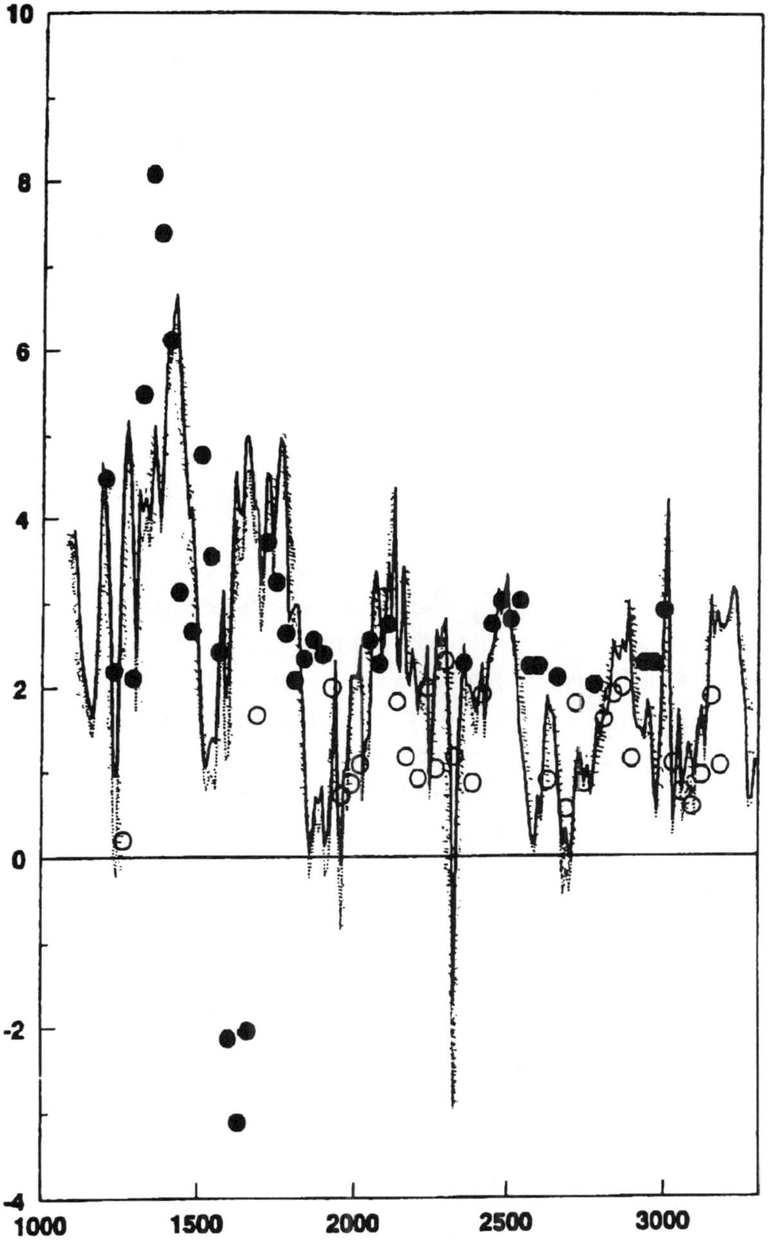

Figure 4.7.5
Calibration with respect to a restricted set of observations: Nominal (dotted line), optimized (continuous line), observation data considered (big dots) and omitted (open big dots)

4.7.3 Conclusions

In this chapter, a variety of calibration problem formulations and their influence on the results were discussed. To illustrate these investigations, a relatively simple but realistic water quality model was chosen. The systematic evaluation of diverse objective function forms and the corresponding solutions, hence, implied a nonnegligible—but still easily manageable—modelling and computational effort.

As has been demonstrated, in general a number of tentative discrepancy measures should be analysed when applying a formally optimized calibration procedure. Moreover, the results should be carefully verified with respect to criteria that may not be properly represented by the optimization problem statement. Special attention must be paid to the calibration of environmental models, as their 'soft' character frequently demands the application of nonstandard discrepancy measures.

Practical knowledge of the model SED (including ex post model runs, applying the formally optimized model parameterizations) made possible the comparative evaluation of the model variants studied. In this way, quantitative and semi-quantitative information was gained with respect to the choice of preferable model variants that led to realistic simulated output. The most preferred model forms can be subsequently used for describing, forecasting, and managing the evolution of the ecological system of Lake Veluwe (and of other lakes with similar characteristics).

Chapter 4.8

Aquifer Model Calibration

4.8.1 Introduction

The opening of the Atlantic Ocean during the Triassic period also resulted in the formation of rift valleys along the eastern border of the North American continental plate. Subsequent infilling of these basins by terrestrial sediments has resulted in the formation of high yielding aquifers of excellent quality. To obtain model parameters that can be subsequently used to characterize these aquifers, the best approach would probably be an intensive drilling program. However, due to the immense size and thickness of the sediments, this cannot be considered a viable strategy; instead, one may turn to the application of a suitable groundwater model and a corresponding automatic calibration procedure.

Inverse modelling—that is, a sequential calibration procedure to obtain improving model parameters using available field data—has been recognized as a suitable approach to aquifer parameter identification for a number of years; see, for example, detailed discussions by Carrera (1984) or Yeh (1986), with extensive reference lists. As illustrated already by the previous examples (in Chapters 4.6 and 4.7), most of these approaches require the use of computationally demanding, iterative solution techniques.

Because of the computational demands referred to, the early attempts at parameter optimization were generally restricted to a very limited number of model parameters, dealing primarily with transmissivity as the aquifer variable. Moreover, most optimization routines currently applied in this field are local in scope, seeking parameterizations only in the neighbourhood of

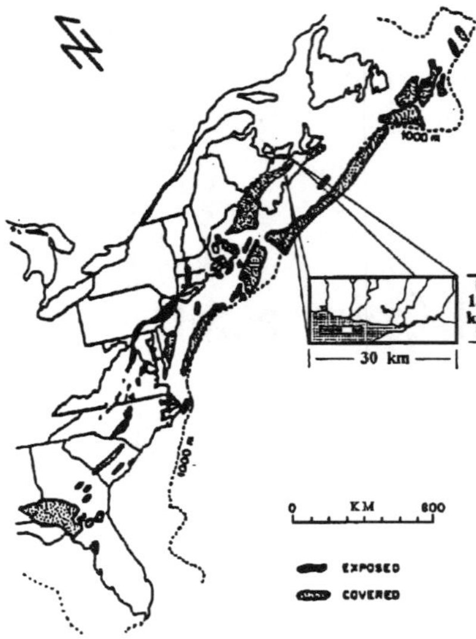

Figure 4.8.1
Location of Triassic-Jurassic rift basins in North America showing the model study area (modified after Forlenza, 1982)

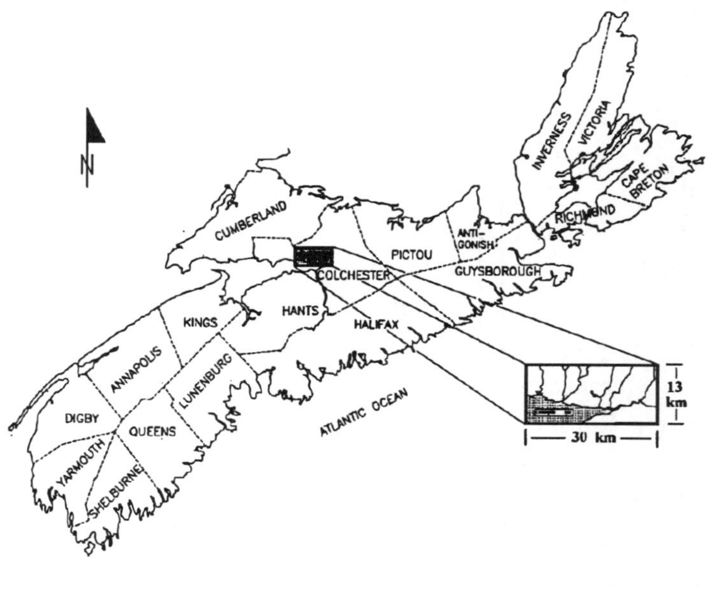

Figure 4.8.2
Location of model study area in Nova Scotia

some 'good' starting point. Among other important methodological issues, this problem has been known to researchers for some time (Wagner and Gorelick, 1987; Gorelick, 1990; De Lange and Pintér, 1991a,b). Recently, Karatzas and Pinder (1992) applied tentative global optimization approaches to find best groundwater model parameterizations.

In this chapter we apply Lipschitz global optimization to fit more than sixty aquifer model parameter simultaneously. Given adequate data, this results in better system description and supports the establishment of more sound groundwater resource management policies. For additional details, consult Finley (1994), and Finley, Pintér and Satish (1994).

4.8.2 Study Area

Fig. 4.8.1 displays a system of rift basins along the eastern seaboard stretching from south of Florida, northward along the coast to Newfoundland, out to the Continental Shelf. In Nova Scotia (Fig. 4.8.2), the sedimentary basin outcrops on the north and south shores of Cobequid Bay and stretches from Truro outward into the Bay of Fundy, forming the Wolfville Formation. Geophysical evidence available has suggested a sediment thickness in this basin of up to 15 km at its deepest point. In the region of the present study, shown in Fig. 4.8.3, the formation has an average thickness of between 1,200 and 1,500 m.

The unknown aquifer parameters for this study are transmissivity, storativity, areal recharge, and boundary flux. Range constraints for these variables were determined using geological data, coupled with a review of existing pump test data: the resulting ranges are given in Table 4.8.1. As will be demonstrated, these very broad ranges preclude the effective use of any 'stand-alone' local optimization routine.

Table 4.8.1
Aquifer parameter ranges

parameter	minimum	maximum
Transmissivity (m/s)	1×10^{-11}	1×10^{-1}
Storativity (–)	5×10^{-6}	3×10^{-2}
Recharge (mm/yr)	60	1200
Boundary flux (m/sec)	4×10^{-18}	3

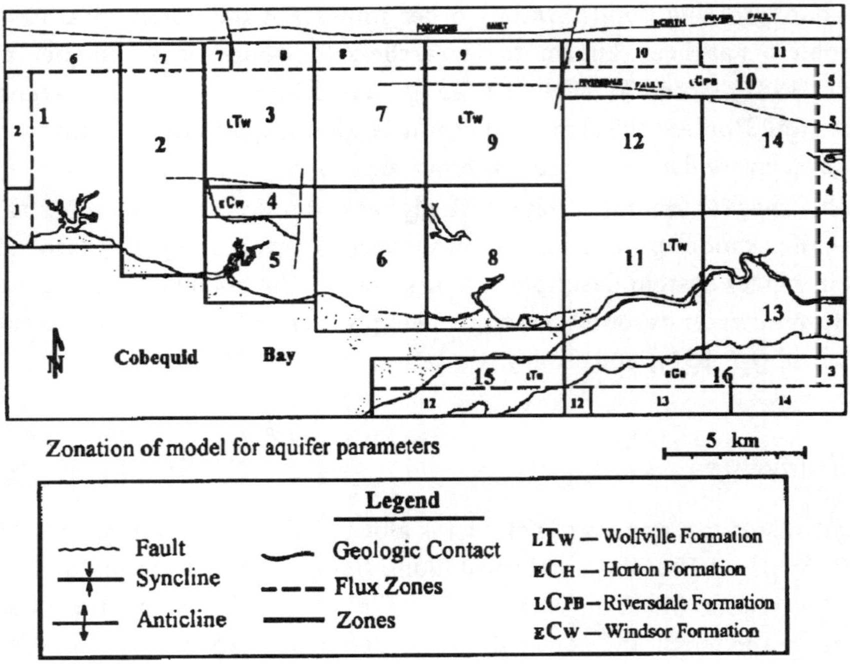

Figure 4.8.3
Zonation of model for parameters
(modified after Donohoe and Wallace, 1982)

4.8.3 Aquifer Model

A model grid was defined for the study area (Fig. 4.8.3) and subsequently used in the groundwater model. The size of this grid, which overlies a selected portion of the Triassic basin, is 13 kilometres N-S and 30 kilometres E-W, hence containing a total of 390 grid squares, each of them having an area equal to 1 km^2. The north boundary is determined by a major fault and the south by Cobequid Bay. The location of rivers on the eastern (North Salmon River) and western (Great Village River) borders define the four boundaries of the modelled area.

Following common groundwater engineering practice, see, for example, Yeh (1986), an aquifer model was adopted for this application. The finite difference model used is based on the alternating direction implicit (ADI) procedure and solves a two-dimensional transient flow system, in an inhomogeneous, anisotropic and confined aquifer (Freeze and Cherry, 1979). The governing equation of this model is written as

$$\frac{\partial}{\partial z}(T\frac{\partial h}{\partial z}) + \frac{\partial}{\partial y}(T\frac{\partial h}{\partial y}) = S\frac{\partial h}{\partial t} - Q, \tag{4.8.1}$$

subject to the following initial and boundary conditions:

$$\begin{aligned} h(z,y,0) &= h_0(z,y), & z,y &\in \Omega & (t=0), \\ h(z,y,t) &= h_1(z,y,t), & z,y &\in \partial\Omega_1 & (t>0), \\ T\frac{\partial h}{\partial n} &= h_2(z,y,t), & z,y &\in \partial\Omega_2 & (t>0). \end{aligned} \tag{4.8.2}$$

In equations (4.8.1) and (4.8.2) the following standard notation is used:

z, y	spatial variables (in two-dimensional aquifer model) [m]
t	time [s]
$h(z,y,t)$	hydraulic head [m]
$T(z,y)$	transmissivity [m/s]
$S(z,y)$	storativity [–]
$Q(z,y,t)$	source/sink term
Ω	flow region
$\partial\Omega$	boundary of the aquifer ($\partial\Omega = \partial\Omega_1 \cup \partial\Omega_2$)
$\partial/\partial n$	normal directional derivative
h_0, h_1, h_2	functions expressing initial and boundary conditions.

4.8.4 Selection of Optimized Parameters

For modelling purposes, the most important aquifer parameters (transmissivity, storativity, areal recharge, and boundary flux) are treated as discretized variables and as such are considered to be constant over a particular area or zone. This implies that the spatially varying parameters will be replaced by their zonal average, based on the assumption that the mean hydraulic gradient does not vary significantly within each zone (Neuman, Fogg and Jacobson, 1980).

The modelling area was subsequently divided into two sets of zones of varying sizes (as shown in Fig. 4.8.3). The first set, concerned with aquifer model parameters (transmissivity, storativity, and recharge), consists of 16 zones. The second set, defined along the model borders, was created specially for the purpose of dealing with the boundary flux, resulting in 14 additional zones. This leads to a total of $16 \times 3 + 14 = 62$ aquifer parameters that need to be calibrated. One can observe that the position of the boundary zones does not conform to the former set: this reduces the interdependence between the corresponding aquifer parameters.

4.8.5 Solution Approach

Optimization techniques which have been applied earlier to solve the inverse (calibration) problem are local in scope, and require calculation of the gradient (Yeh, 1986). In spite of their popularity, there are a number of obvious difficulties inherent in the calculation of the Jacobian matrix. Since the governing equations (4.8.1)–(4.8.2) are used and solved in a discretized version, any derivative determined using these equations must, of itself, be an approximation. In addition to this, insufficient and/or poor quality data will only produce a poor approximation. Consequently, local search algorithms may deliver inaccurate results for yet another reason; this is particularly relevant with large time-steps often used in the numerical approximation. For these reasons, more appropriate is the use of such global optimization algorithms that do not require higher order (gradient or Hessian) information.

Following the model calibration procedure depicted by Fig. 4.5.1, model parameters are selected sequentially; each parameterization—and resulting model output—is then confronted with the set of available measurements, and the error (discrepancy) is calculated. This comparison necessitates a complete run of the system model to be calibrated and—as such, in the case considered here—is computationally very intensive. The circumstances described justify the use of 'black box' Lipschitz global optimization on the parameter range: in the present case, this range is a 62-dimensional interval.

4.8.6 Numerical Results and Discussion

In the following—referring to the related exposition in Chapter 4.6—the calibration error will be defined by the objective function

$$D(\boldsymbol{x}) = \frac{1}{I} \sum_{i=1}^{I} \frac{|h_{ci}(\boldsymbol{x}) - h_{oi}|^2}{V(h_o)}. \tag{4.8.3}$$

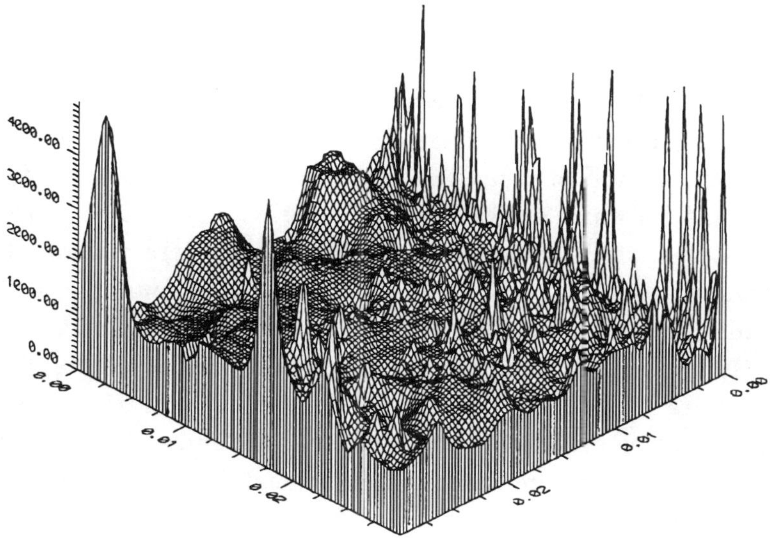

Figure 4.8.4
Aquifer model calibration: 3-D projection of the error function surface

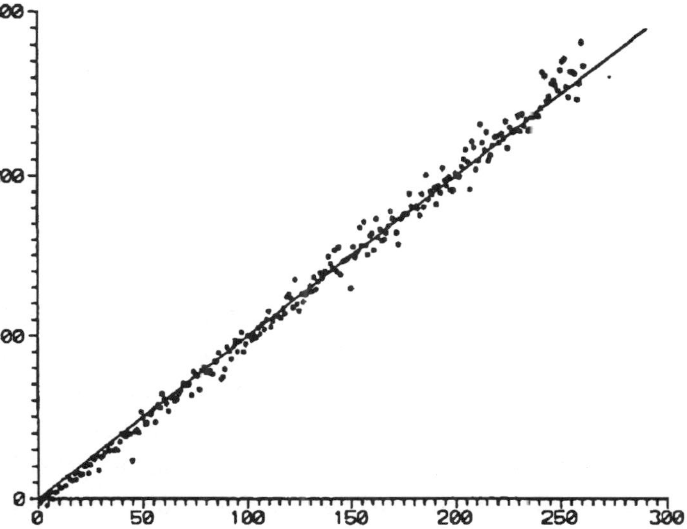

Figure 4.8.5
Aquifer model calibration: Final result
(Horizontal axis: observed values; vertical axis: calculated values)

In (4.8.3), x denotes the vector of optimized model parameters, I is the total number of nodes in the aquifer, $h_{ci}(x)$ is the calculated hydraulic head data (cf. (4.8.1)–(4.8.2)), h_{oi} is the corresponding observed hydraulic head data, and V_{h_o} is the variance of the entire observation set $i = 1, \ldots, I$. In order to illustrate the complicated multiextremal character of the error function $D(x)$, see Fig. 4.8.4 which shows a three-dimensional projection of $D(x)$. In this figure, storativity parameters for Zones 6 and 16 are varied, while all other parameters are kept constant (at their best value found).

An extensive set of calibration results was produced applying global optimization. Only a brief summary of these results is presented here; for additional details, see Finley (1994). The large number of calibration parameters and the computationally intensive character of the embedded groundwater model jointly led to very significant computational times. The global search phase was limited to a predetermined number of function evaluations, in this case 500,000. In comparison, an average number of evaluations for the local search method was 15,000. The corresponding computational time was approximately 9.3 hours on a Digital DEC 3000 workstation (estimated to be equivalent to some 90 hours on a 486DX/66MHz processor based PC).

Having completed the optimization run, the output was examined with respect to the physical constraints and the boundary limits. Fig. 4.8.5 shows the results of the optimized calibration: the solid line on the graph represents an 'ideal' (exact) fit between the calculated and observed data; the vertical distance of a point to this line being a measure of the residual (error) term. For additional numerical detail and interpretation of results, reference is made again to Finley (1994).

4.8.7 Conclusions

The complexity of the 'error surface' (Fig. 4.8.4) and the existence of a large number of local minima has made the optimization of aquifer model parameters extremely difficult. Therefore, the use of local optimization approaches cannot be realistically expected even to approximate the 'best' model parameters. Global optimization procedures, on the other hand, have the capability to provide a better estimate of the 'true' optimum value and corresponding superior parameter estimates with which to characterize the aquifer studied. The importance of this finding cannot be overstated, as an appropriately accurate groundwater system description serves as a basis for better management and control decisions.

Chapter 4.9

Industrial Wastewater Management

4.9.1 Introduction

Major industries typically have a substantial impact on environmental quality. Facing increasing public concern and more stringent environmental legislation, these industries have to find appropriate methods of waste management, while acting under the financial constraints of market realities.

The objective of the Environmentally Sensitive Investment System (ESIS) Project has been to provide quantitative support in this complex decision-making process. Specifically, ESIS assists in selecting wastewater management options that meet technical and economic constraints, as well as environmental regulatory criteria, in an effective manner. With the help of a tool such as ESIS, senior industry and government decision makers, environmental engineers and researchers can systematically analyse and compare strategic investment and operational choices in an interactive, computer-assisted environment.

The ESIS prototype was elaborated for the economically mature Canadian pulp and paper industry, but its broader conceptual relevance is evident with respect to a number of other industries. The study was aimed at the mechanical pulp and paper manufacturing sector: more precisely, to several standard wastewater treatment plant types applied in thermal-mechanical and chemi-thermal-mechanical pulp (TMP/CTMP) mills. Out of a total of some 150 mills across Canada, over 40 belong to this category; they represent about 30 per cent of the total production (newsprint being the major product of this sector). The costs of appropriate waste management—with a view towards needed facility upgrades and improved operational practices—are very significant: typically, in the range of hundreds of thousands to millions of dollars, at each mill site.

The ESIS Project was conducted at Dalhousie University and the Technical University of Nova Scotia (both in Halifax, Nova Scotia); it received support from a consortium of industry, consulting, research and government partners, which represented a range of interests, expertise, and a variety of multidisciplinary professional assistance to the Project. Naturally, in this chapter only the main modelling and optimization features of this work are summarized. For additional details, reference is made to the relevant environmental protection objectives and regulations (Environment Canada, 1990, 1992; Colodey and Wells, 1992), and the corresponding industrial waste management strategies and alternatives (United States Environmental Protection Agency, 1988; Ontario Waste Management Corporation, 1989; Canadian Pulp and Paper Association, 1991; McCubbin Consultants, 1992). Recent theoretical developments in and applications of decision support systems (DSS) to analyse and resolve engineering/economic/environmental issues are discussed, for example, by Glover (1986), Jaumard, Ow and Simeone (1988), Lewandowski and Wierzbicki (1989), Patry and Chapman (1989), Fedra (1991), Lam and Swayne (1991), Pintér (1991a), Greenberg (1993, 1995), Jones (1994), Kularathna and Somlyódy (1994). In particular, the principal ESIS Project documentation (Fels and Lycon, 1992; Pintér, 1993; Pintér, Fels, Lycon, Meeuwig and Meeuwig, 1993; Fels and Pintér, 1994) provide more detailed expositions, also including extensive reference lists.

4.9.2 Environment-Economy Integration in Modelling

Consider Fig. 4.9.1 which, in generic terms, displays the basic links among industrial production, waste management, and the environment. The figure indicates the principal decision alternatives: waste management in the pulp and paper industry can, hence, be considered as one of the specific instances corresponding to this scheme.

Because of the adopted limitations of the ESIS prototype, particular emphasis has been placed on the selection of technologically feasible, environmentally satisfactory, and economically efficient wastewater treatment alternatives. Some other management opportunities, namely,

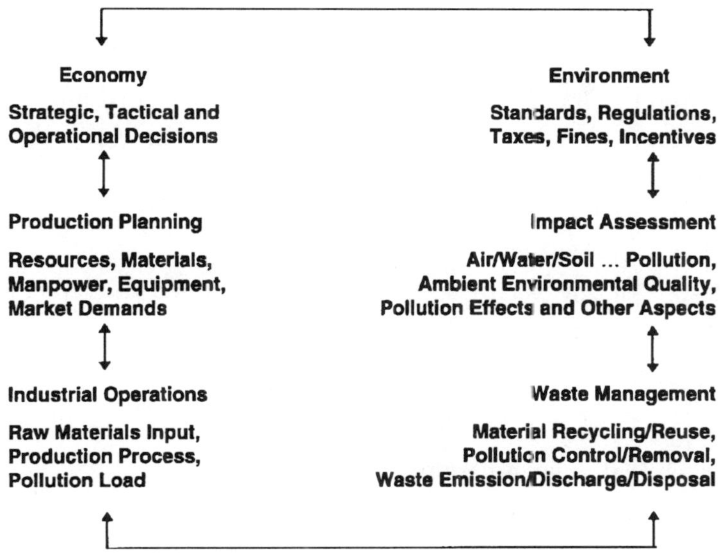

Figure 4.9.1
A generic scheme for environment-economy integration

- production plan (and resulting wastewater load) changes
- nonstandard waste treatment and disposal practices
- adaptively chosen environmental regulations (for example, effluent quality standards)

can, however, be investigated, interactively using the prototype DSS.

4.9.3 Wastewater Treatment Engineering System Model

In the context of the present discussion, 'wastewater' is the total liquid effluent of a pulp mill that enters its treatment system. On the basis of current North American and European experience, including a detailed state-of-the-art survey by McCubbin Consultants (1992), Fels and Lycon (1992) proposed four standard, complete wastewater treatment processes, which were analysed in the ESIS prototype. The following description draws mainly on their report, which can be consulted for additional details.

Wastewater Characteristics

The pulp and paper industry uses copious quantities of water in the processing of wood to make its products. Water is the medium in which the wood fibres are processed; the influent water is subsequently discharged (to a wastewater treatment plant, or without treatment), its volume essentially unchanged but with the addition of various waste components from the wood and the processes employed.

The design of the wastewater treatment systems modelled in ESIS is based on the following (raw) wastewater characteristics:

- flow rate
- total suspended solids (TSS)
- biochemical oxygen demand (BOD)
- acute lethality (toxicity, TOX).

Flow rate is considered the most important characteristic for sizing the wastewater treatment facility. It is usually expressed in terms of m^3/ADMT (air-dry metric ton of production), or in units of m^3/day. In TMP/CTMP mills, flow rates typically vary between 20–100 m^3/ADMT, or (in other terms) between 10,000 and 150,000 m^3/day.

Total suspended solids (TSS) refers to the suspended solids which can be separated out on a glass fibre filter: it consists of biodegradable organic, inert organic, and inorganic components. TSS can be expressed in kg/m^3 or in kg/ADMT: values for raw wastewater in the pulp and paper industry typically range between 0.075–1 kg/m^3, or 6–70 kg/ADMT. The majority (90 to 95 per cent) of these solids is settleable, and can be removed by primary treatment (the corresponding treatment processes will be discussed below).

Biochemical oxygen demand (BOD) is determined by its soluble and biodegradable organic material components. The term BOD refers to the amount of oxygen which is necessary to stabilize a given wastewater (that is, to remove the degradable material) by microbial action. It is expressed in the same units as TSS, and is measured using a 5-day to 7-day oxidation test. (Omitting additional notation, $BOD = BOD_5$ will be used throughout, corresponding to the 5-day test.) BOD in wastewater is commonly reduced by a combination of primary and secondary treatments. Typical ranges for raw wastewater are: 0.2–1.8 kg/m^3, or 10–100 kg/ADMT. It should be noted that chemical oxygen demand (COD)—which is the amount of oxygen necessary to oxidize chemically the organic and inorganic portion of waste—is also a usual water quality indicator, but (in the present context) it is not regulated per se. Typical COD/BOD ratios are about 2 to 3, but this ratio could become greater than 5, after secondary treatment.

Acute lethality (TOX), though not an immediately available basis for design, is the most deleterious currently regulated attribute of wastewater (in the Canadian pulp and paper industry). It is an indication of the effect of the wastewater effluent on a given fish species: usually, salmon or trout are the bioassay population. The test population is directly exposed to the—treated, but undiluted—effluent for 96 hours: the effluent is considered 'acceptable' if 50 per cent of the test population survives. Acceptability of the effluent is denoted by concentration values greater or equal to 100 per cent, meaning that the effluent does not need further treatment, or corresponding dilution. Values for raw wastewater in the pulp and paper industry typically range between 1 to 40 per cent: that is, this wastewater would need to be diluted 2.5 to 100 times to arrive at 'acceptable' quality without additional treatment. Acute lethality can be removed from wastewater only by an appropriate combination of primary and secondary treatment processes.

It is to be emphasized at this point that the wastewater characteristics discussed above are not independent, but their 'exact' interdependence structure is unknown. As TSS and BOD strongly influence the effluent toxicity, and they are significantly easier to determine than the TOX percentage value, on the basis of available measurement data, in Pintér and Lycon (1992a) TOX is approximated by a nonlinear regression model with input variables TSS and BOD. This (sub)model is built into the DSS, to estimate the TOX value of treated wastewater.

The effluent discharged has to meet stated quality standards which, for the Canadian pulp and paper industry in 1993, were expressed by the following values:

maximum TSS – 11.25 kg/ADMT
maximum BO – 7.5 kg/ADMT
minimum TOX – 100%.

Wastewater Treatment Systems

The wastewater treatment systems considered here are determined by a given configuration of unit treatment processes (UTPs) applicable to TMP/CTMP mills. These UTPs can be classified as follows:

Primary treatment

- Bar screen (BAS)
- Primary clarifier (PRC)
- Spill basin (SPB)
- Equalization basin (EQB)

Secondary treatment

- Anaerobic reactor (ANR)
- Activated sludge treatment (AST)
- Aerated stabilization basin (ASB)

Sludge handling devices

- Sludge thickener (SLT)
- Sludge mixer (SLM)
- Sludge dewaterer (SLD).

Fig. 4.9.2 displays the scheme of an investigated wastewater treatment system: a brief description of the UTPs follows.

The bar screen removes the coarsest pollution carried by raw wastewater: it is a fixed part of all the treatment systems investigated, and hence it will not be analysed here in any further detail.

The primary clarifier removes the majority (about 80 to 90 per cent) of settleable solids—and the BOD associated with these solids—from the wastewater, by gravity sedimentation. This removal also implies a corresponding improvement of the toxicity indicator (as approximated by the corresponding submodel description). The main design parameter of the clarifier is the overflow rate, expressed in $m^3/(m^2 day)$, which refers to the rate of daily volumetric loading and the clarifier surface area. Another important decision variable is the effluent sludge consistency (underflow solids concentration, kg/m^3) value which affects the eventual design of the sludge-handling UTPs.

The spill and equalization basins serve similar purposes in a wastewater treatment system: they reduce fluctuations in flow and/or organic loadings by providing additional storage volume. The main design parameters are the volumes (m^3) of these basins.

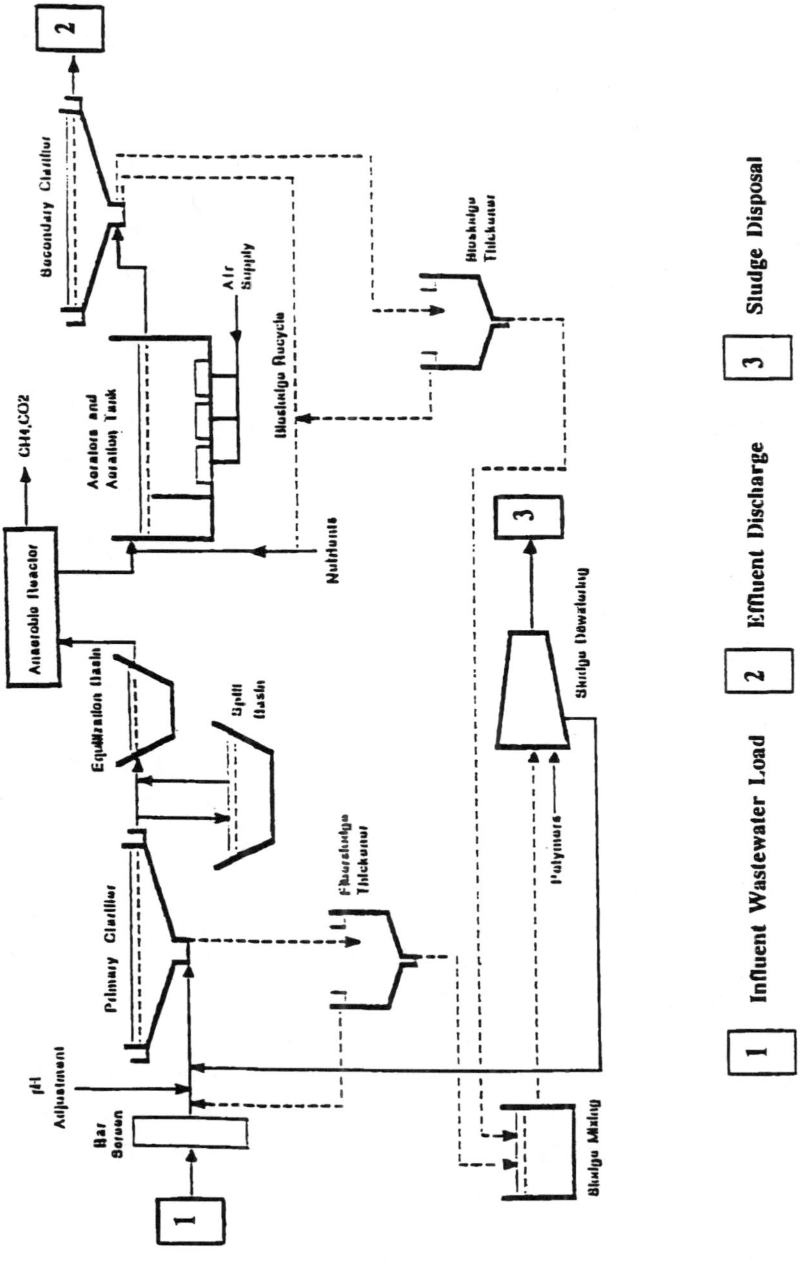

Figure 4.9.2
A wastewater treatment system applicable in mechanical pulp and paper manufacturing

Activated sludge treatment (AST) refers to two coupled treatment units: the aeration tank, equipped with aerators, and the secondary clarifier. In the aeration tank, the organic portion of the wastewater is stabilized using a suspended biomass and added nutrients. The stabilized biomass is then separated from the liquid in the secondary clarifier. The biomass settles at the bottom of the clarifier; a part of it is recycled to the aeration tank to maintain the biomass concentration, and the other part is 'wasted' as secondary sludge. The sludge is eventually separated from the liquid effluent for further processing.

The operation of the AST system is characterized by a kinetic description which depends on the following main design and operational characteristics: reaction temperature ($^\circ$C), soluble organic material (biomass) concentration (kg/m^3), and hydraulic retention time (day) in the aeration tank; overflow rate (m^3/(m^{2}day)) and sludge consistency (underflow solids concentration, kg/m^3) in the secondary clarifier. Activated sludge treatment can remove about 85 to 95 per cent of the BOD, implying also an improvement of the toxicity indicator TOX.

The other forms of secondary treatment, investigated in the frameworks of ESIS, included aerated stabilization basins (ASB) and anaerobic treatment. ASBs function in a similar fashion to the aeration tank in AST, but without recycling the microbial biomass and, consequently, in the presence of considerably lower biomass concentrations. This results in a need for longer residence times (6–10 days, as compared to 10–24 hours for aeration tanks). Because of the longer residence time, the resulting stabilized organic material is digested in the basin itself. In this UTP, residence time is the most important design parameter. The anaerobic reactor (ANR) also uses a microbial biomass to treat the wastewater, but in the absence of oxygen. The stabilized biomass decays anaerobically and biogases (CH_4 and CO_2) are produced. The main design parameter for this system is the oxidizable loading (chemical oxygen demand) rate, expressed in kg COD/(m^{3}day). The anaerobic reactor is useful as a pre-aerobic step for high organic loadings, but, typically, it is not efficient enough to function as a 'stand-alone' secondary treatment system.

The additional UTPs analysed in ESIS are sludge-handling devices which process sludge to the point where it can be disposed of. The sludge thickener is used to partly dewater the sludge arriving from the clarifier(s), resulting in approximately 1 to 4 per cent solids content. This UTP is similar to the secondary clarifier in design: the principal operational parameter is the underflow solids concentration (kg/m^3). The sludge mixer is used to mix primary and secondary sludge in the correct consistencies: this step allows the secondary sludge to be dewatered more easily, in combination with the primary sludge which consists largely of fibres. The final sludge-handling device is the sludge dewaterer which (in the presence of added polymers) dewaters the sludge to a consistency ranging between 20 and 50 per cent solids content. In the case of both the mixing and dewatering units, the main operational parameter is the fraction of the secondary sludge present in the combined sludge.

For each UTP, a corresponding quantitative—deterministic and steady-state—description has been developed (Fels and Lycon, 1992) which can be linked and included into the mathematical system model of complete configurations. The input of this model is, in essence, determined by the pulp and paper manufacturing process, which results in a given production level and corresponding generated raw waste stream. The waste influent is transformed by the treatment system, as a function of the basic design and operational characteristics of the component UTPs. The treated waste stream (effluent and sludge) then affects the spatial and temporal evolution of environmental quality and, hence, has to meet the stated regulatory standards.

Let us remark at this point that complete treatment systems are to be chosen depending on the particular (range of) wastewater load, as their treatment capabilities are different. In the ESIS prototype four standard wastewater treatment configurations, applied in the TMP/CTMP manufacturing practice, are investigated. These systems are variations of a basic configuration, composed of appropriate UTPs for primary treatment, secondary treatment, and sludge handling, as shown by the listings below:

Treatment Plant 1:
BAS, PRC, SPB, EQB | AST | SLT1, SLT2, SLM, SLD

Treatment Plant 2:
BAS, PRC, SPB, EQB | ASB | SLT1, SLD

Treatment Plant 3:
BAS, PRC, SPB, EQB | ANR, AST | SLT1, SLT2, SLM, SLD

Treatment Plant 4:
BAS, PRC, SPB, EQB | ANR, ASB | SLT1, SLD.

Fig. 4.9.2 shows Treatment Plant 3, as an example.

The pollutant removal efficiency, construction, and operational costs of each UTP—consequently, of complete wastewater treatment systems—are governed basically by the choice of several (main) design variables and operational characteristics. Hence, finding 'admissible' decisions (leading to feasible system design and operation) requires the simultaneous choice of these variables; determining the 'best' wastewater treatment option requires systematic search in the decision space of corresponding dimensionality.

4.9.4 Analytical Optimization Model

A General Model Frame for Assisting Industrial Waste Management

Environmentally sensible planning requires the consideration of multiple—typically conflicting and non-commensurable—objectives and requirements. These are related to the technical feasibility, economic efficiency, and environmental performance of strategic, tactical, and operational decisions. In the present context, the underlying inherently 'soft' problem structure excludes the possibility of finding a unique 'best' representation, as reflected by a single mathematical model. Therefore the following, verbally summarized optimization problem statement (4.9.1) is only one possible framework which reflects the basic paradigm. For specific numerical details, see Fels and Lycon (1992), Fels and Pintér (1994).

ESIS Optimization Model Structure (4.9.1)

Primary Objective

- find the most cost-efficient wastewater treatment system.

State Equations and Constraints

- explicit ranges of decision variables: wastewater treatment equipment types and sizes, basic operational limitations of UTPs
- initial conditions: input waste stream characteristics
- analytical description of treatment system: implicit technological constraints and relations among the decision variables
- resulting effluent and solid waste stream

- induced effects of waste on the environment vs. stated environmental quality standards
- implied resource demands and (construction, operation, and maintenance) costs of the wastewater treatment system.

A numerical (programmed) realization of the above optimization model scheme—in the present case—typically involves a few tens of decision variables and constraints, plus hundreds of constants and parameters. In addition to this, the evaluation of each—individually or algorithmically—selected decision alternative necessitates a fair amount of 'black box' operations: here we refer to running a complete cycle of the wastewater treatment system model. Therefore, the corresponding functional relations in model (4.9.1)—first and foremost, the resulting effluent quality—do not lend themselves to a simple and straightforward analysis. That is to say, their mathematical properties—such as convexity, concavity, differentiability, analytical derivatives, etc.—cannot be simply postulated, verified, or exploited. In the terminology of mathematical programming, even the deterministic base model versions of (4.9.1) can merely be classified as (potentially) multiextremal optimization problems defined by Lipschitz-continuous functions, without an obvious possibility of directly justifying a more specific model-class. Another essential problem feature is the inherent uncertainty—imprecise knowledge and/or statistical fluctuations—of some important system attributes. Consequently, one can expect that solving the optimization model variants calls for a nontrivial, computationally intensive procedure.

Mathematical Model Form

The essential structure of problem (4.9.1) can be summarized as

- objective function: minimize wastewater treatment system costs
- explicit ranges (bounds) of principal decision variables
- implicit—possibly 'black box'—functional relationships among decision variables.

In the implementation, the treatment system description is provided algorithmically, by linked UTP subroutines: therefore, the set of implicit, 'black box' relations considered in (4.9.1) can be reduced to the stated environmental quality constraints. Consequently, the following mathematical representation of the model frame (4.9.1) will be considered:

$$\min_{x \in D} f_0(x), \quad D = \{a \leq x \leq b,\ f_i(x) \leq 0,\ \ i = 1, \ldots, I\} \subset \mathbb{R}^n. \tag{4.9.2}$$

With respect to the model form (4.9.2), we shall recall the analytical conditions postulated in Part 2. Namely, we shall suppose that the system design and operational decisions can be represented by real n-vectors belonging to the set D which is the closure of a nonempty, bounded, open set in $\mathbb{R}^n$. Additionally, it is assumed that the functions f_i, $i = 0, 1, \ldots, I$ are all Lipschitz-continuous on D, with corresponding Lipschitz-constants $L_i = L_i(D, f_i)$, $i = 0, 1, \ldots, I$. Finally, we suppose that the objective function f_0 is possibly multiextremal on the feasible set D, and has a finite set $X^\star$ of globally optimal solutions. (Note that, in the present context, the set $X^\star$ is typically expected to be just a single point $x^\star \in \mathbb{R}^n$, corresponding to the unique 'best' wastewater treatment option in any given problem statement.)

Without repeating the earlier related discussion, it is noted that the assumptions with respect to model (4.9.2) are necessary—to make a numerical solution procedure feasible at all—but still sufficiently general to avoid unjustifiable analytical requirements. In particular, ranges are certainly known for the principal system design and operational characteristics. Moreover, the Lipschitz property of functions f_i, $i = 0, 1, \ldots, I$, only means that their dependence on the decision variables is 'regulated' by the inequalities $|f_i(x_1) - f_i(x_2)| \leq L_i \|x_1 - x_2\|$, $x_1, x_2 \in D$. It is also recalled at this point that the (minimal suitable) Lipschitz-constant values $L_i = L_i(D, f_i)$ are unknown: consequently, these need to be estimated, in the course of an appropriate optimization procedure which relies (only) on the Lipschitz structure of the problem.

4.9.5 Solution Approaches

Given the spectrum of targeted practical applications—from obtaining fast, preliminary evaluation of system performance to detailed analysis and 'fine tuning' of wastewater treatment systems—the prototype DSS offers a number of tools that can be used to formulate, analyse, and solve system models of diverse levels and depths. These approaches will be briefly described below; for more details, see Pintér (1993).

Spreadsheet Calculations (Deterministic Simulations)

Spreadsheets are widely applied in numerical systems modelling. With the help of spreadsheets—the entries being alphanumeric values, or results of function evaluation on such arguments—relatively large data sets can be easily manipulated in a sequence of calculations. Specifically, modelling wastewater treatment systems in ESIS, the decision variable entries—together with other model parameters—can be modified in an interactive fashion, and the results of such changes are immediately seen. In this way, the spreadsheet modelling environment enables a fast, preliminary 'trial and error' type evaluation of chosen decision alternatives, in terms of resulting effluent characteristics and financial implications. This method can also be interpreted as an interactive, deterministic simulation environment. Although, in general, this approach does not meet the formal requirements of optimization procedures—for a knowledgeable user—it can provide useful information for further investigations and 'fine tuning'. This 'tuning' can be assisted by a general or site-specific knowledge base and induced semi-heuristic rules ('expert suggestions' to modify the decision) that can be interactively communicated to the user by the system.

Optimization by Pure Random Search

The second solution approach built into ESIS is a simple uniform random search that belongs to the 'folklore' of engineering optimization. (For the underlying concept, the reader is referred to the general exposition in Chapters 1.2 and 3.6.) The search is carried out on the multidimensional interval (box) region defined by the explicit range $[a, b]$ of the decision vector x, see (4.9.2). This method is based on the sequential generation of (pseudo)random vectors x_r, $r = 1, \ldots, R$, where R is the chosen sample size. Following the generation of these randomized decision options, the most cost-effective feasible decision found, $x_R^\star \in D$, can be interpreted as an approximate solution of problem (4.9.2).

As discussed earlier, the main virtue of pure random search lies in its extreme simplicity and robust convergence properties: namely, under mild analytical conditions—certainly satisfied by model (4.9.2)—it is capable of approximating the globally best solution of (4.9.2) in limit, with probability 1. Furthermore, its application needs only a minimal interaction from the user. However, the number R of random search steps, needed to guarantee a good approximation of the 'true' optimum, may become prohibitive—even in decision spaces of modest dimensionality. Besides, the sampling efficiency may be poor on complicated or 'relatively small' (compared to $[a, b]$) feasible sets D. For these reasons, the pure random search strategy is proposed only for the preliminary selection of system designs in ESIS. Note in this context that random search may produce several good alternative solutions which can serve as 'seed points' of smaller search regions in a subsequent 'fine tuning' optimization phase.

Lipschitzian Global Optimization

Advanced mathematical programming methodology can be applied for solving more 'narrow' specifications of the wastewater treatment system design problem, possibly following a spreadsheet analysis (or other deterministic simulation) or a random search phase.

In order to approximate the numerical solution of model (4.9.2), first a penalty function transformation is applied (see, for example, Fiacco and McCormick, 1968; or Fletcher, 1983; recall also Chapter 3.3). Following this approach, the 'complicated' constraints f_i, $i = 1, \ldots, I$, are included into an augmented objective function which has the form

$$f(x, c) = c_0 f_0(x) + \sum_{i=1}^{I} c_i \max\{f_i(x), 0\}. \tag{4.9.3}$$

In the definition of the composite objective function $f(x, c)$, $c_i > 0$ are suitably chosen penalty multipliers, and $c = (c_0, \ldots, c_I)$ is the multiplier vector. Introducing the augmented objective function (4.9.3), (4.9.2) is replaced by the parameterized standard Lipschitzian global optimization problem

$$\begin{aligned} &\min f(x, c). \\ &a \leq x \leq b. \end{aligned} \tag{4.9.4}$$

Apparently, the function f in (4.9.3)–(4.9.4) attempts to balance the aims of reducing the primary objective function value ('minimize total wastewater treatment system costs'), and of satisfying the implicit constraints ('meet environmental quality requirements'). The financial and environmental objectives are, of course, not directly commensurable; however, the greater the multiplier c_i is, the greater importance is associated to f_i in the composite objective function f. Thus, adaptively changing the multiplier vector c, the (approximate) feasibility and optimality of the sequentially generated parametric solutions $x^\star(c)$ can be enforced.

In accordance with the discussion in Chapter 3.3—as an evident consequence of the postulated Lipschitz property of functions f_i, $i = 0, 1, \ldots, I$,—function f is also Lipschitzian. Therefore one can directly apply an adaptive Lipschitzian partition strategy to globally minimize f on the n-dimensional interval region $[a, b]$. Naturally, this global optimization algorithm is, generally speaking, infinite in the theoretical sense. Therefore, instead of strict mathematical optimality, only an approximate solution of model (4.9.2) can be expected, within a finite amount of computational time. This comment has obvious relevance with respect to the particular application in mind, as only a user-dependent fraction of the total computational effort can be spent by the optimizer module of ESIS.

Following, in essence, the implementation discussed in Chapter 3.4, the built-in LGO procedure includes a combination of standard local (convex) and global optimization techniques which can be activated by the user, in interactive or fully automatic modes.

Statistical Uncertainty Analysis and Stochastic Model Extensions

As already discussed, in environmental management problems inherent statistical fluctuations, uncertainties, unforeseeable or unknown factors most often play an important role. In the present case, the description of UTP treatment efficiencies with respect to selected water quality constituents is approximated on the basis of imprecise knowledge, scarce and 'noisy' data: therefore, the empirical regression (sub)models fitted to these data naturally contain random noise. Further uncertainties and fluctuations can be associated with other important system characteristics, such as, for example, the varying influent wastewater load, or the (site-specific, not exactly known) treatment system model parameters. In such cases, it is reasonable to explicitly consider the most relevant stochasticities of the decision problem.

Along the lines of Chapters 3.6 and 3.7, a correct and frequently applied method of stochastic systems modelling and optimization would be to proceed with the search algorithm on a sequence of decision options, while evaluating these alternatives with gradually increasing accuracy. Because of the numerically intensive character of this solution approach, there is no possibility of activating such an 'exact' stochastic optimization algorithm in the real-time ESIS environment. Therefore, on the basis of off-line statistical analysis and corresponding model development, stochastic simulation and/or approximation based techniques can be applied to evaluate preselected or (deterministically) optimized 'good' decision alternatives. Illustrative numerical results with respect to applying this methodology are reported in Pintér and Lycon (1992b).

The solution approaches listed above are capable of providing responses pertinent to a variety of problem statements. The information produced by the solver modules is communicated to the user by a dialogue-based interface, which enables the on-line analysis of results and the adaptive, sequential application of the various solution techniques.

4.9.6 Implementation Aspects

The combination of operations research, artificial intelligence, data management, and visualization concepts and techniques has been realized in ESIS following a modular approach, see Fig. 4.9.3. This generic structure makes possible the adaptive decomposition, substitution, and integration of the various software 'units' that can serve as modules of ESIS.

The principal software components of the prototype DSS were as follows (note that since the completion of the ESIS Project, a number of modifications were implemented; these are not fully reflected by the list below):

- Operating system (Microsoft Windows 3.1 and DOS 6.0)
- Wastewater engineering system module (Microsoft Fortran)
- Spreadsheet module (Microsoft Excel and Lotus 1-2-3 for Windows)
- Optimization module (Microsoft Fortran)
- Stochastic simulation, statistical analysis module (Microsoft Fortran)
- Artificial intelligence/expert system shell (Level 5 Object)
- Help modules (Borland Help Compiler, Microsoft Windows Software Development Kit, Microsoft Visual Basic)
- Graphic modules (CorelDraw, Harvard Graphics)
- Presentation and demonstration modules (Authorware, Level 5 Object).

The prototype implementation of ESIS runs on personal computers (IBM PS/2, Intel 386&387DX or 486DX processor based machines, or compati-

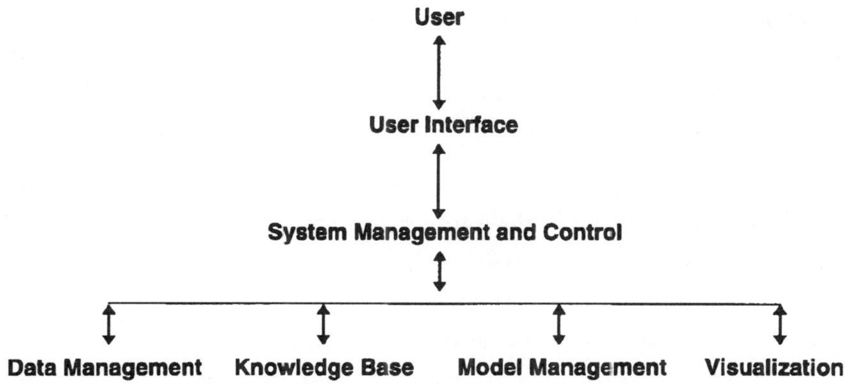

Figure 4.9.3
Decision support system: Principal modules

bles). A workstation development has been proposed for a subsequent stage. Note that, as a consequence of the modular structure, significant portions of the software system developed are directly portable or easily transferable among different hardware platforms, and can be easily adjusted to future application modes.

4.9.7 Illustrative Numerical Results and Discussion

In this section the problem of optimizing a wastewater treatment system is analysed and solved, applying the random search (RS) and Lipschitzian global optimization (LGO) approaches. For the purposes of this illustrative example, Treatment Plan 3 is chosen (recall Fig. 4.9.2); further details and numerical examples are discussed by Pintér (1993).

Selection of Principal Decision Values

The number of decision variables in a detailed wastewater treatment system problem can easily be several tens or even higher. However, to accomplish a thorough global search in higher dimensional spaces—in real time—could be computationally prohibitive. Therefore—as a first step of this illustrative analysis—the most significant decision variables have been selected, on the basis of the Fels and Lycon (1992) report and a subsequent, more detailed, off-line sensitivity study of Lycon (1992). Recalling the component UTPs of Treatment Plant 3: BAS, PRC, SPB, EQB | ANR, AST | SLT1, SLT2, SLM, SLD (Fig. 4.9.2), for the purposes of this example, the following set of principal decision variables is chosen:

- A1 – Primary clarifier base area [m^2]
  (or, alternatively, clarifier overflow rate OR1 [$m^3/(m^2 day)$])
- TO12 – Primary clarifier underflow solids concentration [kg/m^3]
- L9 – Anaerobic reactor loading [kg COD/(m^3day)]
- T9 – Temperature in anaerobic reactor [°C]
- T – Temperature in aeration tank [°C]
- RT41 – Residence time in aeration tank [hour]
- X41 – Biomass concentration in aeration tank [kg/m^3]
- OR4 – Overflow rate of secondary clarifier [$m^3/(m^2 day)$].

The above notation follows that of the Fels and Lycon (1992) report; in the terminology of models (4.9.2)–(4.9.4), the listed variables are the components of vector x. Fixed nominal values are considered for all other design and operational characteristics which are not listed here.

Having reduced the number of principal decision variables (in the present case, to eight), one can expect that the best solution of the derived—possibly multiextremal, 'black box' type—optimization problem (4.9.4) can be reasonably well approximated in the real-time environment of the ESIS prototype.

Pure Random Search Optimization

For running the executable optimizer program, the following basic input information has to be provided (or verified) by the user:

- daily pulp production and raw wastewater load characteristics
- environmental quality criteria (stated effluent standards)
- basic economic indicators (cost factors, total life cycle, discount rate)
- nominal values of fixed design and operational characteristics of wastewater treatment plant, and a number of additional UTP model parameters
- ranges of decision variables.

This input file system information is read by the optimizer, which subsequently writes results to a detailed and a summary output file. The first (optionally generated) output file lists all sequentially produced alternative design options. The second file summarizes the influent and effluent characteristics, the best configuration, and the total system costs, corresponding to the best solution found by random search.

In the present numerical example, 1,000 randomly generated designs are considered: the computational time is approximately 9 seconds (on a 33MHz Intel 386&387DX processor based PC). The most important attributes of the input information, and of the best solution found, are summarized below:

Results of Random Search Optimization Phase

Pulp Production Rate: 900 ADMTD (Air-Dry Metric Ton per Day)

Raw Wastewater Load
- Flow rate = 50000 m^3/day
- TSS = 0.8 kg/m^3 (44.44 kg/ADMT)
- BOD = 1.5 kg/m^3 (83.33 kg/ADMT)
- TOX = 1%

Optimized Decision Variables (rounded values)
- A1 = 3073 m^2 (OR1 = 17.3 $m^3/(m^2 \text{day})$)
- TO12 = 17.63 kg/m^3
- L9 = 7.50 kg COD/(m^3day)
- T9 = 48.8°C
- T = 27.9°C
- RT41 = 11.6 hour
- X41 = 3.69 kg/m^3
- OR4 = 18.8 $m^3/(m^2 \text{day})$

Resulting Effluent Load (rounded values)
- Flow rate = 49630 m^3/day
- TSS = 9.7 kg/ADMT vs. 11.25 kg/ADMT
- BOD = 0.7 kg/ADMT vs. 7.5 kg/ADMT
- TOX = 100% vs. 100 [%] .

Note that the TSS and BOD effluent load values are expressed in relation to the daily production rate, following current Canadian legislative standards.

As can be seen, the best configuration found in 1,000 random search steps meets the effluent standards. (In general, this may not necessarily be the case, but the influent wastewater load is apparently well within the applicability range of Treatment Plant 3.) The financial implications of the incumbent decision variant are summarized as follows:

Treatment Plant Costs (rounded values)

- Equipment costs = 17.258 \$M (million Canadian dollars)
- Yearly operational costs = 2.086 \$M
- Total discounted costs = 32.007 \$M (planning horizon: 10 yrs., discount rate 8%).

In order to improve the cost-efficiency of the design, an automatic LGO search phase was subsequently applied.

Lipschitzian Global Optimization

Although this method can also be used in a 'stand-alone' mode, for illustration it is applied here to improve the results obtained by preliminary random search (RS). For running the executable optimizer program, input information analogous to that in the RS phase has to be provided. Additionally, LGO has a number of procedure parameters which can be set (default values are provided, for the user's convenience). In the present example, the number of alternative decisions evaluated in the (real-time) LGO search phase is set to 5,000. The corresponding runtime, including the local search phase of LGO, is approximately 28 seconds.

The LGO module writes results to a detailed and a summary output file. The first output file can be generated in three versions, depending on the level of detail required by the user. The essential information of the summary output file is shown below, without repeating the daily pulp production rate and raw wastewater load information (identical to the input of RS).

Results of Lipschitz Optimization Phase

Optimized Decision Variables (rounded values)

- A1 = 3118 m^2 (ORI = 17.1 $m^3/(m^2 day)$)
- TO12 = 12.85 kg/m^3
- L9 = 6.80 kg COD/(m^3 day)

- T9 = 48.5°C
- T = 12.9°C
- RT41 = 5.2 hour
- X41 = 2.01 kg/m^3
- OR4 = 23.5 m^3/m^{2}day

Resulting Effluent Load (rounded values)
- Flow rate = 48940 m^3/day
- TSS = 10.19 kg/ADMT vs. 11.25 kg/ADMT
- BOD = 2.67 kg/ADMT vs. 7.5 kg/ADMT
- TOX = 100% vs. 100 [%]

Treatment Plant Costs (rounded values)
- Equipment costs = 14.711 $M
- Yearly operational costs = 1.811 $M
- Total discounted costs = 27.513 $M.

One can observe that some of the optimized parameters have changed quite significantly, in comparison with the result of the RS phase. These modifications led to somewhat higher effluent load—the overall design became more 'tight'—but the treated wastewater still meets the stated environmental protection criteria. At the same time, the estimated total costs of the best configuration found are less by $4.5 million (!) than those of the preliminary design in the RS phase.

Closing the discussion of this example, we emphasize again the illustrative character of the results quoted: the total costs could probably be somewhat further reduced, while satisfaction of the effluent standards could still be maintained. This 'fine tuning' can be done interactively, for example, by changing the parameterization of the optimization procedure. Note also that, with a view towards further analysis of selected configurations, all essential components of the problem statement—daily production rate, raw influent load, decision variable ranges, effluent standards and associated penalty multipliers, cost factors, UTP model descriptions, treatment system configurations, and so forth—can also be varied. Specifically, changing the input parameter files on-line, a sequence of deterministic simulation and optimization runs can be accomplished in a dialogue mode. More substantial modifications, however, necessitate a corresponding software development effort.

Concluding the discussion, one may state that an increasing need for technically sound, economically efficient, and environmentally responsible waste management indicates the potentials of the research work summarized in this chapter. Existing and future application prospects of the DSS include the following:

- site-specific implementations
- refinement of environmental impact analysis
- applications in a regional context, in the presence of multiple pollution sources
- applications to other industries (mining, metallurgy, food processing, etc.) and municipalities.

Chapter 4.10

Multiple Source River Pollution Management

4.10.1 Introduction

The issue of establishing efficient regional water quality management strategies has been the subject of investigations for several decades: consult, for example, Deininger (1965, 1972, 1977), Howe (1971), Dorfman, Jacoby and Thomas (1972), Marks (1975), Bower (1977), Loucks, Stedinger and Haith (1981), Haith (1982), Beck (1985a, 1987a,b), Thomann and Mueller (1987), Loucks and da Costa (1991), Patry and Chapman (1991), Tebbutt (1992), Kularathna and Somlyódy (1994). In spite of the inherent uncertainties related to this problem, mostly deterministic system models have been analysed. Recognizing, however, the significance of partially unknown, uncertain, or statistically fluctuating processes in the environment, and making use of developments in modelling and numerical solution approaches, an increasing number of stochastic models have been introduced in recent years. Beck (1987a) provides a review of descriptive water quality studies, briefly referring also to the issue of decision making under uncertainty. Among the more recent studies in stochastic (river, lake, or subsurface) water quality modelling and management, one may refer, for example, to Brill, Eheart and Liebman (1979), Lohani and Thanh (1979), Padgett and Papadopoulos (1979), Beck and van Straten (1983), Burn and McBean (1985), Fujiwara, Gnanendran and Ohgaki (1986, 1987), Pintér and Somlyódy (1986), Somlyódy and van Straten (1986), Ellis (1987), Kutas and Herodek (1987),

Massmann and Freeze (1987), Wagner and Gorelick (1987), Chen and Papadopoulos (1988), Eheart (1988), Somlyódy and Wets (1988), Tung and Hathhorn (1988), Gorelick (1990), De Lange and Pintér (1991a,b).

Following the detailed analysis and optimization of a single industrial wastewater treatment plant, in this chapter a stochastic framework is presented for exploring cost-effective river basin water quality management strategies in the presence of multiple pollution sources. The approach is based on a stochastic extension of the classical biochemical oxygen demand–dissolved oxygen (BOD-DO) model which leads to a joint chance constraint on water quality. For solving the problem derived, global optimization is combined with Monte Carlo simulation. The methodology is illustrated by numerical examples; additional details can be found in Boon, Pintér and Somlyódy (1989a,b).

4.10.2 Modelling

The model-system applied here for studying river water quality management strategies consists of two main parts, called descriptive and optimization models. First, deterministic model variants will be introduced; this is followed by their stochastic extension.

Descriptive Model

Water quality models are based on the physical and biochemical mass balance of the water system considered. We shall apply here the classical model of Streeter and Phelps (1925) (discussed, e.g., by Dobbins (1964), Orlob (1983), Jolánkai (1992), or Somlyódy and Varis (1992). This model in its simplest form approximates the evolution of dissolved oxygen deficit (D, mg/l) and BOD concentration (L, mg/l) on a river section as a function of their initial values $D(0)$ and $L(0)$, time (t, day), reaeration, oxidation, and decomposition coefficients (k_1, k_2 and k_3, day^{-1}). Assuming steady-state conditions, the coupled differential equations

$$\frac{dD}{dt} = -k_1 D + k_2 L, \qquad (4.10.1)$$
$$\frac{dL}{dt} = -k_3 L$$

lead to the analytical solution

$$D(t) = \frac{k_2}{k_1 - k_2} L(0)[\exp(-k_2 t) - \exp(-k_1 t)] + D(0) \exp(-k_1 t),$$
$$L(t) = L(0)\ \exp(-k_3 t). \tag{4.10.2}$$

Introducing streamflow velocity u (m/s) and longitudinal distance from the beginning of the river section $y = ut$ (m), one can obtain by solving $dD(y)/dy = 0$ for y that the maximal oxygen deficit occurs at location

$$y^c = \frac{u \ln A}{k_2(f-1)}, \tag{4.10.3}$$

where

$$f = k_1/k_2, \quad A = f[1 - (f-1)D(0)/L(0)].$$

Furthermore, the critical oxygen deficit value at location y^c equals

$$D^c = \frac{k_2}{k_1 - k_2} L(0)\{A^{-1/(f-1)} - A^{-B}\} + D(0)A^{-B}, \tag{4.10.4}$$

where

$$B = f(f-1).$$

Obviously, the values y^c and D^c are of primary importance in the context of water quality management, as they characterize the least favourable location and in-stream water quality along a river section (in terms of dissolved oxygen deficit).

As is well known in environmental engineering, the derivation of equations (4.10.1)–(4.10.4) is based on a number of strong simplifying assumptions. This point will be addressed, to a certain extent, by introducing stochastic model extensions. Note also that more sophisticated basic descriptive model versions are also known: consult Orlob (1983), Barnwell and Brown (1987), Jolánkai (1992), or Somlyódy and Varis (1992). Assuming that such extended models can be appropriately supported by site-specific data, the methodology presented here is applicable, mutatis mutandis.

Management Model

Given the analytical solutions (4.10.3)–(4.10.4) and the DO saturation level D^s, the maximal oxygen deficit uniquely corresponds to the minimal DO level, $D^{\min}$:

$$D^{\min} = D^s - D^c.$$

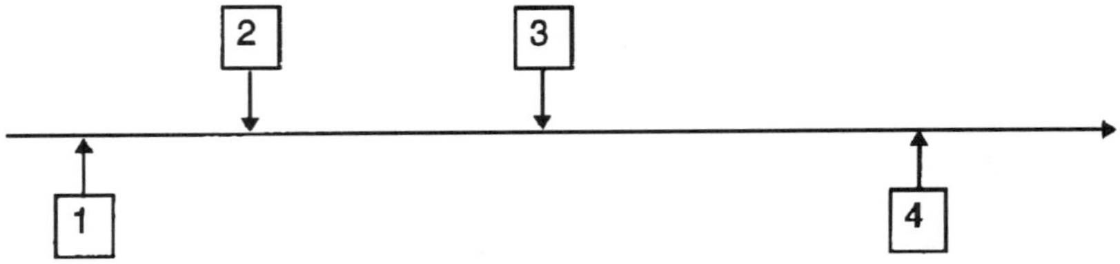

Figure 4.10.1
Multiple point-source pollution sites along a river section

Assume now that we investigate a river section with multiple point source polluters (effluent discharges) located sequentially: for illustration, see such a scheme in Fig. 4.10.1. Assigning minimum DO water quality standards to all river sections of interest—each, by supposition, having a single significant point-source pollution discharge—the effects of site-specific wastewater treatment can be directly traced. Let us introduce index $i = 1, \ldots, I$ for subsequent river sections and, simultaneously, for the river water quality monitoring sites. Denote by x_i the treatment efficiency (BOD load removal rate) to be chosen at pollution discharge site i. Further let $D_i^{\min}$, D_i^s and D_i^0 be the minimum occurring DO level, the saturation level, and the stated DO standards for river section $i = 1, \ldots, I$. Then, assuming simple linear additivity and complete mixing in the load mass-balance relations, the overall effect of wastewater treatment can be determined from the following relations:

$$\begin{aligned} L_{0i} &= [L_{ri}Q_{ri} + (1 - x_i)L_i q_i]/(Q_{ri} + q_i) && i = 1, \ldots, I, \\ D_i^c &= F_i(D_{0i}, L_{0i}, k_1, k_2) && i = 1, \ldots, I, \\ D_i^{\min} &= D_i^s - D_i^c && i = 1, \ldots, I. \end{aligned} \tag{4.10.5}$$

In (4.10.5),

D_{0i}, L_{0i} initial DO deficit and BOD values on reach i,

L_{ri}, Q_{ri} BOD concentration and streamflow in river at the beginning of reach i (note that L_{ri} depends on $L_{0i-1} := L_{0,i-1}$, as expressed by (4.10.2); $L_{00} := L_{0,0}$ is the background BOD load in the river)

L_i, q_i input BOD concentration and discharge flow at section i

D_i^c critical deficit as a function $F_i(\cdot)$ of D_{0i}, L_{0i}, k_1 and k_2 (see the expressions (4.10.3) and (4.10.4).

In this way, the first set of equations expresses mixing at the beginning of reach i (based on the pollutant load carried by the river and the effluent discharge). The second set of equations leads to determining the critical oxygen level on each river section. Finally, the last set of equations simply expresses the water quality, in terms of the minimal DO level.

Observe that D_{0i} and L_{0i}, $i = 1, \ldots, I$, can be sequentially determined by applying (4.10.2), given the location of all wastewater discharge sites, and accepting a deterministic description. Under these assumptions, the water quality implicitly depends on the decision variables x_i. Assuming now that, at each wastewater treatment discharge site, a corresponding total cost value $C_i(x_i)$ can be associated, a basic water quality management model can be written in a summary form as

$$\text{minimize} \sum_{i=1}^{I} C_i(x_i) \tag{4.10.6}$$

$$D_i^{\min}(x) \geq D_i^0, \qquad i = 1, \ldots, I,$$

$$x = (x_1, \ldots, x_I), \quad x_{i,\min} \leq x_i \leq x_{i,\max}, \quad i = 1, \ldots, I.$$

(The last row represents explicit technological bounds.) Note that, frequently, the cost functions C_i are convex, thus allowing piecewise linearization. Based on this approximation and simpler, linearized constraint sets, many water quality model variants have been traditionally solved via linear programming, see, e.g., Dorfman, Jacoby and Thomas (1972), Loucks, Stedinger and Haith (1981), and Haith (1982). By contrast, model (4.10.1)–(4.10.6) is genuinely nonlinear (in x_i, $i = 1, \ldots, I$).

Observe additionally that (4.10.1)–(4.10.6) can be interpreted as a basic deterministic version of more realistic problem formulations. Specifically, streamflow, temperature, and related physical, biochemical processes are all subject to fluctuations; therefore, at least the implicit constraint set in (4.10.5) should be replaced by some appropriate stochastic model component. Note also at this point that the modelling of the individual wastewater treatment plants can be done in more sophisticated ways, following, e.g., a similar approach to that of the previous chapter. This would lead to possibly complex 'black box' functions (in more variables) replacing both the objective function and the constraint set in (4.10.6).

Stochastic Model Extension

All statistically varying model components, referred to above, can—at least in principle—be modelled by introducing a corresponding random variable (r.v.), with an associated probability distribution function (p.d.f.). Typically —and also in the context of this study—the p.d.f.'s are then estimated from actual observations and using data from the literature. Furthermore, from the available modelling tools of stochastic programming (recall the brief review provided in Chapter 3.5), we choose the well-known chance constrained model. More precisely, the following probabilistic constraint will be introduced:

{the joint probability of satisfying given minimum DO level constraints, for all river sections, is 'sufficiently high'}. (4.10.7)

In (4.10.7) the minimum DO level can be analytically determined, assuming that given is a joint random realization of the river streamflow, temperature and load conditions (initial discharge loads, and treatment efficiencies). In this way, via Monte Carlo simulation, statistical performance estimates can be produced for each decision variant considered. We emphasize here that, although (mostly for numerical reasons) separate chance constraints are often used, in the present context the use of a joint water quality constraint is more realistic. This—together with the implicit nonlinearities of the underlying model (4.10.1)–(4.10.6)—implies a significant increase of model complexity and necessitates the application of suitable global optimization techniques.

4.10.3 Solution Method

The stochastic optimization model, including the requirement (4.10.7), can be formulated as

$$\text{minimize} \sum_{i=1}^{I} C_i(x_i) + w[\alpha - P(\cdot)]_+ \tag{4.10.8}$$

$$P(D_i^{\min}(x) \geq D_i^0,\ i = 1, \ldots, I) \geq \alpha$$

$$x = (x_1, \ldots, x_I), \quad x_{i,\min} \leq x_i \leq x_{i,\max}, \quad i = 1, \ldots, I.$$

The notation $D_i^{\min}$ and D_i^0, $i = 1, \ldots, I$, was introduced earlier; additionally, the symbol $P(\cdot)$ refers to the probability of the embedded random event $(\cdot)$ in the second row of model (4.10.8); $0 < \alpha < 1$ is a prescribed reliability level; $w > 0$ is a given multiplier for 'penalizing' unfavourable deviations from the reliability level prescribed; and $[h]_+ = \max(h, 0)$.

As can be seen—in spite of the relatively simple model setup—the stochastic optimization model (4.10.8) can be expected to have a fairly complicated structure. Although the primary objective function term $\sum_i C_i(x_i)$ often may be approximated linearly, and the treatment efficiency bounds define a simple rectangular region, the probabilistic constraint in (4.10.8) is highly nonlinear and its analytical form is, in general, unknown. (Recall that the probabilistic constraint involves the numerical evaluation of $D_i^{\min}$ by following the calculations (4.10.2)–(4.10.4), for each wastewater treatment system studied, and—within that context—for each random realization of the modelled water quality description.) For this reason, Lipschitzian optimization is applied to search for the globally best wastewater treatment configuration.

Note that, for streamlining the optimization procedure, the interval feasible region can be reduced on the basis of the following observation: if, for each plant, the maximal possible efficiency were taken at all previous plants, then this obviously yields a lower bound of the removal rate to be selected there. This simple observation may lead to more or less significant reductions of the feasible region.

In the numerical examples four treatment plants were to be chosen: experience has shown that the best solution could be approximated in a few hundred iterations (involving the sequential selection of wastewater treatment configurations and their statistical performance evaluation by Monte Carlo simulation). The results were finally checked via ex post performance analysis, in the course of which a more detailed stochastic simulation study was accomplished, with respect to the best solution found.

4.10.4 Illustrative Results and Discussion

We shall consider a set of numerical examples related to water quality management of a small river in Hungary. River Zala is a major pollution load source to Lake Balaton (discussed in the subsequent chapter). Observation data concerning river conditions were collected weekly between 1975 and 1986. A tentative wastewater treatment plant/discharge configuration (recall Fig. 4.10.1) will be analysed, to illustrate the approach presented above.

Stochastic Model Input

- streamflow (m^3/s): modelled by a lognormal distribution (mean value = 7.63; standard deviation = 5.45)
- water temperature (C°): seasonality and streamflow-dependent, its approximate p.d.f. is derived applying (off-line) regression analysis
- river DO: water temperature-dependent; again, its p.d.f. is derived from regression analysis
- stream velocity, river depth (m), reaction parameters k_1, k_2, k_3, and saturation DO level: all of them streamflow- and/or temperature-dependent, approximated by hydrological and empirical equations (for details, see Boon, Pintér and Somlyódy, 1989a).

Deterministic Model Parameters and Background Information

- joint probability bound of meeting the critical DO deficit standards for all river sections: varied systematically between 0.50 and 0.98
- initial conditions, based on the following (illustrative) set of municipal wastewater discharges:

Table 4.10.1
Input data in numerical examples

river section	BOD-load (kg/d) (untreated wastewater)	wastewater discharge flow (m^3/s)	DO (mg/l) standard	removal efficiency initial bounds
1	15000	0.434	4.0	[0, 0.97]
2	15000	0.434	6.0	[0, 0.98]
3	6000	0.170	6.0	[0, 0.97]
4	21000	0.608	6.0	[0, 0.97]

- wastewater treatment costs: piecewise linear approximation is applied (corresponding to treatment phases of 30, 85, 95, 98 and 99 per cent removal rate); in the numerical examples only investment (construction) costs are considered.

A sequence of stochastic optimization runs to solve model (4.10.8) was performed, systematically increasing the probability bound in meeting the joint chance constraint. For comparison, corresponding deterministic computations were also completed: in these runs, that streamflow value was taken (as main hydrologic input) which is exceeded with the probability given in the joint chance constraint of the corresponding stochastic model version. In each deterministic run, the streamflow value referred to is used in combination with the other corresponding (deterministic) model input parameters.

For illustration, see Table 4.10.2, which summarizes the solution of a single example run in some detail. Note that costs are given in millions of Hungarian forints (Ft; at the time of the study, the exchange rate 1 Ft $\approx$ 2 U.S. cents was applied). Hence, the estimated total construction costs of the four wastewater treatment plants are over 40 million US dollars. (For comparison, recall the numerical example in Chapter 4.8.)

Table 4.10.2
Comparison of optimal configurations found in (adjusted) deterministic and stochastic model versions

	Deterministic Model Results probability of flow rate exceedance: 0.9			Stochastic Model Results reliability level: 0.9		
river section	lower bound vector x	optimal vector $x^\star$	total costs	lower bound vector x	optimal vector $x^\star$	total costs
1	0.623	0.846		0.620	0.850	
2	0.639	0.799	2356	0.684	0.777	2356
3	0.127	0.689		0.105	0.656	
4	0.717	0.830		0.711	0.837	

Next, in Table 4.10.3, we present some more results in summarized form. These correspond to solving model (4.10.8), for different settings of the reliability level α in the joint chance constraint.

Table 4.10.3
Comparison of (adjusted) deterministic and stochastic model versions

Deterministic Model Results		Stochastic Model Results	
probability of flow rate exceedance	total costs	joint probability bound	total costs
0.500	2164	0.500	2231
0.700	2287	0.700	2260
0.800	2301	0.800	2302
0.900	2356	0.900	2356
0.950	2418	0.950	2504
0.970	2483	0.970	2625
0.975	2512	0.975	2701
0.980	2516	0.977*	4342**

*the maximum reliability level which can be attained
**the corresponding total cost

From Table 4.10.3 one can see that, in this particular numerical example, the deterministic results approximate very well the stochastic ones up to a certain threshold (probability) level; then the results tend to deviate in a rapidly increasing manner. In the present example, this phenomenon seems to be the consequence of the interdependence structure of randomly varying streamflow rates, water temperature, and background DO level. It is expected that the difference between deterministic and stochastic solutions increases if the dependence becomes stronger. (A comprehensive sensitivity study—including also this aspect—was reported in Boon, Pintér and Somlyódy, 1989a.)

Summing up the results, one may conjecture that, on the one hand, easier and faster obtained deterministic results may provide realistic information under problem-specific circumstances. On the other hand, assuming that proper, statistically sound background information is available, the stochastic modelling approach yields a more detailed and realistic description. In several cases—that typically include the most important ones, from a management point of view—this will lead to significantly different results from the deterministic ones. The above finding is in accordance with those in most of the (earlier) references on stochastic water quality modelling.

Apparently, the approach presented in this chapter can be developed further in several directions. Practically important generalizations include, inter alia, the following:

- consideration of more refined descriptive (hydrometeorological, hydraulic, biochemical, etc.) submodels
- corresponding generalization of the base management model(s)
- formulation, solution, and comparison of different stochastic model variants, applying global optimization.

For additional details on these issues, the reader is referred again to Boon, Pintér and Somlyódy (1989a), and also to the related discussion in Chapters 3.5 to 3.7.

Chapter 4.11

Lake Eutrophication Management

4.11.1 Introduction

As already discussed in previous chapters, due to the worldwide danger of environmental degradation, water quality management has attracted increasing attention in the last decades. A large number of descriptive (simulation, assessment) and management (scenario analysis, optimization) models have been developed and applied to assist decision-making. The interrelated (sub)models frequently have very different temporal and/or spatial scaling and underlying information structure, so that their proper combination poses a significant challenge. Moreover, even recognizing the inherent stochasticity of the management problems studied—and analysing them in detail by descriptive submodels—most optimization models are formulated as deterministic: this transition requires a carefully balanced approach.

In this chapter a stochastic programming model is presented for studying a lake eutrophication management problem. (More details related to this problem can be found, for example, in Somlyódy and Pintér, 1984; Somlyódy and van Straten, 1986; Somlyódy and Wets, 1988.) The methodological findings of this study are of direct relevance to many other complicated stochastic optimization problems, such as, for example, reservoir system design and operation, (stochastic) facility location and operation, or multi-item dynamic production and inventory control (see the earlier quoted references on these subjects).

4.11.2 Lake Eutrophication: Management Alternatives

Man-made eutrophication in lakes is considered to be one of the major water quality problems worldwide. It is caused primarily by increasing municipal and industrial wastewater discharges and the intensive use of chemicals in agriculture, which frequently lead to a nutrient 'overdose' in the recipient water body. The typical symptoms of eutrophication—algal blooms, water colouration, floating water plants, organic debris, unpleasant taste and odour—frequently lead to serious limitations on (recreational or other) water use.

Both the main causes of artificial eutrophication (increasing amounts of nutrient load reaching the lake) and the possible water quality control alternatives physically originate from and are connected to the region surrounding the lake. Their impact on lake water quality is the result of interdependent—physical, chemical, and biological—in-lake processes. Thus, eutrophication management requires an integrated analysis of the watershed, including all relevant natural phenomena/processes and human activities.

In the study summarized in this chapter the region of Lake Balaton (Hungary) will be analysed. The most important tributaries and water quality management options are shown in Fig. 4.11.1. For modelling reasons, the lake (total area, 596 km^2, watershed area, 5,776 km^2) is subdivided into four basins: the water quality in these basins shows marked differences, with a well-defined longitudinal profile. (Let us note that the pollution control example, discussed in Chapter 4.10, was related to Zala: this river flows from the north into Basin 1 of the lake, via the larger pre-reservoir.)

Because of the obvious stochasticity of the problem studied—driven primarily by varying hydrological and meteorological factors—even to a deterministic management option (for example, a fixed configuration of wastewater treatment plants), the lake response will be stochastic. This fact can be illustrated, for example, by Fig. 4.11.2, in which empirical water quality density functions (each based on 1,000 Monte Carlo simulations) are displayed, corresponding to a 'cheaper' and a 'more expensive' water quality management option. In this figure, the resulting water quality of each sub-basin is characterized by the value of the annual peak chlorophyll concentration $(\text{Chl-a})_{\max}$, $[mg/m^3]$, a frequently used eutrophication indicator.

The relevance of random factors, influencing the impact of decision alternatives, implies that eutrophication management problems should be nat-

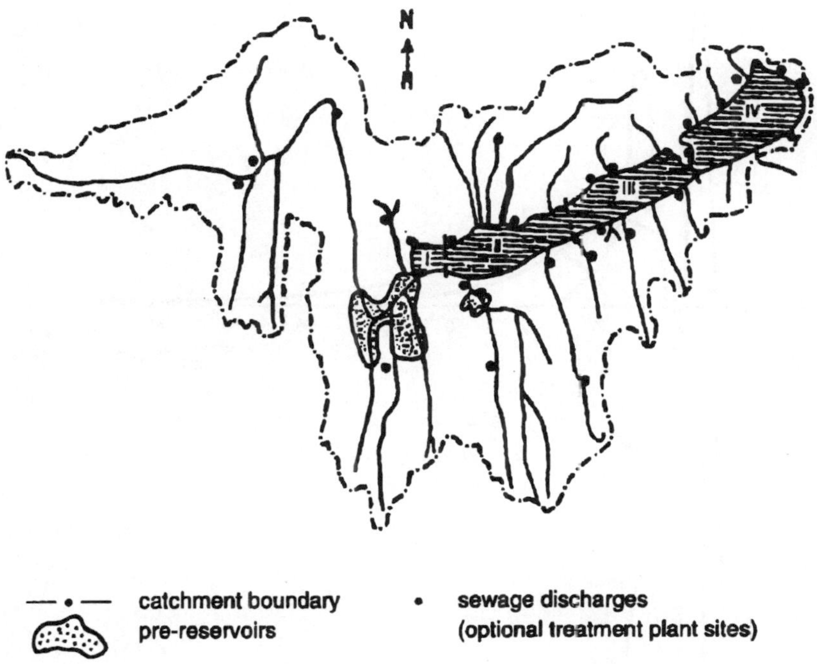

Figure 4.11.1
Lake Balaton (Hungary): Water quality management options

urally formulated applying stochastic models. Before discussing this point, in the next section we shall consider briefly the submodels and research methodology applied.

4.11.3 Decomposition and Aggregation

Descriptive and optimization models often have a different information structure: while the first model-type is, as a rule, applicable to analyse in detail any single scenario (decision alternative), the second model-type is capable of evaluating sequentially a set of alternatives, based on their aggregated characteristics. The coordinated application of these submodels necessitates the consideration of related theoretical and technical issues, see the exposition in Chapters 3.5–3.7, and the ecological modelling literature referred to earlier in Chapters 4.5–4.7.

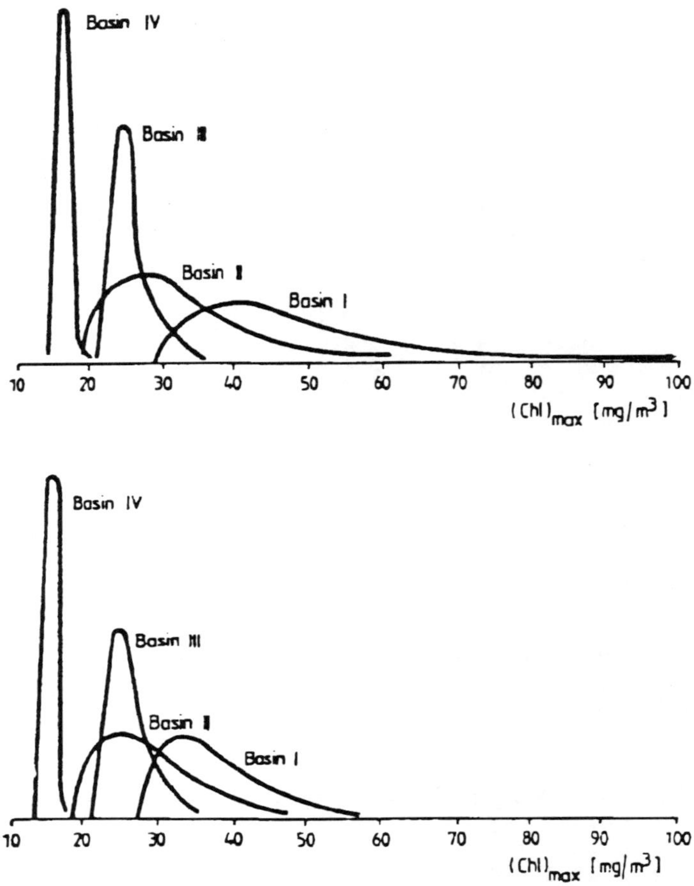

Figure 4.11.2

Density functions of (Chl-a)$_{max}$ realizations resulting from two different water quality management configurations

An appropriate system modelling approach is based on the principle of decomposition and aggregation: for a detailed discussion, see Somlyódy (1982b), or Somlyódy and Varis (1992). In the first phase of applying this methodology, a complex system is decomposed into 'homogeneous' parts which can be investigated in detail: these parts (modules) form a hierarchy which has a tree structure. In the second phase—at each hierarchical level—the information is sequentially aggregated for further analysis, while neglecting details, unnecessary at the present level of hierarchy. The de-

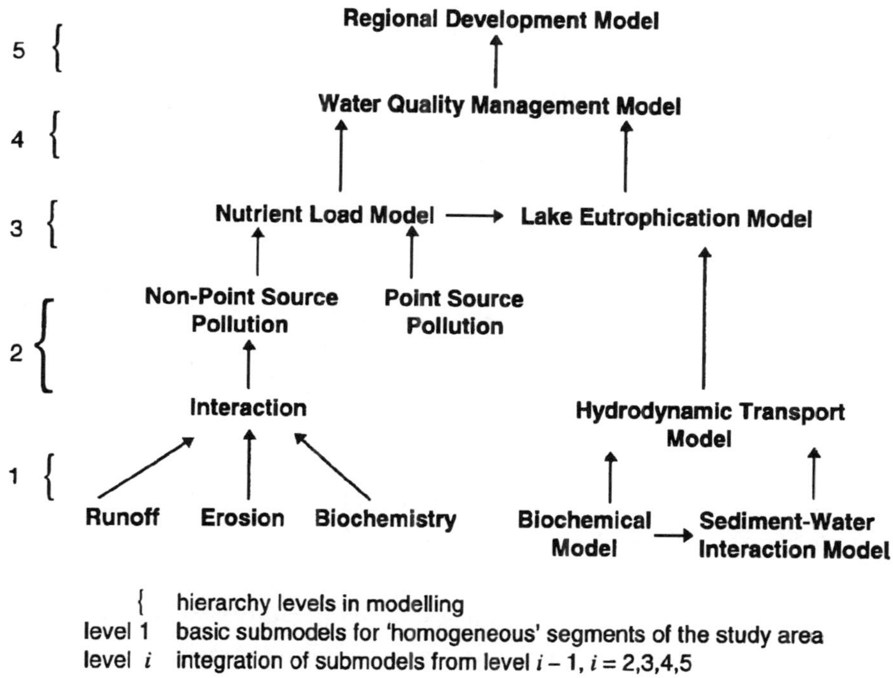

Figure 4.11.3
Decomposition and aggregation in lake water quality management: Submodels and their interrelations

composition/aggregation procedure is obviously of iterative character, with adaptive feedback possibilities. As a result of this modelling approach—instead of constructing a huge, hardly tractable model—a coherent system of detailed and aggregated submodels is obtained.

The application of this method to shallow lake eutrophication management is illustrated by Fig. 4.11.3, which shows the principal submodels of the Lake Balaton water quality management study. On the top of this hierarchical structure (root level) is located a regional development model, which summarizes the most important features of the man-made activities and natural processes considered. The lower levels provide additional details as indicated by the figure. With respect to details of these submodels, consult Somlyódy and Pintér (1984), or Somlyódy and van Straten (1986). Note also at this point the possible connection to this scheme of several submodels (describing the fate of micropollutants, or sediment-water interaction) discussed in Chapters 4.6 and 4.7, and of the wastewater treatment models presented in Chapters 4.9 and 4.10.

4.11.4 A Stochastic Modelling Framework

The following notation is introduced (on the basis of Pintér and Somlyódy, 1986):

$x \in \mathbb{R}^n$	feasible decision alternatives (water quality management options), represented as vectors of the n-dimensional real Euclidean space
$\omega \in \Omega$	elementary random events and their set which are considered in the stochastic lake eutrophication management model; the realizations $y = y(\omega) \in \mathbb{R}^\ell$ of the random factors are ℓ-vectors
$(\Omega, \mathcal{A}, P)$	the underlying probability space
$C^{(i)}(x, \omega)$	(scalar random variable) water quality resulting, for a given decision x and a random outcome ω of the environmental conditions, in lake segment (sub-basin) i, $C^{(i)}(x, y) : \mathbb{R}^n \times \mathbb{R}^\ell \to \mathbb{R}^1$, $i = 1, \ldots, I$
$C(x, \omega)$	water quality vector defined by components $C^{(i)}(x, \omega)$, $i = 1, \ldots, I$
$C_g^{(i)}, C_g^{(i)}(\omega)$	(scalar) target water quality, respectively given by a deterministic target level or by a probability distribution function (p.d.f), $C_g^{(i)}(y) : \mathbb{R}^\ell \to \mathbb{R}^1$ for $i = 1, \ldots, I$
$C_g, C_g(\omega)$	the corresponding target lake water quality vector, and vector random variable
F	a real-valued functional, measuring the undesirable deviations from the target lake water quality (its concrete form will be discussed below)
$A(x) : \mathbb{R}^n \to \mathbb{R}^m$	a vector-function, expressing technological constraints related to water quality management alternatives
$b \in \mathbb{R}^m$	right-hand side constant vector of technical constraints
$K(x) : \mathbb{R}^n \to \mathbb{R}^1$	a scalar function, representing the summarized financial (resource) implications of decision x
$k \in \mathbb{R}^1$	right-hand side parameter in the resource constraint
$\underline{x}, \overline{x} \in \mathbb{R}^n$	lower and upper bounds with respect to x.

Applying the above notation, the following general model-type can be formulated for lake eutrophication management:

$$\min F\{C(x,\omega), C_g(\omega)\} \tag{4.11.1}$$

$$A(x) \leq b \tag{4.11.2}$$

$$K(x) \leq k \tag{4.11.3}$$

$$\underline{x} \leq x \leq \overline{x}. \tag{4.11.4}$$

The technical and resource constraints (4.11.2) and (4.11.3) can often be reasonably well approximated by piecewise linear, convex functions (more details on this point are provided, for example, in Somlyódy and van Straten, 1986). Consequently, (4.11.2) and (4.11.3) can be substituted by linear constraints of the form

$$Ax \leq b \tag{4.11.5}$$

$$K^T x \leq k, \tag{4.11.6}$$

with a corresponding $(m \times n)$-matrix A and n-vector K.

Evidently, the functional F in (4.11.1) can be defined in a number of ways. First of all, it is assumed that a proper weighting of the sub-basins can be expressed by some scalar factors w_i such that

$$w_i \geq 0, \quad i = 1, \ldots, I, \quad \text{and} \quad \sum_{i=1}^{I} w_i = 1.$$

This conceptually straightforward technique originates from multiobjective decision analysis considering commensurable objectives, see, for example, Hwang and Masud (1980), Sawaragi, Nakayama and Tanino (1985), Romero (1986), or Steuer (1986). Next, F is defined in the specialized form

$$F = \sum_{i=1}^{I} w_i F_i, \quad \text{where } F_i = F_i\{C^{(i)}(x,\omega), C_g^{(i)}(x,\omega)\}. \tag{4.11.7}$$

Naturally, the component functionals F_i can still be defined in many different forms. Let us emphasize that, generally speaking, these formulations are not equivalent and reflect more or less different considerations on the stochastic factors. Principles of formulating stochastic programming models with numerous examples are given, for example, in the works of Ermoliev (1976), Yudin (1979), Dempster (1980), Kall and Prékopa (1980), Wets (1983), Arkin, Shiriaev and Wets (1986), Klein Haneveld (1986), Frauendorfer (1992), Wets (1994); see also Chapter 3.5. For instance, Somlyódy (1983) considered the following objective function components:

$$F_i\{C^{(i)}(\boldsymbol{x},\omega),C_g^{(i)}\} = E(C^{(i)}(\boldsymbol{x},\omega)-C_g^{(i)})^w \tag{4.11.8}$$

and

$$\begin{aligned}F_i\{C^{(i)}(\boldsymbol{x},\omega),C_g^{(i)}\} &= E(C^{(i)}(\boldsymbol{x},\omega)-C_g^{(i)})\\ &\quad + wD(C^{(i)}(\boldsymbol{x},\omega)-C_g^{(i)}).\end{aligned} \tag{4.11.9}$$

In (4.11.9), E and D respectively denote the (existing) mathematical expectation and standard deviation of the involved random variables, and $w > 0$ is a model parameter.

In Somlyódy and Pintér (1984), objective functions with the following components were investigated:

$$F_i\{C^{(i)}(\boldsymbol{x},\omega),C_g^{(i)}\} = E(C^{(i)}(\boldsymbol{x},\omega)-C_g^{(i)})^2, \tag{4.11.10}$$

$$F_i\{C^{(i)}(\boldsymbol{x},\omega),C_g^{(i)}(\omega)\} = \int_{-\infty}^{\infty} |H_x^{(i)}(z)-H_g^{(i)}(z)|\, dH_g^{(i)}(z), \tag{4.11.11}$$

$$F_i\{C^{(i)}(\boldsymbol{x},\omega),C_g^{(i)}(\omega)\} = \int_{-\infty}^{\infty} (H_x^{(i)}(z)-H_g^{(i)}(z))^2\, dH_g^{(i)}(z), \tag{4.11.12}$$

where

$$H_x^{(i)}(z) = P(C^{(i)}(\boldsymbol{x},\omega) < z) \tag{4.11.13}$$

and

$$H_g^{(i)}(z) = P(C_g^{(i)}(\omega) < z) \tag{4.11.14}$$

are the p.d.f.'s of the resulting and target water qualities per basin (recall Fig. 4.11.1). Note that the objective function component (4.11.11) is based on Fisher–von Mises statistic, while (4.11.12) is based on Cramér–von Mises statistic: recall the related discussion in Chapter 4.3, or consult Sherman (1957), Ermakov (1975). Note that both of these statistics give a characterization of the 'overall' deviation of two p.d.f.'s in a more complete manner than, for example, the frequently applied Kolmogorov–Smirnov type statistics; besides, they have also less inherent subjectivity than, for example, the more commonly applied chi-square statistics. Note also that Somlyódy and

Wets (1988) formulated another stochastic model for lake eutrophication management which leads to a two-stage stochastic programming problem; the numerical solution of that model is described by King (1988).

4.11.5 Solution Method

It is apparent from the model variants presented above that, for an arbitrary decision alternative, the result of that particular decision depends also on a number of statistically varying—hydrological, meteorological, biochemical, ecological—factors. Therefore the solution method is based on a combination of optimization and Monte Carlo simulation: the sequentially generated decisions are evaluated with gradually increasing accuracy (for details of the theoretical background, recall again Chapters 3.6–3.7).

As all considered versions of the optimization problem (4.11.1)–(4.11.4) are in principle deterministic nonlinear (or linear) programming problems—except the numerically very relevant fact that the objective function values can only be estimated via computationally expensive simulation cycles leading to inherent inaccuracies—an efficient nonlinear programming method could be applied to solve them. Note, however, that some of the objective function forms—notably, for example, (4.11.11) and (4.11.12)—do not lend themselves easily to analytical conclusions concerning their convex or nonconvex (multiextremal) structure. Therefore—at least in a strict mathematical sense—a stochastic global optimization approach is advisable. At the time of the original study (Somlyódy and Pintér, 1984), a relatively simple multistart approach (recall Chapter 1.2) was used. That is, launching the search from a set of initial 'expert guesses' of the solution, and also from other randomly generated starting points, the nonlinear optimization package MINOS, originally developed by Murtagh and Saunders (1983, 1987), was applied.

4.11.6 Summary of Results and Conclusions

The numerical results for model variants with different objective function types (4.11.8)–(4.11.12) were discussed in detail elsewhere (Somlyódy and Pintér, 1984; Somlyódy and van Straten, 1986; Somlyódy and Wets, 1988). Therefore only a qualitative summary of the results is provided here. Independently of the special form of the objective function, the results basically follow the same tendency: if the amount of available financial resources (parameterized by k in (4.11.3)) is relatively small, then only municipal and industrial wastewater treatment plants are present in the optimal solution, and the more costly 'natural filter' pre-reservoirs are absent (recall Fig. 4.11.1). On the other hand, as the statistical variability of the resulting water quality can be substantially decreased only by these reservoirs, beyond a certain threshold value of available financial resources the reservoirs begin to play an important role: they gradually substitute the wastewater treatment plants in the optimal solution, whenever this is possible (and sufficient). The basically analogous behaviour of the various models indicates their coherence in this respect. This point deserves attention as, in particular, it shows that properly aggregated, simpler models (such as, for example, the linear programming model variant, presented in Somlyódy, 1983) can be capable of representing the main features of the genuinely nonlinear, stochastic problem studied here. The different stochastic programming versions, of course, can (and did) yield a fair amount of additional information, besides basically verifying the findings of the LP model.

Note also that—while following the general tendency mentioned—non-negligible differences among the 'optimal' solutions could also be observed, depending on the form of the objective function. For example, one of the sub-basins clearly had a dominating role in the composed objective function value (4.11.7), when using (4.11.8), (4.11.9), or (4.11.10) for F_i, while, when applying the statistical criteria (4.11.11) and (4.11.12), this feature was much less evident. This finding clearly shows the importance of proper stochastic formulations of optimization problems, given a complex underlying random structure—as in the present case.

The Lake Balaton eutrophication management study has indicated a number of challenging theoretical and numerical issues (besides its important practical findings, which led to a strategic water quality management approach implemented in the watershed). Among these, the following deserve special attention:

- complex stochastic systems are to be studied, applying interactive decision support tools, rather than 'purely' automatic optimization procedures
- the comparative evaluation of tentative model variants can effectively assist the determination of additional modelling and data requirements,

in accordance with the decomposition-aggregation principle

- efficient global optimization methodology is needed, in combination with statistical modelling and simulation techniques to provide robust, well-established solutions to the selected model variants.

Chapter 4.12

Risk Management of Accidental Water Pollution

4.12.1 Introduction

In this chapter a fairly general risk assessment and management approach is proposed for analysing and controlling accidental—as opposed to chronic, long-term—environmental pollution events. This concept is illustrated by a simplified case study, describing (hypothetical) point-source toxic pollution of a large European river (River Danube) and its estimated effect on the downstream bank-filtered well system. A numerical example indicates the viability of the suggested approach, and shows the potentials of applying advanced optimization techniques in risk management.

Managing accidental environmental pollution has become an increasingly important issue of our era; see, for example, World Health Organization (1985), Kleindorfer and Kunreuther (1987), Richardson (1988), Cooke and Vogt (1990), Chibber and Apostolakis (1992), Cooke (1992). Many environmental hazards result from chemical accidents, such as, for example, collision of ships, followed by oil spills, explosion or valve failure at a chemical storage facility, collision of transportation vehicles, etc. Typically, such events may cause significant environmental degradation, and may have serious health-related consequences; moreover, their mitigation and the correcting actions taken are, as a rule, very costly and demanding.

The review of Benedek (1988) summarizes the principal characteristics of chemical accidents occurring in recent years along the River Danube. Without a reliable early warning system and proper emergency technologies, these accidents may, inter alia, endanger operating waterworks. In addition, the planned construction of the Gabčikovo–Nagymaros barrage system on the river—which has been subject to heated debate during (and since) the time of this case study—also underlines the need for investigating safety measures to be taken (or modified, if necessary). Analogous comment would, naturally, apply to other large-scale engineering structures on the river.

Below an attempt will be made to demonstrate that the conceptual (and quantitative) approach of risk analysis may contribute to more 'traditional' environmental hazard management. (The exposition draws on Cooke and Pintér, 1989; Pintér, Benedek and Darázs, 1990; and Chapter 3.5.) This is done by introducing a general framework for modelling environmental risk, and then applying it in a simplified case study. Throughout this chapter a somewhat 'technocratic' standpoint will be taken when concentrating on the modelling and numerical solution aspects; therefore, other important (societal, political, legislative, etc.) details are, by necessity, mentioned only very briefly.

4.12.2 Risk Assessment and Management: Principles, Models, Solution Methods

Basic Notation and Model Forms

Decision problems in the sciences, engineering, and economics have been analysed applying optimization theory/mathematical programming for some five decades. In spite of the variety of existing concepts and techniques, there exist relatively few applications in assessing and managing environmental risks. This is probably due to several interconnected factors, namely, difficulties in quantitative problem description, modelling, data collection, and choosing appropriate solution methods. Below we shall highlight stochastic modelling techniques which can be applied for structuring and solving problems in environmental risk management. Related additional discussions can be found, for example, in Rowe (1977), Henley and Kumamoto (1981), Kaplan and Garrick (1981), Cohen (1984), Cooke and Pintér (1989); recall also Chapter 3.5.

For modelling purposes, consider an environmental system S with 'components' $c_1, \ldots, c_N$ (for example—in water resources—flood control reservoirs, water supply networks, wastewater treatment plants, and their physical, economic, and societal environment may form the system S in question). Assume that the failure probability $p_n(x_n)$ of component c_n can be expressed as some function of a decision vector x_n, $n = 1, \ldots, N$. It is also assumed that the system S has $k = 1, \ldots, K$ accident scenarios, with consequences $y_k = y_k(x)$ and probabilities $P_k = P_k(x) = P_k(p_1(x_1), \ldots, p_N(x_N))$. The risk function associated with this problem-type is some given function $R(x) = R(\{y_k(x), P_k(x)\}, k = 1, \ldots, K)$ which describes scenario probabilities and consequences as depending on the decision $x = \{x_1, x_2, \ldots, x_N\}$.

In view of the typical scarcity of resources available for system maintenance and accident management, reasonable decision strategies may be sought, following, for example, one of the schemes below:

$$\begin{aligned}&\{\text{minimize system operations and maintenance costs}\}\\&\{\text{constraints on 'acceptable' risks}\}\\&\{\text{technical constraints on decision variables}\}\end{aligned} \tag{4.12.1}$$

or, alternatively,

$$\begin{aligned}&\{\text{minimize risk}\}\\&\{\text{resource constraints}\}\\&\{\text{technical constraints}\}.\end{aligned} \tag{4.12.2}$$

Subsequently, we shall formalize these qualitative problem statements. First, it will be assumed that the risk–descriptor (negative utility, or loss) relationship can be given as a scalar function $U(R(x))$ of the decision vector x, and that—besides the technical, economic, etc. constraints—the total system costs can also be expressed by a function $C = C(x)$. Then a simple model form corresponding to (4.12.1) can be given as

$$\begin{aligned}&\text{minimize } C(x)\\&U(R(x)) \leq U_{\max}\\&g_m(x) \leq 0, \qquad m = 1, \ldots, M.\end{aligned} \tag{4.12.3}$$

($U_{\max}$ is the maximally acceptable loss, $g_m(x)$, $m = 1, \ldots, M$ are the technical constraint functions.)

Similarly, a possible symbolic formulation of (4.12.2) is

$$\begin{aligned} &\text{minimize } U(R(x)) \\ &C(x) \leq C_{\max} \\ &g_m(x) \leq 0, \qquad m = 1, \ldots, M. \end{aligned} \tag{4.12.4}$$

($C_{\max}$ is the admissible cost level.)

Note at this point that $U(R(x))$ is often expressed in the—positive or negative sense—expected utility functional form (cf., e.g., Berger, 1985; or French, 1986):

$$U(R(x)) = \sum_{k=1}^{K} y_k(x) p_k(x), \tag{4.12.5}$$

but this is neither essential nor always appropriate. Observe also that although the notation introduced suggests a (partial) discretization—namely, of probabilities and consequences of hazardous events—the possibility of analogous continuous problem description is evident.

Stochastic Model Extensions

Although random factors of the above problem statement have been considered by introducing component failure and scenario occurrence probabilities, it has been tacitly assumed that all relevant model parameters are known 'exactly'. This assumption may hold for 'well-defined' technical systems, but is certainly not a priori valid in the context of environmental risk management. For example, the parameters y_k (consequences), p_n and P_k (probabilities) themselves can frequently be regarded as partially unknown or uncertain. In stochastic models this fact is taken into account explicitly, handling relevant model uncertainties as random variables.

Consider, for example, model (4.12.3): if we accept that risks can be given only in a statistical sense, then (at least) the respective risk constraints are to be replaced by some statistically meaningful criteria. Examples are (recall the discussion in Chapter 3.5) the following:

$$E\{U(R(x))\} \leq U_{\max}, \tag{4.12.6}$$

{the constraints related to acceptable risks are met on average};

$$P\{U(R(x)) \leq U_{\max}\} \geq 1 - \alpha \qquad (0 < \alpha < 1 \text{ model parameter}), \tag{4.12.7}$$

{the probability of satisfying the risk constraints is 'sufficiently near' to one};

$$E\{U(R(x)) \mid U(R(x)) > U_{\max}\} \leq U_{\max} + \beta \tag{4.12.8}$$
$$(U_{\max}\,,\ \beta > 0 \text{ are model parameters}),$$

{the extent of unfavourable deviations from 'acceptable' risks is 'sufficiently small', on average}.

Similarly, if total system costs and resource demands may also statistically vary, then objective functions of the type

$$\min E\{C(x)\}, \tag{4.12.9}$$

{minimize expected total costs};

$$\min\ \alpha\ E\{C(x)\} + (1-\alpha)D\{C(x)\}, \tag{4.12.10}$$

{minimize a weighted combination of the mean and standard deviation of costs}

have a statistically well-defined interpretation.

Obviously, these model forms are not equivalent, and they serve only for illustration. In the next section a simplified case study will be analysed in a stochastic modelling framework, using the (simplest) model component (4.12.9).

4.12.3 An Illustrative Case Study

All industries which, in their production process, use, consume, and/or discharge hazardous—toxic, explosive, inflammable, etc.—materials, can be considered potentially dangerous to the environment. We shall analyse below water quality degradation, resulting water supply shortage, and other damages (primarily, the financial consequences) which might be induced by accidental point-source toxic pollution of the River Danube. The Danube is one of the largest rivers in Europe, with streamflow rates of 1,000 to 6,000 m^3/s on its Hungarian section. The industrial plant taken as our example produces viscose fibres and other plastics; it has a list of stored raw materials and intermediates. For illustration we chose carbon disulfide (CDS) from this list, as it is stored in fairly large quantities, which in the case of a possible accident may escape, together with the fire extinguishing agent, through the drain network of the plant towards the Danube. Note here that CDS is soluble in water to 2000 mg/l; furthermore, that it reacts with diamines to form even more noxious compounds. The industrial plant is situated in the impoundment section of the Nagymaros barrage site, and many water intake works (bank-filtered wells) might be endangered by an accidental CDS release. As seen from the above, we shall investigate here exclusively the short-term negative effects of a single polluting material. Furthermore, due to the lack of reliable data and related background information, simplifications and 'educated guesses' need to be applied during the model formulation. In spite of the strong simplifications made, we attempt to present an example which already reflects realistic problem complexity and highlights the potential of the global optimization approach in environmental risk management.

Scenario Outline

Let us suppose that the following chain of events may take place:

- Several (1, 2, 3 or 4) storage tanks containing CDS fail by accident: as a consequence, CDS is discharged into the river in a (partially) uncontrollable manner.

- The CDS load disperses in the river: the spatial and temporal pattern of its distribution depends on the pollutant flux and on the actual river hydrologic/hydraulic conditions.

- Given the CDS load and its distribution process, depending on the topographical location of the water supply works (bank well-systems), a number of these waterworks may be forced to close. This action implies severe adverse effects: water supply shortages—or possibly even toxic water supplies—for a short period. Corresponding measures (resources/costs) needed for mitigating the risk can be temporary water supply from emergency reserves, special water treatment, rehabilitation of the well-system (and its filtration layer), or the eventual close-down of some irreversibly affected wells.

4.12.4 Quantitative Analysis

Storage Tank Failure Probabilities

Obviously, the estimation of these is one of the crucial elements in the present illustrative case study; however, data are scarce and failure rate estimates (from reference data sets or from expert opinions) might vary to a great extent, cf. the related discussions in Kleindorfer and Kunreuther (1987), Chibber and Apostolakis (1992), or Cooke (1992). For illustration, we shall choose here a simple model in which it is supposed that the subsequent (theoretical) tank failure times—that is, the time period between two consecutive failure events—follow an exponential probability distribution function with known parameters $a_n > 0$, $n = 1, \ldots, N$ (in the example $N = 4$). In addition, we shall assume that if tank n is found defective on a routine maintenance test, then its corresponding part (say, its valve) is immediately repaired or replaced. Supposing that the probability of a tank being operated is independent of the testing schedule, the probability of a failure event (unavailability on demand) of tank n is expressed by

$$p_n(T_n) = \frac{1}{T_n} \int_0^{T_n} (1 - \exp(-a_n t))\, dt \approx \frac{a_n T_n}{2}, \qquad (4.12.11)$$

where T_n denotes the testing (that is, regular checking) interval for tank n. (This approximation is acceptable for small values of a_n which are typical in the present context.) Now, it is supposed that one can control, to some extent, the (un)availability of tanks by an appropriate choice of the testing intervals T_n. Again, for simplicity, we may assume that the testing costs linearly depend on the test frequency, that is, they are inversely proportional to the values T_n:

$$C_n(T_n) := \frac{e_n}{T_n} \qquad (e_n > 0 \text{ constant cost factors}). \qquad (4.12.12)$$

Note that, typically, such testing will interrupt the normal operational procedure, thus 'continuous' hands-on monitoring is simply not possible, in the present case.

Coupling the relations (4.12.11) and (4.12.12), we can approximate the maintenance/testing costs as a function of the corresponding failure probability:

$$C_n(p_n) = \frac{e_n a_n}{2 p_n}. \qquad (4.12.13)$$

Note that this simple model will allow us to 'optimize' the choice of the testing schedule, as all possible consequences studied below are explicitly dependent on the failure probabilities. For lack of a similar description, one is confined to a 'pure' scenario analysis which, per se, does not allow one to consider formally optimized actions.

Below we shall apply an additional simplification, by which the same testing schedule (T) and the same failure rates (a) are taken for all ($n = 1, \ldots, 4$) tanks. Denote the number of tanks failing during any time period T by $P_k(T)$, $k = 0, 1, \ldots, 4$. At this point, we shall take into account the 'domino effect', quite typical in the accident-type analysed: if it is strong, then $P_k(T) \approx P_1(T)$, $k = 2, 3, 4$. Consequently, it can be modelled, for example, in the following simple way:

$$P_1(T) = \frac{aT}{2}, \qquad P_{k+1} = P_k(T) q_k, \qquad k = 1, 2, 3. \tag{4.12.14}$$

In (4.12.14) the factors q_k are near to 1, reflecting the strength of the 'domino effect'. Note that the discrete probability distribution $\{P_k(T)\}$ is an explicit function of the time-horizon considered. For simplicity, we shall suppress T (assuming a prefixed horizon, say, a year) and the notation P_k will stand for $P_k(T)$, when this does not lead to ambiguity.

The Amount and Rate of Toxic Discharge

As mentioned earlier, tank failures may result in an uncontrolled toxic pollution load entering the river via the internal drains of the factory—supposing that appropriate measures cannot be taken. (We shall not investigate here in any detail the possibility of in-plant emergency measures.) For illustrative purposes, the following simple model is applied: all tanks have the same capacity $V_{\max} = 37.5$ metric tons; moreover, if a tank by accident discharges its actual content, then this (random) content V, for each tank independently, can be approximated by a corresponding (truncated) normal distribution with mean value $E(V) = V_{\max}/2 = 18.75$ and standard deviation $D(V) = 7.30$. (Note that according to this approximation the interval $[0, 37.5]$ covers, with probability 0.99, the range of the normal variate $\mathcal{N}(18.75, 7.30)$.)

From the above assumptions it also follows that in the case of simultaneous accidents of several tanks, the summed accidental discharge from k

tanks is, approximately, also normally distributed with mean value $kE(V)$ and standard deviation $\sqrt{k}D(V)$. A similar model may describe the discharge time: it is supposed that the toxic pollutant load leaves the factory at a uniform rate during a time period which follows (truncated) normal p.d.f. with mean value 6.50 hr and standard deviation 2.17 hr. (Again, this implies that the interval [1 hr, 12 hr] covers 99 per cent of the range of the modelled random variable $\mathcal{N}(6.50, 2.17)$.)

Consequently, the primary pollution process, modelled by a flux of (randomly realized) intensity I, equals

$$I = \{\text{total pollution load}\}/\{\text{discharge time}\} = \text{TPL}/\text{DT} \quad (\text{kg/s}), \tag{4.12.15}$$

where both TPL and DT are respectively parameterized normal variates.

Pollutant Dispersion in the River

As seen from the above description, the pollution load is influenced by a number of random (accidental) circumstances; further significant uncertainties originate from the actual hydrologic and hydraulic conditions of the recipient river.

The mixing of point-source pollutants with the bulk of water in a stream can be described by the general transport equation (cf., e.g., Rich, 1973):

$$\begin{aligned}\frac{\partial c}{\partial t} &+ \frac{\partial}{\partial x}(v_x c) + \frac{\partial}{\partial y}(v_y c) + \frac{\partial}{\partial z}(v_z c) \\ &= \frac{\partial}{\partial x}(D_x \frac{\partial c}{\partial x}) + \frac{\partial}{\partial y}(D_y \frac{\partial c}{\partial y}) + \frac{\partial}{\partial z}(D_z \frac{\partial c}{\partial z}),\end{aligned} \tag{4.12.16}$$

where t denotes time (s); x, y, and z are the axes of the coordinate system in which the pollution dispersion process is described (the movement of the pollution flux is measured in meters, m); $c(\text{g/m}^3)$ is concentration of the pollutant; v_x, v_y, and $v_z(\text{m/s})$ are coordinate-wise velocities of the river; finally, D_x, D_y, and $D_z(\text{m}^2/\text{s})$ are diffusion coefficients.

The general (three-dimensional) form of the transport equation (4.12.16) usually has to be simplified, in order to obtain numerically tractable specifications (which are also properly supported by data). The special form of equation (4.12.16) applied here is derived by integrating it with respect to water depth h(m): this leads to the equation of turbulent dispersion in two dimensions:

$$\frac{\partial}{\partial t}(h\bar{c}) + \frac{\partial}{\partial x}(h\bar{v}_x\bar{c}) + \frac{\partial}{\partial y}(h\bar{v}_y\bar{c}) = \frac{\partial}{\partial x}(hD_x^\star\frac{\partial \bar{c}}{\partial x}) + \frac{\partial}{\partial y}(hD_y^\star\frac{\partial \bar{c}}{\partial y}). \quad (4.12.17)$$

In equation (4.12.17) depth-wise average values of the previously introduced quantities c, v_x and v_y are given; further, the notation $D_x^\star$ and $D_y^\star$ stands for the respective dispersion coefficients.

On the basis of the above description, the numerical integration of equation (4.12.17) can be accomplished, under various initial conditions (namely, pollution intensity (kg/s), streamflow rate (m^3/s), respective water depth h(m), velocity $\bar{v}_x$(m/s), $\bar{v}_y \approx 0$ (m/s), and corresponding dispersion coefficients $D_x^\star(m^2/s)$, $D_y^\star \approx 0(m^2/s)$. The results of a detailed (off-line) numerical study served as a basis for the following conclusions (see Fig. 4.12.1):

- In the present risk management context, the pollution process is properly characterized by the resulting maximal pollutant concentration $c_{\max}$.
- $c_{\max}$ is a monotonically decreasing function of the distance between the pollution discharge and the downstream observation point: the form of this function can be numerically approximated (but it is not known analytically).
- $c_{\max}$ is, to a reasonable approximation, linearly proportional to the pollution intensity and inversely proportional to the actual streamflow rate. Further, its value is only slightly dependent on the pollution load duration (at least on the time range [1 hr, 12 hr] investigated here).
- Large-scale engineering construction works may significantly affect the quantitative characteristics of the above analysis. For example, side effects of the barrage system mentioned would include the following (as estimated from our off-line numerical model):
    - the dispersion coefficient D_y decreases by more than an order of magnitude
    - the average velocity of water v_x decreases by a factor of about 4
    - the (average) water depth may grow significantly
    - the overflow above the barrage increases volatilization of the pollutant.

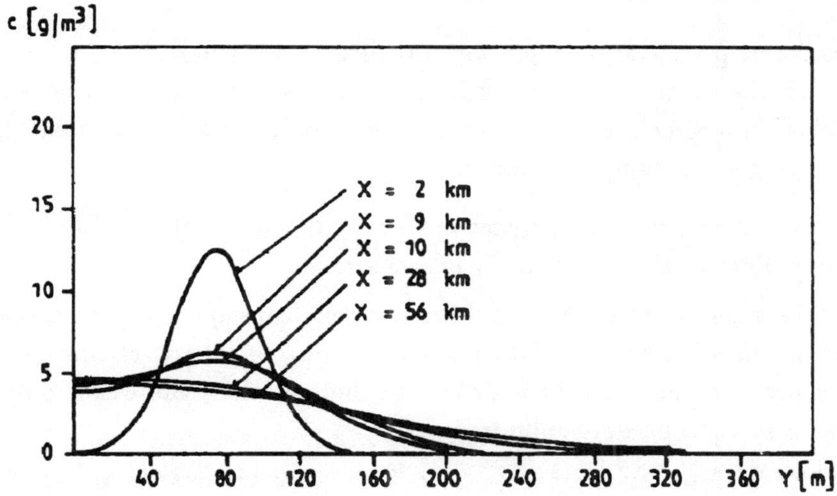

Downstream distance x(km)	Maximal pollutant concentration $c_{max}(g/m^3)$
2	12.65
9	6.02
10	5.74
28	4.59
31	4.53
37	4.39
42	4.26
47	4.12
56	3.94

Figure 4.12.1
Longitudinal (X, km) and cross-sectional (Y, m) evolution of c_{max}

Now, the findings listed allow for the numerical generation of c_{max}. That is, for a hypothetical pollution event and corresponding river hydrologic and hydraulic state, the concentration maximum value can be numerically traced to a reasonable distance (up to several tens of kilometres). This, in turn, makes it possible to analyse the (hypothetical) accident consequences with respect to the downstream waterworks affected.

Pollution Transport in the Filtration Zone

The transport processes through the bank filter layers will be briefly reviewed below. In the present framework it is not necessary to go into details of physical and biogeochemical transport phenomena. Instead, we shall make the following simplifying assumptions:

- The river bed has no significant bottom sediment (that could be organically polluted) above the sandy gravel strata.
- For the sake of safety in modelling accident consequences, no transformation (adsorption) of CDS is supposed during its travel through the filter layer to the wells; in addition, at the section of the wells complete dissolution can be duly supposed.
- On the basis of experience (gained from observations separately from this study), one can estimate that in the case of horizontal wells, approximately 2 days, and in the case of vertical wells, some 4 days travelling time can be expected in the filter layers.

Note here that in a more detailed study much more precise site-specific information would be desirable. In the absence of such information, we shall confine ourselves to a tentative analysis of the consequences of an accidental water pollution event.

Pollution Effects and Consequences

The calculations outlined yield maximum CDS concentrations which affect the water supply well-system. Namely, if the (site-dependent) value $c_{\max}$ exceeds a certain threshold, then the wells in question have to be closed down, since their water cannot be utilized, at least for drinking purposes. In other words, given an accidental CDS load, the estimation of its spatial distribution process leads in a straightforward manner to estimating the need for i) early warning of those water supply works which are potentially affected, and ii) corresponding mitigation actions.

In our example, only the direct cost implications of accidental pollution will be considered, that is, additional (health, etc.) effects will not be directly included. In this way, with any accident scenario, a corresponding estimated monetary loss will be simply associated. The risk of a particular maintenance policy will be defined as the expected total loss (R) which might be anticipated (in the long run taking, say, a period of several years):

$$R = R(T) = \sum_{k=1}^{K} P_k E(L_k), \qquad P_k = P_k(T), \qquad L_k = L_k(T). \quad (4.12.18)$$

In (4.12.18), T symbolizes the operations policy (in our simple example, the regular testing/maintenance interval is the sole decision variable), $\{P_k\}$ represents the aggregated statistical model of the accident occurrences, and $\{L_k\}$ represents the consequences. Hence (4.12.18) shows the expected losses, as a function of the accident probabilities and related consequences indexed by the values $k = 1, \ldots, K$.

Let us consider now the testing cost factors. Recall (cf. (4.12.12)) that, for simplicity, it has been assumed that the operating policy costs depend only on T according to

$$C(t) = \frac{e}{T}, \qquad (e > 0). \quad (4.12.19)$$

In addition, we shall define total damage as the sum of close-down costs (CDC) and rehabilitation costs (RC), per waterworks affected. These will be calculated as follows:

$$CDC = \{\text{Number of close-down days}\} \cdot \{\text{50 percent daily water supply rate}\} \cdot \{\text{unit cost of emergency water supply}\}; \quad (4.12.20)$$

$$RC = \{\text{Number of pollution travel days}\} \cdot \{\text{daily amount of safety water recharge needed}\} \cdot \{\text{unit cost of water recharge}\}. \quad (4.12.21)$$

Note that—in the present example—fixed costs of an early warning system or stand-by emergency technologies will not be subject to optimization. Consequently, in our greatly simplified example, only the testing interval will be chosen according to the optimization model below:

$$\min \frac{e}{T} + \sum_{k=1}^{K} P_k E(CDC_k + RC_k), \quad P_k = P_k(T) \quad (4.12.22)$$
$$T_{\min} \leq T \leq T_{\max},$$

in which the expected value term $E(\cdot)$ is an implicit 'black box' function of T. The index k in CDC_k and RC_k is understood for all possible scenarios, that is, for all meaningful collections of waterworks which can be affected by the corresponding accidental water pollution events considered.

4.12.5 Numerical Example and Discussion

We shall consider the following data and accident occurrence model:

failure rate: $a = 0.002/\text{yr}$

testing interval range: $T_{\min} = 1$ day ≈ 0.00274 yr, $\quad T_{\max} = 1$ yr;

accident occurrence model: $N = 4$ storage tanks

$$P_1(t) = a\,\frac{T}{2}, \qquad P_{k+1}(T) = P_k(T)q_k, \qquad k = 1, 2, 3,$$
$$q_1 = 0.8, \qquad q_2 = 0.9, \qquad q_3 = 0.95;$$

testing procedure costs: 10,000 Fts/test (1 Ft $\approx$ 2 cents U.S., at time of study).

In this simplified example, we shall assume that if a single storage tank fails, then only the first three waterworks are affected, while in the case of two, three, or four tanks failing, the first five, six, or all waterworks are respectively affected. The estimated costs are shown in Table 4.12.1.

Taking first $T = T_{\min}$ (the minimal testing interval considered), and evaluating a set of corresponding randomly generated accident scenarios, the estimated objective function value in (4.12.22) equals

$$\begin{aligned} 3.65+ &(2.74 \cdot 98.2 + 2.19 \cdot 102.5 + 1.97 \cdot 106.8 + 1.87 \cdot 473.5) \cdot 10^{-6} \\ &\approx 3.652\text{MFt/yr}. \end{aligned}$$

The result shows that in this case testing costs are predominant: namely, 'very frequent testing is far too costly'. Take now at the other extreme $T = T_{\max}$ (the maximal testing interval), then (under the same assumptions) the estimated costs are

$$\begin{aligned} 0.01+ &(1 \cdot 98.2 + 0.8 \cdot 102.5 + 0.72 \cdot 106.8 + 0.684 \cdot 473.5) \cdot 10^{-3} \\ &\approx 0.591\text{MFt/yr}. \end{aligned}$$

Table 4.12.1
Distances and accident costs per affected waterworks

waterworks no.	distance from pollution discharge (kms)	total costs of possible accident (CDC+RC, million Fts)
1	2	19.2
2	9	68.3
3	10	10.7
4	28	1.0
5	31	3.3
6	37	4.3
7	42	2.6
8	47	78.4
9	50	23.3
10	56	21.3
11	56	53.3
12	56	78.4
13	61	1.3
14	64	49.5
15	64	9.2
16	64	29.1
17	68	20.3

Note that in this case the expected damage costs are predominant: 'too infrequent testing might prove dangerous'. Taking now, say, $T = 0.25$ yr, one obtains that the expected total costs and damages are

$$0.04 + (2.5 \cdot 98.2 + 2 \cdot 102.5 + 1.8 \cdot 106.8 + 1.71 \cdot 473.5) \cdot 10^{-4} \approx 0.185 \text{MFt/yr}.$$

Hence, the last testing policy seems definitely better than the two chosen previously. By simple univariate Lipschitz global optimization, one can obtain in a few iterations that the best maintenance policy results by choosing $T \approx 0.13$ (that is, some 7.69 tests/year): the expected approximate cost (of testing and estimated potential damages) equals 0.152 MFt/yr.

In conclusion, let us recall first that in the above illustrative example, a number of greatly simplifying assumptions were made. In reality, a much more precise description would be necessary when, for example, expected damages are calculated; one has to consider a proper statistical description of the actual storage tank contents, a detailed analysis of the in-factory emergency actions and their possible results. Furthermore, an in-depth statistical investigation of the environmental conditions is also desirable: this has to include a streamflow-pollution dispersion study, proper account of the infiltration process, and a site-specific analysis of the emergency actions at the waterworks. In this way, a sound risk assessment can be provided which, in turn, makes it possible to suggest efficient management policies.

Note finally that (univariate) global optimization may seem like 'overkill' to solve the strongly simplified model discussed here. Evidently, the model should be refined in many respects. For instance, if—instead of a single maintenance policy and failure event model—the system components (in the example, storage tanks) have a corresponding description, then this leads directly to multidimensional optimization problems. Such, more realistic—typically multivariate and stochastic 'black box'—model formulations can be handled similarly to the case studies presented in the preceding chapters of Part 4.

Afterword

In Part 4, several prospective applications of Lipschitz and continuous global optimization have been discussed in some detail. With the purpose of highlighting a more extensive 'palette' of potential application areas—and to stimulate interest in them—a collection of research subjects is listed in the following pages. The few supporting references cited should serve only as pointers, especially since not all of them are equally elaborate in terms of mathematical modelling and optimization aspects and needs. Naturally, the subject listing is not 'complete' in any reasonable sense, and it serves only for illustration. Nevertheless, it is hoped that the reader will become (even more) convinced concerning the potentials of global optimization in the most diverse fields.

The apparent need for robust and efficient global optimization procedures should motivate theoretical research and foster work on good implementations. Conversely, important new theoretical results and sophisticated software will certainly find their way to important applications which then can be analysed and solved more effectively than before. Theory and applications are meant to complement and stimulate each other, for the benefit of society.

Some Further Research Perspectives

The research areas are listed alphabetically; the subsequent pointer references are provided in chronological order

- Aerospace engineering,
  - Dempster and Coupé (1982)
  - Adeli and Cheng (1993)
- Air pollution control / Acid rain abatement
  - Fortin and McBean (1983)
  - Ellis, McBean and Farquhar (1985, 1986)
  - Fronza and Melli (1985)
  - Pintér (1987a)
  - McBean and Ponnambalam (1989)
  - Chang and Delleur (1992)
- Analysis and simulation of random fields
  - Fenton and Griffith (1993)
- Artificial intelligence
  - Jaumard, Ow and Simeone (1988)
  - Triantaphyllou (1994)
- Automatic assembly design
  - Boothroyd, Poli and Murch (1982)
- Chemical engineering
  - Li, Pardalos and Levine (1992)
  - Pintér, Fels, Lycon, Meeuwig and Meeuwig (1993)

Said and Kerr (1993)
Syaiful, Wilhelm and Delmas (1993)

- Chromatography
  Wegscheider, Lankmayr, and Budna (1982)
  Debets, Bajema, and Doornbos (1983)
  Wegscheider, Lankmayr, and Otto (1983)
- Communication systems (analysis and design)
  Erup and Harris (1988)
  Jeruchim, Shanmugan, Biglieri, and Balaban (1988)
- Computational chemistry
  Shalloway (1992)
  Le Grand and Merz (1993)
  McDonald and Floudas (1994)
  Mockus (1994)
  Pardalos and Xue (1994)
- Control engineering
  Adamczyk, Zohdy and Khan (1993)
  Tang (1993)
- Crystallography
  Rous (1993)
- Electromechanical design
  Di Barba and Savini (1993)
- Engineering design (mechanical, automated)
  Wilde (1978)
  Johnson (1979)
  Jaumard, Lu and Sriskandarajah (1988)
  Papalambros and Wilde (1988)
  Hansen, Jaumard and Lu (1989a,b,c, 1991b)
  Papalambros (1993)
  Wilde (1993)
- Environmental resources and quality monitoring networks
  Pintér and Somlyódy (1987)
  Meyer and Brill (1988)
  Deininger and Lee (1989)

- Equilibrium problems and energy functions in mechanics and engineering
  Motreanu and Panagiotopoulos (1995)
- Exploration and exploitation of natural resources
  Medova (1981)
  Flåm and Pintér (1985)
  Gassmann (1989)
- Fuzzy clustering, pattern recognition, location-allocation
  Laporte (1988)
  Pintér and Pesti (1991)
  Tuy and Al-Khayyal (1992)
  Al-Sultan and Selim (1993)
  Galvão (1993)
  Ghost and Harche (1993)
  Carrizosa, Conde Muñoz-Marquez and Puerto (1995)
  Pardalos (1995)
- Hierarchical (multilevel) decision problems (in engineering, management)
  Vicente and Calamai (1994)
- Hydrologic process analysis and forecasting
  Szöllősi-Nagy (1982)
  Bras and Rodriguez-Iturbe (1985)
  Hipel (1985)
- Integrated circuit testing; electric fault location
  Miranda and Lopez-Barrio (1993)
- Large combinatorial optimization problems
  Cooper (1976)
  Golden and Alt (1979)
  Lawler, Lenstra, Rinnooy Kan, and Schmoys (1985)
  Laporte (1988)
  Vizvári and Yilmaz (1994)
  Feo and Resende (1995)
- Manufacturing, production management
  Kusiak (1986)
  Nitti and Speranza (1987)
  Bard, Venkatraman, and Feo (1993)

Chen, Hearn, and Lee (1994)

- Measuring techniques, signal generation and processing
  Csendes, Daróczy and Hantos (1992)
  Mescheryakov and Rovkij (1992)
  Chen and Hodgkiss (1993)

- Medium and large-scale modelling of environmental pollution effects
  United States Environmental Protection Agency (1980)
  Eliassen and Saltbones (1983)
  Alcamo, Hordijk, Kamari, Kauppi, Posch and Runca (1985)
  Shaw (1986)
  McBean (1987)
  Hettelingh (1989)
  Gorelick (1990)
  De Lange and Pintér (1991a,b)
  Jolánkai (1992)
  Kularathna and Somlyódy (1994)

- Microelectronics, computer aided VLSI circuit design
  Muroga (1982)
  Hu and Kuh (1987)
  Boese, Kahng and Tsao (1993)

- Multilevel inventory systems
  González, Rofman, and Sagastizábal (1995)

- Network flow problems
  Guisewite and Pardalos (1991)
  Guisewite (1995)

- Neurobiological modelling
  White, Manis and Young (1992)
  Evans, Kember and Major (1993)

- Number theory
  Finch (1992)

- Numerical approximation problems (in electric power system planning)
  Fülöp (1992)
  Fülöp and Prill (1992)

- Parallel computational environments (in the context of global optimization)
  - Dert (1986)
  - Sergeyev and Strongin (1989)
  - Zenios (1989)
  - Gergel, Sergeyev and Strongin (1992)
  - Strongin and Sergeyev (1992)
  - Gergel and Sergeyev (1994)
  - Sergeyev and Grishagin (1994a,b)

- Parameterization of (engineering, biological, statistical, etc.) models
  - Wingo (1984)
  - Suki, Csendes and Daróczy (1990)
  - Huang, Wang and Kuo (1993)
  - Gourdin, Hansen and Jaumard (1994)

- Reliability engineering, safety studies
  - Cooke and Pintér (1989)
  - Chibber and Apostolakis (1992)
  - Cooke (1992)
  - Gutjahr (1994)

- Risk analysis and its applications (in agriculture, or in other areas)
  - Schweigman, Joosten, Klein Haneveld, Kijne, and Snijders (1985)
  - Hendrix (1989)

- Robust calibration of environmental models
  - Hendrickson, Sorooshian and Brazil (1988)
  - Pintér (1990c)
  - Chang and Delleur (1992)
  - Porto and Lopes (1993)

- Robust (tolerance, centered, etc.) design, in engineering and production
  - Halász (1982)
  - Thach (1988)
  - Csendes (1990)
  - Hansen (1991)
  - Hendrix and Pintér (1991)
  - Zabinsky, Graesser, Tuttle and Kim (1992)

Csendes and Pintér (1993b)
He and Polak (1993)
Hendrix, Mecking and Hendriks (1993)
Kristinsdottir, Zabinsky, Csendes, and Tuttle (1993)
Mockus (1994)
Phong, Tao and An (1995)

- Software testing
  Barbagallo and Reccioni (1992)

- Standardized methodology for evaluation of global optimization software
  Floudas and Pardalos (1990)
  Pintér (1990b)
  Jackson, Boggs, Nash, and Powell (1991)
  Schoen (1993)
  Moshirvaziri (1994)
  Romeijn and Smith (1994)

- Statistical quality control
  Seidel (1991)

- Surface reconstruction, image reconstruction and interpretation
  Wang, Xing, and Zhuang (1993)

- Systems of nonlinear equations (especially, recent developments)
  Forster (1992, 1995)
  Thach (1993)
  Diener (1995)

- Time series analysis and forecasting
  Hipel and McLeod (1987)
  Mockus (1994)

As even this purely illustrative listing indicates, seemingly unrelated, 'distant' problem-types can be modelled and solved, applying various global optimization techniques. Naturally, the list of pointers could be much extended: the references on the subsequent pages provide additional information for the interested reader.

References

Aalderink, R. H., Lijklema, L., Breukelman, J., van Raaphorst, W., and Brinkman, A. G. (1984). Quantification of wind induced resuspension in a shallow lake. *Water Science and Technology* 17, 903–914.

Aarts, E. H. L., and Korst, J. H. M. (1988). *Simulated Annealing and Boltzmann Machines*. Wiley, Chichester.

Abraham, B. and Ledolter, J. (1983). *Statistical Methods for Forecasting*. Wiley, New York.

Adamczyk, B., Zohdy, M. A., and Khan, A. A. (1993). Global optimization of stochastic multivariable functions. *Proceedings of the 1993 American Control Conference*, San Francisco, Calif., pp. 2339–2344. IEEE Service Center, Piscataway, NJ.

Adeli, H., and Cheng, N.-T. (1993). Integrated genetic algorithm for optimization of space structures. *Journal of Aerospace Engineering* 6, 315–328.

Alcamo, J., Hordijk, L., Kamari, J., Kauppi, P., Posch, M., and Runca, E. (1985). Integrated analysis of acidification in Europe. *Journal of Environmental Management* 21, 47–61.

Alexandrov, A. D. (1950). On surfaces which can be represented by the difference of convex functions. *Dokladi Akademii Nauk SSSR* 72, 613–616. (In Russian.)

Al-Khayyal, F. (1987). An implicit enumeration procedure for the general linear complementarity problem. *Mathematical Programming Study* 31, 1–20.

Al-Khayyal, F., and Falk, J. E. (1986). Jointly constrained biconvex programming. *Mathematics of Operations Research* 8, 273–286.

Allgower, E. L., and Georg, K. (1983). Predictor-corrector and simplicial methods for approximating fixed points and zero points of nonlinear mappings. In: *Mathematical Programming. The State of the Art,* (ed. A. Bachem, M. Grötschel, and B. Korte), pp. 15–56. Springer, Berlin.

Allgower, E. L., and Georg, K. (1990). *Numerical Continuation Methods*. Springer, Berlin.

Al-Sultan, K. S., and Selim, S. Z. (1993). Global algorithm for the fuzzy clustering problem. *Pattern Recognition* 26, 1357–1361.

Aluffi-Pentini, F., Parisi, V., and Zirilli, F. (1985). Global optimization and stochastic differential equations. *Journal of Optimization Theory and Applications* 47, 1–16.

Apostolakis, G. (1985). The broadening of failure rate distributions in risk analysis: How good are the experts? *Risk Analysis* 5, 89–95.

Apostolakis, G. (1988). Expert judgment in probabilistic safety assessment. In: *Accelerated Life Testing and Experts: Opinions in Reliability,* (ed. C. Clarotti and D. Lindley). *Proceedings of the International School of Physics Enrico Fermi,* Amsterdam.

Archetti, F., and Betrò, B. (1978). A priori analysis of deterministic strategies for global optimization problems. In: *Towards Global Optimisation,* Vol. 2, (ed. L. C. W. Dixon and G. P. Szegö), pp. 31–48. North-Holland, Amsterdam.

Archetti, F., and Betrò, B. (1979). A probabilistic algorithm for global optimization. *Calcolo* 61, 335–343.

Archetti, F., and Betrò, B. (1980). Stochastic models and optimization. *Bolletino della Unione Matematica Italiana* 51, 17-A, 295–301.

Archetti, F., Betrò, B., and Steffe, S. (1975). A theoretical framework for global optimization via random sampling. Technical Report, CNR, University of Pisa.

Archetti, F., and Schoen, F. (1984). A survey on the global optimization problem: General theory and computational approaches. *Annals of Operations Research* 1, 87–110.

Ariyawansa, K. A., and Templeton, J. G. C. (1983). On statistical control of optimization. *Optimization* 14, 393–410.

Ashton, A. H., and Ashton, R. H. (1985). Aggregating subjective forecasts: Some empirical results. *Management Science* 31, 1499–1508.

Avriel, M. (1976). *Nonlinear Programming: Analysis and Methods.* Prentice-Hall, Englewood Cliffs, NJ.

Baba, N. (1981). Convergence of a random optimization method for constrained optimization problems. *Journal of Optimization Theory and Applications* 33, 451–461.

Bachem, A., Grötschel, M., and Korte, B., eds. (1983). *Mathematical Programming: The State of the Art.* Springer, Berlin.

Banks, H. T., and Fitzpatrick, B. G. (1990). Statistical methods for model comparison in parameter estimation problems for distributed systems. *Journal of Mathematical Biology* 28, 501–527.

Barbagallo, R., and Reccioni, M. C. (1992). A global optimization approach to software testing. In: *Recent Advances in Global Optimization,* (ed. C. A. Floudas and P. M. Pardalos), pp. 616–633. Princeton University Press.

Bard, J. F., Venkatraman, K., and Feo, T. A. (1993). Single machine scheduling with flow time and earliness penalties. *Journal of Global Optimization* 3, 289–309.

Baritompa, W. (1993). Customized methods for global optimization – A geometric viewpoint. *Journal of Global Optimization* 3, 337–358.

Baritompa, W. (1994). Accelerations for a variety of global optimization methods. *Journal of Global Optimization* 4, 37–45.

Baritompa, W., and Cutler, A. (1994). Accelerations of global optimization covering methods using second derivatives. *Journal of Global Optimization* 4, 329–341.

Barnwell, T. O., and Brown, L. C. (1987). *The Enhanced Stream Water Quality Models QUAL2e and QUAL2e–UNCAS: Documentation and User Manual.* U.S. Environmental Protection Agency, Athens, Georgia.

Bartell, S. M., Cale, W. G., O'Neill, R. V., and Gardner, R. H. (1988). Aggregation error: Research objectives and relevant model structure. *Ecological Modelling* 41, 157–168.

Basso, P. (1982). Iterative methods for the localization of the global maximum. *SIAM Journal on Numerical Analysis* 19, 781–792.

Basso, P. (1985). Optimal search for the global maximum of a function with bounded seminorm. *SIAM Journal on Numerical Analysis* 22, 888–903.

Beck, M. B. (1985a). *Water Quality Management: A Review of the Development and Application of Mathematical Models.* Springer, Berlin.

Beck. M. B. (1985b). Lake eutrophication: Identification of tributary nutrient loading and sediment resuspension dynamics. *Applied Mathematics and Computation* 17, 1–26.

Beck, M. B. (1987a). Water quality modelling: A review of the analysis of uncertainty. *Water Resources Research* 23, 1393–1442.

Beck, M. B., ed. (1987b). *Systems Analysis in Water Quality Management.* Pergamon Press, Oxford.

Beck, M. B., and van Straten, G., eds. (1983). *Uncertainty and Forecasting in Water Quality.* Springer, Berlin.

Benedek, P. (1988). River Danube pollution and its risk assessment. In: *Risk Assessment of Chemicals in the Environment*, (ed. M. L. Richardson). The Royal Society of Chemistry, London.

Benson, H. P. (1982). On the convergence of two branch and bound algorithms for nonconvex programming problems. *Journal of Optimization Theory and Applications* 36, 129–134.

Benson, H. P. (1995). Concave minimization: Theory, applications and algorithms. In: *Handbook of Global Optimization* (ed. R. Horst and P. M. Pardalos), pp. 43–148. Kluwer, Dordrecht.

Benson, H. P., Erenguc, S. S., and Horst, R. (1990). A note on adapting methods for continuous optimization to the discrete case. *Annals of Operations Research* 25, 243–252.

Berger, J. O. (1985). *Statistical Decision Theory and Bayesian Analysis* (2nd edn.) Springer, Berlin.

Betrò, B. (1984). Bayesian testing of nonparametric hypotheses and its application to global optimization. *Journal of Optimization Theory and Applications* 42, 31–50.

Betrò, B. (1991). Bayesian methods in global optimization. *Journal of Global Optimization* 1, 1–14.

Betrò, B., and Schoen, F. (1987). Sequential stopping rules for the multistart algorithm in global optimization. *Mathematical Programming* 38, 271–286.

Betrò, B., and Schoen, F. (1992). Optimal and suboptimal stopping rules for the multistart algorithm in global optimization. *Mathematical Programming* 57, 445-458.

Bodin, L. D., Golden, B. L., Assad, A., and Ball, M. (1983). Routing and scheduling of vehicles and crews: The state of the art. *Computers and Operations Research* 10, 69–211.

Boender, C. G. E. (1984). The generalized multinomial distribution: A Bayesian analysis and applications. Ph.D. Dissertation. Erasmus University, Rotterdam.

Boender, C. G. E., and Rinnooy Kan, A. H. G. (1987). Bayesian stopping rules for multistart global optimization methods. *Mathematical Programming* 37, 59-80.

Boender, C. G. E., and Rinnooy Kan, A. H. G. (1991). On when to stop sampling for the maximum. *Journal of Global Optimization* 1, 331–340.

Boender, C. G. E., Rinnooy Kan, A. H. G., Stougie, L., and Timmer, G. T. (1982). A stochastic method for global optimization. *Mathematical Programming* 22, 125–140.

Boender, C. G. E., and Romeijn, E. H. (1995). Stochastic methods. In: *Handbook of Global Optimization* (ed. R. Horst and P. M. Pardalos), pp. 829–869. Kluwer, Dordrecht.

Boese, K. D., Kahng, A. B., and Tsao, C.-W. A. (1993). Best-so-far vs. where you are: New perspectives on simulated annealing for CAD. *Proceedings of the European Design Automation Conference*, Hamburg, Germany, pp. 78–83. IEEE Service Center, Piscataway, NJ.

Bogomolov, N. A., and Karmanov, V. G. (1977). Convergence of a random search method for searching stationary points of the general nonlinear programming problem. *Bulletin of the Moscow State University, Ser. 15. Computational Mathematics and Cybernetics* 1, 20–26. (In Russian.)

Bohachevsky, I. O., Johnson, M. E., and Stein, M. L. (1986). Generalized simulated annealing for function optimization. *Technometrics* 28, 209–217.

Boon, J. G., Pintér, J., and Somlyódy, L. (1989a). River basin water quality management: A stochastic approach. Research Report, Department of Nature Conservation, Agricultural University, Wageningen.

Boon, J. G., Pintér, J., and Somlyódy, L. (1989b). A new stochastic approach for controlling point source river pollution. in: *Proceedings of the 3rd Scientific Assembly of the International Association for Hydrologic Sciences, IAHS Publication 180*, pp. 241–249, IAHS, London.

Boothroyd, G., Poli, C., and Murch, L. E. (1982). *Automatic Assembly.* Marcel Dekker, New York.

Bopp, A. E. (1985). On combining forecasts: Some extensions and results. *Management Science* 31, 1492–1508.

Born, J., and Bellman, K. (1982). Numerical adaptation of parameters in simulation models using evolution strategies. In: *Modelling and Simulation of Molecular Genetic Information Systems,* (ed. K. Bellman), pp. 291–320. Akademie-Verlag, Berlin.

Bower, B. T., ed. (1977). *Regional Residuals Environmental Quality Management Modeling.* Johns Hopkins University Press, Baltimore.

Box, G. E. P., and Jenkins, G. M. (1970). *Time Series Analysis: Forecasting and Control.* Holden Day, San Francisco.

Braess, D. (1986). *Nonlinear Approximation Theory.* Springer, Berlin.

Branin, F. H. (1972). Widely convergent methods for finding multiple solutions of simultaneous nonlinear equations. *IBM Journal of Research and Development* 16, 504–522.

Bras, R. L., and Rodriguez-Iturbe, I. (1985). *Random Functions and Hydrology.* Addison-Wesley, Reading, Massachusetts.

Bratley, P., Fox, B. L., and Schrage, L. (1983). *A Guide to Simulation.* Springer, New York.

Breiman, L., and Cutler, A. (1993). A deterministic algorithm for global optimization. *Mathematical Programming* 58, 179–199.

Bremmerman, H. A. (1970). Method of unconstrained global optimization. *Mathematical Biosciences* 9, 1–15.

Brent, R. P. (1973). *Algorithms for Minimization Without Derivatives.* Prentice-Hall, Englewood Cliffs, NJ.

Briggs, J. (1992). *Fractals – The Patterns of Chaos.* Simon and Schuster, New York.

Brill, E. D., Eheart, J. W., and Liebman, J. C. (1979). Discussion: Stochastic programming model for water quality management in a river. *Journal of Water Pollution Control Federation* 51, 29–58.

Bromberg, M., and Chang, T.-S. (1992). One dimensional global optimization using linear lower bounds. In: *Recent Advances in Global Optimization,* (ed. C. A. Floudas and P. M. Pardalos), pp. 200–220. Princeton University Press.

Brooke, A., Kendrick, D., and Meeraus, A. (1988). *GAMS: A User's Guide.* Scientific Press, Palo Alto.

Brooks, S. H. (1958). Discussion of random methods for locating surface maxima. *Operations Research* 6, 244–251.

Brubaker, S. (1973). Environmental matrix. Resources for the Future, Inc. Reprint No. 107. Washington, D.C.

Brucker, P. (1978). On the complexity of clustering problems. In: *Optimization and Operations Research.* (ed. M. Beckmann and H. P. Kunzi), pp. 45–54. *Lecture Notes in Economics and Mathematical Systems* 157. Springer, Berlin.

Burn, D. H., and McBean, E. A. (1985). Optimization modeling of water quality in an uncertain environment. *Water Resources Research* 21, 934–940.

Butz, A. R. (1968). Space filling curves and mathematical programming. *Information and Control* 12, 319–330.

C. Fodor, J. (1983). Extremal distributions and their application in mathematical programming. Dr. rer. nat. Dissertation. Eötvös University (ELTE), Budapest. (In Hungarian.)

C. Fodor, J., and Pintér, J. (1988). Extreme order statistics for optimum estimation in 'hard' MP problems. In: *Transactions of the Tenth Prague Conference on Information Theory, Statistical Decision Functions, Random Processes*, pp. 321–328. Academia Publishing House, Prague.

Canadian Pulp and Paper Association (1991). *A Statement by the Pulp and Paper Industry: Solid Waste Management.* CPPA, Montreal.

Carrera, J. (1984). Estimation of aquifer parameters under transient and steady-state conditions. Ph.D. Dissertation. The University of Arizona, Tucson.

Carrizosa, E., Conde, E., Muñoz-Marquez, M., and Puerto, J. (1995). Planar point-objective location problems with nonconvex constraints: A geometrical construction. *Journal of Global Optimization* 6, 77–86.

Casti, J. L. (1990). *Searching for Certainty.* Morrow and Co., New York.

Caswell, H. (1988). Theory and models in ecology: A different perspective. *Ecological Modelling* 43, 33–44.

Chang, F., and Delleur, J. W. (1992). Systematic parameter estimation of a watershed acidification model. *Hydrological Processes* 6, 29–44.

Charnes, A., and Cooper, W. W. (1962). Chance constraints and normal deviates. *Journal of the American Statistical Association* 57, 134–148.

Chen, G. C., and Hodgkiss, W. S. (1993). VLF source localization with a freely drifting acoustic sensor array. *IEEE Journal of Oceanic Engineering* 18, 209–223.

Chen, H. D., Hearn, D. W., and Lee, C. Y. (1994). A new dynamic programming algorithm for the single item capacitated dynamic lot size model. *Journal of Global Optimization* 4, 285–300.

Chen, K.-W., and Papadopoulos, A. S. (1988). A nonparametric method for estimating the joint probability density of BOD and DO. *Ecological Modelling* 41, 183–191.

Chibber, S., and Apostolakis, G. (1992). A taxonomy of issues related to the use of expert judgement in probabilistic safety studies. *Reliability Engineering and System Safety* 38, 27–47.

Chichinadze, V. K. (1967). Random search to determine the extremum of a function of several variables. *Engineering Cybernetics* 5, 115–123.

Chichinadze, V. K. (1969). The Ψ-transform for solving linear and nonlinear programming problems. *Automatica* 5, 347–355.

Clarotti, C., and Lindley, D., eds. (1988). *Accelerated Life Testing and Expert Opinions in Reliability. Proceedings of the International School of Physics Enrico Fermi,* Amsterdam.

Clough, D. J. (1969). An asymptotic extreme value sampling theory for the estimation of a global maximum. *Canadian Operational Research Society Journal* 7, 102–115.

Cochrane, K. L., Ashton, P. J., Jarvis, A. C., Twinch, A. J., and Zohary, T. (1987).

An ecosystem model of phosphorus cycling in a warm monomictic, hypertrophic impoundment. *Ecological Modelling* 37, 207–233.

Cohen, A. V. (1984). Overview and definition of risk: Its assessment and management. *Environment International* 10, 359–366.

Cohon, J. l. (1978). *Multiobjective Programming and Planning.* Academic Press, New York.

Colodey, A. G., and Wells, P. G. (1992). Effects of pulp and paper mill effluents on estuarine and marine ecosystems in Canada: A review. *Journal of Aquatic Ecosystem Health* 1, 201–226.

Cooke, R. M. (1990). Statistics in expert resolution: A theory of weights for combining expert opinions. In: *Statistics in Science: The Foundations of Statistical Methods in Biology, Physics and Economics,* (ed. R. M. Cooke and D. Costantini). Kluwer, Dordrecht.

Cooke, R. M. (1992). *Experts in Uncertainty: Expert Opinion and Subjective Probability in Science.* Oxford University Press.

Cooke, R. M. (1994). Parameter fitting for uncertain models: Modelling uncertainty in smaller models. *Reliability Engineering and System Safety* 44, 89–102.

Cooke, R. M., Mendel, M., and Thys, W. (1988). Calibration and information in expert resolution. *Automatica* 24, 87–94.

Cooke, R., and Pintér, J. (1989). Optimization in risk management. *Civil Engineering Systems* 6, 122–128.

Cooke, R., and Vogt, F. (1990). PARFUM: Parameter fitting for uncertain models: Concepts and code for accident consequence modelling. Research Report, Department of Mathematics and Informatics, Delft University of Technology.

Cooper, D. F. (1976). Heuristics for scheduling resource-constrained projects: An experimental investigation. *Management Science* 22, 1186–1194.

Cormack, R. M. (1971). A review of classification. *Journal of the Royal Statistical Society, Ser. A*, 134, 321–367.

Cosby, B. J., Hornberger, G. M., Galloway, J. N., and Wright, R. F. (1985a). Modelling the effects of acid deposition: Assessment of a lumped parameter model of soil water and streamwater chemistry. *Water Resources Research* 21, 51–63.

Cosby, B. J., Wright, R. F., Hornberger, G. M., and Galloway, J. N. (1985b). Modelling the effects of acid deposition: Estimation of long-term water quality responses in a small forested catchment. *Water Resources Research* 21, 1591–1601.

Csendes, T. (1988). Nonlinear parameter estimation by global optimization – efficiency and reliability. *Acta Cybernetica* 8, 361–370.

Csendes, T. (1989). An interval method for bounding level sets of parameter estimation problems. *Computing* 41, 75–86.

Csendes, T. (1990). Interval method for bounding level sets: Revisited and tested with global optimization problems. BIT 30, 650–657.

Csendes, T. (1993). Computational methods for solving parameter estimation problems. Ph.D. Dissertation, József Attila University (JATE), Szeged, Hungary. (In Hungarian.)

Csendes, T., Daróczy, B., and Hantos, Z. (1986). Nonlinear parameter estimation by global optimization: Comparison of local search methods in respiratory system modelling. In: *System Modelling and Optimization,* (ed. A. Prékopa, B. Straziczky and J. Szelezsán), pp. 188–192. *Lecture Notes in Control and Information Sciences* 84. Springer, Berlin.

Csendes, T., Daróczy, B., and Hantos, Z. (1992). Optimal measuring signal generation. *Journal of Global Optimization* 2, 225–226.

Csendes, T., and Pintér, J. (1993a). The impact of accelerating tools on the interval subdivision algorithm for global optimization. *European Journal of Operational Research* 65, 314–320.

Csendes, T., and Pintér, J. (1993b). A new interval method for locating the boundary of level sets. *International Journal of Computer Mathematics* 49, 53–59.

Csiszár, I. (1975). I-divergence geometry of probability distributions and minimization problems. *Annals of Probability* 3, 146–158.

Cunge, J. A., Holly, F. M., and Verwey, A. (1980). *Practical Aspects of Computational River Hydraulics.* Pitman, London.

Daellenbach, H. G., George, J. A., and Nickle, D. C. (1983). *Introduction to Operations Research Techniques,* (2nd edn.). Allyn and Bacon, Newton, Mass.

Danilin, Y. M. (1971). Estimation of the efficiency of an absolute-minimum finding algorithm. *USSR Computational Mathematics and Mathematical Physics* 11, 261–267.

Danilin, Y. M., and Piyavskii, S. A. (1967). On an algorithm for finding the absolute minimum. In: *Theory of Optimal Solutions,* pp. 25–37. Institute of Cybernetics (IK AN USSR), Kiev. (In Russian.)

Debets, H. J. G., Bajema, B. L., and Doornbos, D. A. (1983). A critical evaluation of quality criteria for the optimization of chromatographic multicomponent separations. *Analytica Chimica Acta* 151, 131–141.

DeGroot, M. H. (1974). Reaching a consensus. *Journal of the American Statistical Association* 69, 118–121.

DeGroot, M. H., and Fienberg, S. (1983). The comparison and evaluation of forecasters. *The Statistician* 32, 12–22.

DeGroot, M. H., and Fienberg, S. (1986). Comparing probability forecasters: Basic binary concepts and multivariate extensions. In: *Bayesian Inference and Decision Techniques,* (ed. P. Goel and A. Zellner). Elsevier, London.

De Haan, L. (1970). *On Regular Variation and Its Application to the Weak Convergence of Sample Extremes.* North-Holland, Amsterdam.

De Haan, L. (1981). Estimation of the minimum of a function using order statistics. *Journal of the American Statistical Assocation* 76, 467–469.

Deininger, R. A. (1965). Water quality management: The planning of economically optimal pollution control systems. Ph.D. Thesis, Northwestern University.

Deininger, R. A. (1972). Minimum cost regional pollution control systems, In: *Proceedings of the International Symposium on Modeling Techniques in Water Resources Systems,* Toronto, Ontario, pp. 352–361.

Deininger, R. A., ed. (1977). *Systems Analysis for Water Quality Management (Proceedings of the WHO Symposium, Budapest, 1975).* School of Public Health, The University of Michigan, Ann Arbor.

Deininger, R. A., and Lee, B. H. (1989). Monitoring strategies for water distribution systems. Working Paper, School of Public Health, University of Michigan, Ann Arbor.

De Lange, W. J., and Pintér, J. (1991a). Groundwater quality assessment and management: A stochastic modelling approach. Research Report 91.009, Institute for Inland Water Management and Waste Water Treatment, Lelystad.

De Lange, W. J., and Pintér, J. (1991b). Groundwater quality assessment and management: Illustrative numerical results. Research Report 91.031, Institute for Inland Water Management and Waste Water Treatment, Lelystad.

Delattre, M., and Hansen, P. (1980). Bicriterion cluster analysis. *IEEE Transactions on Pattern Analysis and Machine Intelligence* PAMI-2/4, 277–291.

Delft Hydraulics (1990). IMPAQT – A mathematical model for simulation of the fate of micropollutants in aquatic systems. User's manual. Delft Hydraulics, Delft.

Dempster, M. A. H., ed. (1980). *Stochastic Programming.* Academic Press, London.

Dempster, M. A. H., and Coupé, G. M. (1982). Investigation of the stability of satellite large angle attitude manoeuvres using non-linear optimization methods. In: *Proceedings of the IFAC/ESA Symposium on Automatic Control in Space,* ESTEC, Noordwijk.

Dempster, M. A. H., Fisher, M. L., Jansen, L., Lageweg, B. J., Lenstra, J. K., and Rinnooy Kan, A. H. G. (1981). Analytical evaluation of hierarchical planning systems. *Operations Research* 29, 707–716.

Denisov, D. V. (1978). On the method of random search for solving constrained optimization problems. *USSR Computational Mathematics and Mathematical Physics* 18, 1103–1111. (In Russian.)

Dennis, J. E., and Schnabel, R. B. (1983). *Numerical Methods for Nonlinear Equations and Unconstrained Optimization.* Prentice-Hall, Englewood Cliffs, NJ.

Dert, C. (1986). A parallel algorithm for global optimization. M.Sc. Thesis, Erasmus University, Rotterdam, and University of Colorado, Boulder.

De Vries, D. J. (1987). IMPAQT – A mathematical model for simulation of the fate of micropollutants in aquatic systems. Model description. Delft Hydraulics Report TOW–IW T250, Delft.

Devroye, L. P. (1978). Progressive global random search of continuous functions. *Mathematical Programming* 15, 330–342.

Devroye, L. P. (1979). A bibliography of random search. Technical Report, SOCS, McGill University, Montreal.

Dewdney, A. K. (1989). *The Turing Omnibus.* Computer Science Press, Rockville, Md.

Di Barba, P., and Savini, A. (1993). Global optimization methodologies and application in electromechanics. *International Journal of Applied Electromagnetics in Materials* 4, 93–98.

Diener, I. (1986). Trajectory nets connecting all critical points of a smooth function. *Mathematical Programming* 36, 340–352.

Diener, I. (1987a). On non-uniqueness of nonlinear L_2-approximation. *Journal of Approximation Theory* 51, 54–67.

Diener, I. (1987b). On the global convergence of path-following methods to determine all solutions to a system of nonlinear equations. *Mathematical Programming* 39, 181–188.

Diener, I. (1990). *Einführung in die Methoden der globalen Optimierung* (2nd edn.) Lecture Notes, Institute for Applied Mathematics, University of Göttingen.

Diener, I. (1995). Trajectory methods in global optimization. In: *Handbook of Global Optimization* (ed. R. Horst and P. M. Pardalos), pp. 649–668. Kluwer, Dordrecht.

Dixon, L. C. W., and Szegö, G. P., eds. (1975, 1978). *Towards Global Optimisation.* Vols. 1–2. North-Holland, Amsterdam.

Dobbins, W. W. (1964). BOD and oxygen relationships in streams. *Journal of the Sanitary Engineering Division, Proceedings of the ASCE* 90, 53–78.

Donohoe, Jr., H. V., and Wallace, P. I. (1982). *Geological Map of the Cobequid Highlands, Colchester, Cumberland and Pictou Counties, Nova Scotia.* Nova Scotia Department of Mines and Energy, Halifax.

Dorfman, R., Jacoby, H. D., and Thomas, H. A., eds. (1972). *Models for Managing Regional Water Quality.* Harvard University Press.

Du, D.-Z. (1995). Minimax and its applications. In: *Handbook of Global Optimization* (ed. R. Horst and P. M. Pardalos), pp. 339–367. Kluwer, Dordrecht.

Eheart, J. W. (1988). Effects of streamflow variation on critical water quality for multiple discharges of decaying pollutants. *Water Resources Research* 24, 1–8.

Eliassen, A., and Saltbones, J. (1983). Modelling of long-range transport of sulphur over Europe: A two-year model run and some model experiments. *Atmospheric Environment* 17, 1457–1473.

Ellis, J. H. (1987). Stochastic water quality optimization using imbedded chance constraints. *Water Resources Research* 23, 2227–2238.

Ellis, J. H., McBean, E. A., and Farquhar, G.J . (1985). Chance-constrained/stochastic linear programming model for acid rain abatement. I: Complete colinearity and noncolinearity. *Atmospheric Environment* 19, 925–937.

Ellis, J. H., McBean, E. A., and Farquhar, G. J. (1986). Chance-constrained/stochastic linear programming model for acid rain abatement. II: Limited colinearity. *Atmospheric Environment* 20, 501–511.

Environment Canada (1990). Cleaning up pollution in the pulp and paper industry: An overview of the federal regulatory strategy. Working document. Ottawa, Canada.

Environment Canada (1992). Regulations for the Canadian pulp and paper industry. Proposed legislation. Ottawa, Canada.

Ermakov, S. M. (1975). *Monte Carlo Methods and Related Problems.* Nauka, Moscow. (In Russian.)

Ermoliev, Yu. M. (1976). *Stochastic Programming Methods.* Nauka, Moscow. (In Russian.)

Ermoliev, Yu. M. (1988). Facility location problems. In: *Numerical Techniques for Stochastic Optimization*, (ed. Yu. M. Ermoliev and R. J.-B. Wets), pp. 413–435. Springer, Berlin.

Ermoliev, Yu. M., and Wets, R. J.-B., eds. (1988). *Numerical Techniques for Stochastic Optimization.* Springer, Berlin.

Erup, L., and Harris, R.A. (1988). On numerical optimization of communications system design. *IEEE Journal on Selected Areas in Communications* 6, No. 1, 106–125.

Evans, J. D., Kember, G. C., and Major, G. (1993). Techniques for obtaining analytical solutions to the multicylinder somatic shunt cable model for passive neurones. *Biophysical Journal* 63, 350-365.

Everaars, C. T. H., Hemker, P. W., and Stortelder, W. J. H. (1995). Manual of spIds. Technical Report, Department of Numerical Mathematics, CWI, Amsterdam.

Evtushenko, Yu. G. (1971). Algorithm for finding the global extremum of a function (case of a non-uniform mesh). *USSR Computational Mathematics and Mathematical Physics* 11, No. 6, 1390–1403.

Evtushenko, Yu. G. (1985). *Numerical Optimization Techniques.* Springer, Berlin.

Falk, J. E., and Hoffman, K. L. (1976). A successive underestimating method for concave minimization problems. *Mathematics of Operations Research* 1, 251–259.

Falk, J. E., and Soland, R. M. (1969). An algorithm for separable nonconvex programming problems. *Management Science* 15, 550–569.

Fedorov, V.V., ed. (1985). *Problems of Cybernetics: Models and Methods in Global Optimization.* USSR Academy of Sciences, Moscow. (In Russian.)

Fedra, K. (1991). Smart software for water resources planning and management. In: *Decision Support Systems in Water Resources Planning,* (ed. P. Loucks and J. R. Da Costa). NATO ASI Series, Vol. G 26, Springer, Berlin.

Feller, W. (1971). *An Introduction to Probability Theory and Its Applications.* Wiley, New York.

Fels, M., and Lycon, D. S. (1992). ESIS Project: Waste Water Treatment Modules. Technical Report, Department of Chemical Engineering, Technical University of Nova Scotia, Halifax, N.S.

Fels, M., and Pintér, J. (1994). Optimized design of wastewater treatment systems: Application to the mechanical pulp and paper industry. Working Paper, Department of Chemical Engineering, Technical University of Nova Scotia, Halifax, N.S. (To appear.)

Fenton, G. A., and Griffith, D. V. (1993). Statistics of block conductivity through a simple bounded stochastic medium. *Water Resources Research* 29, 1825–1830.

Feo, T. A., and Resende, M. G. C. (1995). Greedy randomized adaptive search procedures. *Journal of Global Optimization* 6, 109–133.

Ferrell, W. R. (1985). Combining individual judgments. In: *Behavioral Decision Making,* (ed. G. Wright), pp. 111–145. Plenum, New York.

Fiacco, A. V., and McCormick, G. P. (1968). *Nonlinear Programming: Sequential Unconstrained Minimization Techniques*. Wiley, New York.

Fields, D. E., and Miller, C. W. (1988). A methodology for deriving model input parameters from a set of environmental data. *Ecological Modelling* 40, 155–159.

Finch, S. (1992). A convex maximization problem. *Journal of Global Optimization* 2, 419.

Finley, J. R. (1994). An aquifer model study of the Wolfville Formation using co-kriging and optimization techniques. M.Sc. Thesis. Technical University of Nova Scotia, Halifax, N.S.

Finley, J. R., Pintér, J., and Satish, M. (1994). Aquifer model calibration applying global optimization. In: *Proceedings of the 3rd IASTED Conference* (Washington, D.C., September 1994). Working Paper 94–05, Department of Industrial Engineering, Technical University of Nova Scotia, Halifax, N.S.

Fishman, G. S. (1978). *Principles of Discrete Event Simulation.* Wiley, New York.

Flåm, S. D., and Pintér, J. (1985). Selecting oil exploration strategies: Some stochastic programming formulations and solution methods. Research Report CMI 852611–1. Christian Michelsen Institute, Fantoft, Bergen.

Fletcher, R. (1983). Penalty functions. In: *Mathematical Programming: The State of the Art,* (ed. A. Bachem, M. Grötschel, and B. Korte). Springer, Berlin.

Fletcher, R., and Reeves, C. M. (1964). Function minimization by conjugate gradients. *Computer Journal* 7, 149–154.

Flippo, O. E., and Rinnooy Kan. A. H. C. (1993). Decomposition in general mathematical programming. *Mathematical Programming* 60, 361–382.

Florian, M., ed. (1984). *Transportation Planning Models.* North-Holland, Amsterdam.

Floudas, Ch. A., and Pardalos, P. M. (1990). *A Collection of Test Problems for Constrained Global Optimization Algorithms.* Springer, Berlin.

Floudas, Ch. A., and Pardalos, P. M., eds. (1992). *Recent Advances in Global Optimization.* Princeton University Press.

Floudas, Ch. A., and Visweswaran, V. (1995). Quadratic optimization. In: *Handbook of Global Optimization* (ed. R. Horst and P. M. Pardalos), pp. 217–269. Kluwer, Dordrecht.

Forgó, F. (1972). Cutting plane methods for solving nonconvex quadratic problems. *Acta Cybernetica* 1, 171–192.

Forgó, F. (1978). *Nonconvex and Discrete Programming.* Közgazdasági és Jogi Könyvkiadó, Budapest. (In Hungarian.)

Forgó, F. (1988). *Nonconvex Programming.* Akadémiai Kiadó, Budapest.

Forlenza, M. F. (1982). Fluvial sedimentology of the Middle and Upper Triassic Wolfville Redbeds, Hants County, Nova Scotia. M.Sc. Thesis. University of Massachusetts.

Forster, W., ed. (1980). *Numerical Solution of Highly Nonlinear Problems.* North-Holland, Amsterdam.

Forster, W. (1992). Some computational methods for systems of nonliner equations and systems of polynomial equations. *Journal of Global Optimization* 2, 317–356.

Forster, W. (1995). Homotopy methods. In: *Handbook of Global Optimization* (ed. R. Horst and P. M. Pardalos), pp. 669–750. Kluwer, Dordrecht.

Fortin, M., and McBean, E. A. (1983). A management model for acid rain abatement. *Atmospheric Environment* 17, 2331–2336.

Fourer, R., Gay, D. M., and Kernighan, B. W. (1987). AMPL: A mathematical programming language. Computing Science Technical Report No. 133, AT&T Bell Laboratories, Murray Hill, NJ.

Fourer, R., Gay, D. M., and Kernighan, B. W. (1993). *AMPL: A Modelling Language for Mathematical Programming.* The Scientific Press, San Francisco.

Frauendorfer, K. (1992). *Stochastic Two-Stage Programming.* Springer, Berlin.

Freeze, R. A., and Cherry, J. A. (1979). *Groundwater.* Prentice-Hall, Englewood Cliffs, N.J.

French, S. (1985). Group consensus probability distributions: A critical survey. In: *Bayesian Statistics,* (ed. J. M. Bernardo, M. H. DeGroot, D. V. Lindley, and A. E. M. Smith). Elsevier, London.

French, S. (1986). *Decision Theory: An Introduction to the Mathematics of Rationality.* Ellis Horwood, Chichester.

Fronza, G., and Melli, P. (1984). Assignment of emission abatement levels by stochastic programming. *Atmospheric Environment* 18, 531–535.

Fujii, V., Ichida, K., and Ozasa, M. (1986). Maximization of multivariable functions using interval analysis. In: *Interval Mathematics 1985* (ed. K. L. E. Nickel). Springer, Berlin.

Fujiwara, O., Gnanendran, S. K., and Ohgaki, S. (1986). River quality management under stochastic streamflow. *Journal of the Environmental Engineering Division, Proc. ASCE,* 112, 185–198.

Fujiwara, O., Gnanendran, S. K., and Ohgaki, S. (1987). Chance constrained model for river water quality management. *Journal of the Environmental Engineering Division, Proc. ASCE,* 1018–1031.

Fülöp, J. (1988). Reverse convex programming over a polytope. Working Paper, Computer and Automation Institute of the Hungarian Academy of Sciences, Budapest.

Fülöp, J. (1990). A finite cutting plane method for solving linear programs with an additional reverse convex constraint. *European Journal of Operational Research* 44, 395–405.

Fülöp, J. (1991). A finite cutting plane method for facial disjunctive programs. *ZOR – Methods and Models of Operations Research* 35, 1–13.

Fülöp, J. (1992). A semi-infinite programming method for approximating load duration curves by polynomials. *Computing* 49, 201–212.

Fülöp, J., and Prill, M. (1992). On the minimax approximation in the class of the univariate piecewise constant functions. *Operations Research Letters* 12, 307–312.

Galambos, J. (1977). Bonferroni inequalities. *Annals of Probability* 5, 577–581.

Galambos, J. (1978). *The Asymptotic Theory of Extreme Order Statistics.* Wiley, New York.

Galperin, E. A. (1985). The cubic algorithm. *Journal of Mathematical Analysis and Applications* 112, 635–640.

Galperin, E. A. (1988). Precision, complexity, and computational schemes of the cubic algorithm. *Journal of Optimization Theory and Applications* 57, 223–238.

Galperin, E. A. (1993). The fast cubic algorithm. *Computers and Mathematics with Applications* 25, 147–160.

Galperin, E. A., and Zheng, Q. (1987). Nonlinear observation via global optimization methods: Measure theory approach. *Journal of Optimization Theory and Applications* 54, 63–92.

Galvão, R. D. (1993). The use of Lagrangean relaxation in the solution of uncapacitated facility location problems. *Location Science* 1, 57–79.

Garcia, C. B., and Gould, F. J. (1980). Relations between several path following algorithms and local and global Newton methods. *SIAM Review* 22, 263–274.

Garcia, C. B., and Zangwill, W. I. (1981). *Pathways to Solutions, Fixed Points and Equilibria.* Prentice-Hall, Englewood Cliffs, NJ.

Gardner, R. H., Cale, W. G., and O'Neill, R. V. (1982). Robust analysis of aggregation error. *Ecology* 63, 1771–1779.

Garfinkel, R. S., and Nemhauser, G. L. (1972). *Integer Programming*. Wiley, New York.

Gassmann, H. I. (1989). Optimal harvest of a forest in the presence of uncertainty. *Canadian Journal of Forest Research* 19, 1267–1274.

Gassmann, H. I., and Ziemba, W. T. (1986). A tight upper bound for the expectation of a convex function of a multivariate random variable. *Mathematical Programming Study* 27, 39–53.

Gaviano, M. (1975). Some general results on the convergence of random search algorithms in minimisation problems. In: *Towards Global Optimisation,* Vol. 1, (ed. L. C. W. Dixon and G. P. Szegö), pp. 149–157. North-Holland, Amsterdam.

Ge, R. P., and Qin, Y. F. (1987). A class of filled functions for finding global minimizers of a function of several variables. *Journal of Optimization Theory and Applications* 54, 241-252.

Genest, Ch., and Zidek, J. V. (1986). Combining probability distributions: A critique and an annotated bibliography. *Statistical Science* 1, 114–148.

Gergel, V. P., and Sergeyev, Ya. D. (1994). Sequential and parallel global optimization algorithms using second derivatives. Research Report 7–94, Department of Mathematics, University of Calabria, Italy.

Gergel, V. P., Sergeyev, Ya. D., and Strongin, R. G. (1992). A parallel global optimization method and its implementation on a transputer system. *Optimization* 26, 261–275.

Gergel, V. P. (1995). A global optimization algorithm for multivariate functions with Lipschitz first derivatives. Technical Report, Software Department, University of Nizhni Novgorod. (Submitted.)

Ghost, A., and Harche, F. (1993). Location-allocation models in the private sector: Progress, problems and prospects. *Location Science* 1, 81–106.

Glover, F. (1978). Integer programming and combinatorics. Chapter II, Section 2, in: *Handbook of Operations Research,* Vol. 1, (eds. J. J. Moder and S. E. Elmaghraby). Van Nostrand Reinhold, New York.

Glover, F. (1986). Future paths for integer programming and links for artificial intelligence. *Computers and Operations Research* 13, 533–549.

Gnedenko, B. V. (1943). Sur la distribution limite du terme maximum d'une série aléatoire. *Annals of Mathematics* 44, 423–453.

Goicoechea, A., Hansen, D. R., and Duckstein, L. (1982). *Multiobjective Decision Analysis with Engineering and Business Applications.* Wiley, New York.

Golden, B. L., and Alt, F. B. (1979). Interval estimation of a global minimum for large combinatorial problems. *Naval Research Logistics Quarterly* 26, 69–77.

Golden, B. L., and Assad, A. (1986). Perspectives in vehicle routing: Exciting new developments. *Operations Research* 34, 803–810.

Goldstein, A. A., and Price, J. F. (1971). On descent from local minima. *Mathematics of Computation* 25, 569–574.

Gomulka, J. (1975). Remarks on Branin's method for solving nonlinear equations. In: *Towards Global Optimisation,* Vol. 1, (ed. L. C. W. Dixon and G. P. Szegö), pp. 29–54. North-Holland, Amsterdam.

Gomulka, J. (1978). Two implementations of Branin's method: Numerical experience. In: *Towards Global Optimisation,* Vol. 2, (ed. L. C. W. Dixon and G. P. Szegö), pp. 151–163. North-Holland, Amsterdam.

González, R., Rofman, E., and Sagastizábal, C. (1995). Global optimization of arborescent multilevel inventory systems. *Journal of Global Optimization* 6, 269–292.

Gordon, A. D. (1981). *Classification Methods for the Exploratory Analysis of Multivariate Data.* Chapman and Hall, New York.

Gorelick, S. M. (1990). Large scale nonlinear deterministic and stochastic optimization: Formulations involving simulation of subsurface contamination. *Mathematical Programming* 48, 19–39.

Gourdin, E., Hansen, P., and Jaumard, B. (1994). Finding maximum likelihood estimators for the three-parameter Weibull distribution. *Journal of Global Optimization* 5, 373–397.

Greenberg, H. J. (1993). A bibliography for the development of an intelligent mathematical programming system. Working Paper, Department of Mathematics, University of Colorado, Denver. (To appear.)

Greenberg, H. J. (1995). Mathematical programming models for environmental quality control. Working Paper, Department of Mathematics, University of Colorado, Denver. (To appear.)

Griewank, A. O. (1981). Generalized descent for global optimization. *Journal of Optimization Theory and Applications* 34, 11–39.

Griewank, A. (1994). Computational differentiation and optimization. In: *Mathematical Programming – State of the Art*, (ed. J. R. Birge and K. G. Murty), pp. 102–131. The University of Michigan, Ann Arbor.

Guddat, J., Guerra Vazguez, F., and Jongen, H.,Th. (1990). *Parametric Optimization: Singularities, Pathfollowing, and Jumps*. Wiley and Teubner, New York.

Guisewite, G. M., and Pardalos, P. M. (1991). Algorithms for the single-source uncapacitated minimum concave-cost network flow problem. *Journal of Global Optimization* 1, 245–265.

Guisewite, G. M. (1995). Network problems, In: *Handbook of Global Optimization* (ed. R. Horst and P. M. Pardalos), pp. 609–648.

Gumbel, E. J. (1958). *Statistics of Extremes*. Columbia University Press, New York.

Gutjahr, W. J. (1994). A global optimization problem in series-parallel networks with maximum reliability. *Journal of Global Optimization* 5, 403–404.

Haith, D. A. (1982). *Environmental Systems Optimization*. Wiley, New York.

Halász, E. (1982). Design centering and tolerancing, considering environmental effects via a new type of minimax optimisation. *IEE Proceedings* 129, Pt. G, No. 4, 134–138.

Hall, C. A. S. (1988). An assessment of several of the historically most influential theoretical models used in ecology and of the data provided in their support. *Ecological Modelling* 43, 5–31.

Hall, N., ed. (1991). *Exploring Chaos*. W. W. Norton and Co., New York.

Hansen, E. R. (1979). Global optimization using interval analysis: The one-dimensional case. *Journal of Optimization Theory and Applications* 29, 331–344.

Hansen, E. R. (1980). Global optimization using interval analysis: The multidimensional case. *Numerische Mathematik* 34, 247–270.

Hansen, E. R. (1991). Bounding the set of solutions of a perturbed global optimization problem. *Journal of Global Optimization* 1, 359–374.

Hansen, E. R. (1992). *Global Optimization Using Interval Analysis*. Marcel Dekker, New York.

Hansen, P., and Delattre, M. (1980). Complete-link cluster analysis by graph coloring. *Journal of the American Statistical Association* 73, 397–403.

Hansen, P., and Jaumard, B. (1987). Minimum sum of diameters clustering. *Journal of Classification* 4, 215–226.

Hansen, P., and Jaumard, B. (1988). Maximum sum of splits clustering. Research Report 88–07, GERAD, HEC, Montréal.

Hansen, P., and Jaumard, B. (1995). Lipschitz optimization. In: *Handbook of Global Optimization*, (ed. R. Horst and P. M. Pardalos), pp. 407–493. Kluwer, Dordrecht.

Hansen, P., Jaumard, B., and Lu, S. H. (1989a). Some further results on monotonicity in globally optimal design. *Transactions of the ASME, Journal of Mechanisms, Transmissions, and Automation in Design* 111, 345–352.

Hansen, P., Jaumard, B., and Lu, S. H. (1989b). A framework for algorithms in globally optimal design. *Transactions of the ASME, Journal of Mechanisms, Transmissions, and Automation in Design* 111, 353–360.

Hansen, P., Jaumard, B., and Lu, S. H. (1989c). An automated procedure for globally optimal design. *Transactions of the ASME, Journal of Mechanisms, Transmissions, and Automation in Design* 111, 361–367.

Hansen, P., Jaumard, B., and Lu, S. H. (1991a). On the number of iterations of Piyavskii's global optimization algorithm. *Mathematics of Operations Research* 16, 334–350.

Hansen, P., Jaumard, B. and Lu, S.-H. (1991b). An analytical approach to global optimization. *Mathematical Programming* 52, 227–254.

Hansen, P., Jaumard, B., and Lu, S. H. (1992a). Global optimization of univariate Lipschitz functions: I. Survey and properties. *Mathematical Programming* 55, 251–272.

Hansen, P., Jaumard, B., and Lu, S. H. (1992b). Global optimization of univariate Lipschitz functions: II. New algorithms and computational comparison. *Mathematical Programming* 55, 273–292.

Hansen, P., Jaumard, B., and Lu, S. H. (1992c). On the use of estimates of the Lipschitz constant in global optimization. *Journal of Optimization Theory and Applications* 75, 195–200.

Hantos, Z., Daróczy, B., Csendes, T., Suki, B., and Nagy, S. (1990). Modelling of low-frequency pulmonary impedance in the dog. *Journal of Applied Physiology* 68, 849–860.

Hardy, J. W. (1975). An implemented extension of Branin's method. In: *Towards Global Optimisation,* Vol. 1, (ed. L. C. W. Dixon and G. P. Szegö), pp. 117–142. North-Holland, Amsterdam.

Hartigan, J. A. (1975). *Clustering Algorithms.* Wiley, New York.

He, L. and Polak, E. (1993). Multistart method with estimation scheme for global satisfycing problems. *Journal of Global Optimization* 3, 139–156.

Hemker, P. W., and Kok, J. (1993). A project on parameter identification in reaction kinetics. Technical Report NM–R9300, Department of Numerical Mathematics, CWI, Amsterdam.

Hendrickson, J. D., Sorooshian, S., and Brazil, L. E. (1988). Comparison of Newton-type and direct search algorithms for calibration of conceptual rainfall-runoff models. *Water Resources Research* 24, 691–700.

Hendrix, E. M. T. (1989). Risk and uncertainty in OR models which can be used in agriculture. Technical Note 89–01, Department of Mathematics, Agricultural University, Wageningen.

Hendrix, E. M. T., Mecking, C. J., and Hendriks, Th. H. B. (1993). A mathematical formulation of finding robust solutions for a product design problem. Technical Note 93–02, Department of Mathematics, Agricultural University, Wageningen.

Hendrix, E. M. T., and Pintér, J. (1991). An application of Lipschitzian global optimization to product design. *Journal of Global Optimization* 1, 389–401.

Henley, E., and Kumamoto, H. (1981). *Reliability Engineering and Risk Assessment.* Prentice-Hall, Englewood Cliffs, N.J.

Hestenes, M. R. (1980). *Conjugate Direction Methods in Optimization.* Springer, New York.

Hettelingh, J.-P. (1989). Uncertainty in Modeling Environmental Systems: The Generalization of a Watershed Acidification Model for Predicting Broad Scale Effects. Ph.D. Dissertation, Free University, Amsterdam.

Hillestad, R. J., and Jacobsen, S. E. (1980). Reverse convex programming. *Applied Mathematics and Optimization* 6, 63–78.

Hillier, F. S., and Lieberman, G. J. (1986). *Introduction to Operations Research* (4th edn.). Holden Day, Oakland.

Himmelblau, D. M. (1972a). *Applied Nonlinear Programming.* McGraw-Hill, New York.

Himmelblau, D. M. (1972b). A uniform evaluation of unconstrained optimization techniques. In: *Numerical Methods for Nonlinear Optimization,* (ed. F. A. Lootsma), pp. 69–97. Academic Press, London.

Hipel, K. W. (1985). Time series analysis in perspective. *Water Resources Bulletin* 21, 609–624.

Hipel, K. W., and McLeod, A. I. (1987). *Time Series Modelling for Water Resources and Environmental Engineers.* Elsevier, Amsterdam.

Hiriart-Urruty, J.-B. (1995) Conditions for global optimality. *Handbook of Global Optimization* (ed. R. Horst and P. M. Pardalos), pp. 1–26. Kluwer, Dordrecht.

Hoeffding, W. (1963). Probability inequalities for sums of bounded random variables. *Journal of the American Statistical Association* 58, 13–30.

Hoffman, K. L. (1981). A method for globally minimizing concave functions over convex sets. *Mathematical Programming* 20, 22–32.

Hogan, A. J., Morris, J. G., and Thompson, H. E. (1981). Decision problems under risk and chance constrained programming: Dilemmas in the transition. *Management Science* 27, 698–716.

Holling, C. S., ed. (1978). *Adaptive Environmental Assessment and Management.* IIASA–Wiley, Chichester.

Hornberger, G. M., and Cosby, B. J. (1985). Selection of parameter values in environmental models using sparse data: A case study. *Applied Mathematics and Computation* 17, 335–355.

Hornberger, G. M., and Spear, R. C. (1981). An approach to the preliminary analysis of environmental systems. *Journal of Environmental Management* 12, 7–18.

Horst, R. (1976). An algorithm for nonconvex programming problems. *Mathematical Programming* 10, 312–321.

Horst, R. (1980). A note on the convergence of an algorithm for nonconvex programming problems. *Mathematical Programming* 19, 237–238.

Horst, R. (1984a). On the convexification of nonconvex programming problems. *European Journal of Operational Research* 15, 382–392.

Horst, R. (1984b). On the global minimization of concave functions: Introduction and survey. *Operations Research Spektrum* 61, 195–205.

Horst, R. (1986). A general class of branch-and-bound methods in global optimization, with some new approaches for concave minimization. *Journal of Optimization Theory and Applications* 51, 271–291.

Horst, R. (1988). Deterministic global optimization with partition sets whose feasibility is not known. Application to concave minimization, DC-programming, reverse convex constraints and Lipschitzian optimization. *Journal of Optimization Theory and Applications* 58, 11–37.

Horst, R. (1990). Deterministic methods in constrained global optimization: Some recent advances and new fields of application. *Naval Research Logistics* 37, 433–471.

Horst, R., Nast, M., and Thoai, N. V. (1994). A new *LP*-bound in multivariate Lipschitz optimization: Application to unconstrained and linearly constrained problems and to systems of inequalities. Research Report 94–08, Department of Mathematicss, University of Trier. (To appear.)

Horst, P., and Pardalos, P. M., eds. (1995). *Handbook of Global Optimization.* Kluwer, Dordrecht.

Horst, R., and Thoai, N. V. (1988). Branch-and-bound methods for solving Lipschitzian equations and inequalities. *Journal of Optimization Theory and Applications* 58, 139–146.

Horst, R., and Thoai, N. V. (1993). A decomposition approach for the global minimization of biconcave functions over polytopes. Research Report 93–23, Department of Mathematics, University of Trier. (To appear.)

Horst, R.. and Thoai, N. V. (1994). Constraint decomposition algorithms in global optimization. *Journal of Global Optimization* 5, 333–348.

Horst, R., Thoai, N. V., and Tuy, H. (1989) On an outer approximation concept in global optimization. *Optimization* 20, 255–264.

Horst, R., and Tuy, H. (1987). On the convergence of global methods in multiextremal optimization. *Journal of Optimization Theory and Applications* 54, 253–271.

Horst, R., and Tuy, H. (1990). *Global Optimization: Deterministic Approaches.* Springer, Berlin. (2nd edn., 1993.)

Howe, C. W. (1971). *Benefit-Cost Analysis for Water System Planning.* American Geophysical Union, Washington, D.C.

Hu, T. C., and Kuh, E. S., eds. (1987). *VLSI Circuit Layout: Theory and Design.* IEEE Press, New York.

Huang, Z.-K., Wang, S.-D., and Kuo, T.-S. (1993). Multi-modal parameter identification by automata approach. *Journal of the Chinese Institute of Engineers*, Series A, 16, 603–613.

ten Hulscher, D., Bak, C., and Pintér, J. (1991). Calibration of the model system DELWAQ-IMPAQT for the lake Ketelmeer. Research Report 91.001, Institute for Inland Water Management and Waste Water Treatment, Lelystad.

Hwang, C. L., and Masud, A. S. M. (1980). *Multiple Objective Decision Making: Methods and Applications.* Springer, Berlin.

Ibaraki, T., ed. (1987). Enumerative approaches to combinatorial optimization. *Annals of Operations Research*, Vols. 10–11.

Incerti, S., Parisi, V., and Zirilli, F. (1979) A new method for solving nolinear simultaneous equations. *SIAM Journal on Numerical Analysis* 16, 779–789.

Ivanov, V. V. (1972). On optimal algorithms of minimization in the class of functions with the Lipschitz condition. *Information Processing* 2, 1324–1327.

Jackson, R. H. F., Boggs, P. T., Nash, S. C., and Powell, S. (1991). Guidelines for reporting results of computational experiments: Report of the ad hoc committee. *Mathematical Programming* 49, 413–425.

Jacobsen, S. K., and Madsen, O. B. G. (1980). A comparative study of heuristics for a two level routing-location problem. *European Journal of Operational Research* 5, 378–387.

Jaumard, B., Lu, S.-H., and Sriskandarajah, C. (1988). Design parameter selection and ordering of part-orienting devices. Research Report G–88–39, GERAD, HEC, Montréal.

Jaumard, B., Ow, P. S., and Simeone, B. (1988). A selected artificial intelligence bibliography for operations researchers. *Annals of Operations Research* 12, 1–50.

Jeruchim, M. C., Shanmugan, K. S., Biglieri, E., and Balaban, P. (1988). Computer-aided modeling, analysis, and design of communication links: Introduction and issue overview. *IEEE Journal on Selected Areas in Communications* 6, No. 1, 1–4.

Johnson, R. C. (1979). *Optimum Design of Mechanical Elements.* Wiley, New York.

Jolánkai, G. (1992). Hydrological, chemical and biological processes of contaminant transformation and transport in river and lake systems. IHP–IV Projects H–3.2, International Hydrological Programme, UNESCO, Paris.

Jones, C. V. (1994). Visualization and optimization. Working Paper, Faculty of Business Administration, Simon Fraser University, Burnaby, B.C. (To appear.)

Jones, D. R., Perttunen, C. D., and Stuckman, B. E. (1993). Lipschitzian optimization without the Lipschitz constant. *Journal of Optimization Theory and Applications* 79, 157–181.

Jongen, H. Th., Jonker, P., and Twilt, F. (1979). On Newton flows in optimization. *Methods of Operations Research* 31, 345–359.

Jongen, H. Th., Jonker, P., and Twilt, F. (1983, 1986). *Nonlinear Optimization in* $I\!R^n$, Vols. 1 and 2. Peter Lang, Frankfurt a. M.

Jørgensen, S. E., ed. (1983). *Applications of Ecological Modelling in Environmental Management.* Elsevier, Amsterdam.

Kalantari, B. (1986). Quadratic functions with exponential number of local maxima. *Operations Research Letters* 5, 47–49.

Kalantari, B., and Rosen, J. B. (1987). Algorithms for global minimization of linearly constrained concave quadratic functions. *Mathematics of Operations Research*, 12, 544–561.

Kall, P. (1976). *Stochastic Linear Programming.* Springer, Berlin.

Kall, P. (1987). Stochastic programming with recourse: Upper bounds and moment problems. Research Report, Institute for Operations Research, University of Zurich.

Kall, P., and Prékopa, A., eds. (1980). *Recent Results in Stochastic Programming. Lecture Notes in Economics and Mathematical Systems* 179. Springer, Berlin.

Kaplan, S., and Garrick, B. (1981). On the quantitative definition of risk. *Risk Analysis* 1, 11–27.

Kaplinskij, A. I., and Propoj, A. I. (1993). Global optimization techniques based on the potential theory. *Avtomatika i Telemekhanika* 7, 55–65. (In Russian.)

Karatzas, G. P., and Pinder, G. F. (1992). Combination of groundwater simulation with an outer approximation method for global optimization. In: *Finite Elements in Water Resources,* Vol. 1, pp. 337–344. *Proceedings of the 9th International Conference on Computational Methods in Water Resources* (Denver, Colorado, June 1992). Computational Mechanics Publications, Southampton.

Kaufman, L., and Rousseeuw, P. (1987). Clustering by means of medoids. Research Report 87–03, Department of Mathematics and Informatics, Delft University of Technology.

Kaufman, L., and Rousseeuw, P. (1990). *Finding Groups in Data: An Introduction to Cluster Analysis.* Wiley, New York.

Kelley, J. E. (1960). The cutting-plane method for solving convex programs. *SIAM Journal* 8, 703–712.

Khachiyan, L. G. (1979). A polynomial algorithm in linear programming. *Soviet Mathematics Doklady* 20, 191–194.

Kiefer, J. (1953). Sequential minimax search for a maximum *Proceedings of the American Mathematical Society* 4, 502–506.

King, A. (1988). An implementation of the Lagrangian finite generation method. In: *Numerical Techniques for Stochastic Optimization* (ed. Yu. M. Ermoliev and R. J.-B. Wets). Springer, Berlin.

Klafszky, E., Mayer, J., and Terlaky, T. (1989). Linearly constrained estimation by mathematical programming. *European Journal of Operational Research* 34, 254–267.

Klastorin, T. D. (1985). The p-median problem for cluster analysis: A comparative test using the mixture model approach. *Management Science* 31, 84–95.

Kleijnen, J. P. C., van Ham, G., and Rotmans, J. (1990). Techniques for sensitivity analysis of simulation models: A case study of the CO_2 greenhouse effect. Research Report FEW 448, Tilburg University.

Klein Haneveld, W. K. (1986). *Duality in Stochastic Linear and Dynamic Programming.* Springer, Berlin.

Kleindorfer, P. R., and Kunreuther, H. C., eds. (1987). *Insuring and Managing Hazardous Risks: From Seveso to Bhopal and Beyond.* Springer, Berlin.

Klepper, O. (1989). A model of carbon flows in relation to macrobenthic food supply in the Oosterschelde estuary (S.W. Netherlands). Ph.D. Dissertation, Agricultural University, Wageningen.

Klepper, O., and Hendrix, E. M. T. (1994). A method for robust calibration of ecological models under different types of uncertainty. *Ecological Modelling* 74, 161–182.

Kneese, A. V., Ayres, R. U., and d'Arge, R. C. (1970). *Economics and the Environment: A Materials Balance Approach.* Johns Hopkins University Press, Baltimore.

Kneese, A. V., and Bower, B. T. (1968). *Managing Water Quality: Economics, Technology, Institutions.* Johns Hopkins University Press, Baltimore.

Konno, H., and Kuno, T. (1995). Multiplicative programming problems. In: *Handbook of Global Optimization,* (ed. R. Horst and P. M. Pardalos), pp. 369–405. Kluwer, Dordrecht.

Korotkich, V. V. (1989). Multilevel dichotomy algorithm in global optimization. In: *Proceedings of the 14th IFIP Conference on Systems Modelling and Optimization,* pp. 161–169. Springer, Berlin.

Kovács, L. B. (1969). *Combinatorial Methods in Discrete Programming.* Bolyai Mathematical Society, Budapest. (In Hungarian.)

Kristinsdottir, B. P., Zabinsky, Z. B., Csendes, T., and Tuttle, M. E. (1993). Methodologies for tolerance intervals. *Interval Computations* 3, 133–147.

Kularathna, M., and Somlyódy, L. (1994). River basin water quality management models: A state-of-the-art review. Working Paper 94–3, International Institute for Applied Systems Analysis, Laxenburg, Austria.

Kumar, P. R., and Varaiya, P. (1986). *Stochastic Systems Estimation, Identification and Adaptive Control.* Prentice-Hall, Englewood Cliffs, NJ.

Kushner, H. J. (1964). A new method of locating the maximum point of an arbitrary multipeak curve in the presence of noise. *Transactions of ASME, Series D, Journal of Basic Engineering* 86, 97–105.

Kushner, H. J., and Clark, D. S. (1978). *Stochastic Approximation Methods for Constrained and Unconstrained Systems.* Springer, New York.

Kushner, H. J., and Dupuis, P. G. (1992). *Numerical Methods for Stochastic Control Problems in Continuous Time.* Springer, New York.

Kusiak, A., ed. (1986). *Flexible Manufacturing Systems: Methods and Studies.* North-Holland, Amsterdam.

Kutas, T., and Herodek, S. (1987). Effects of load reductions on the water quality of a large shallow lake. *Ecological Modelling* 39, 85–99.

Lahey Computer Systems, Inc. (1994a). *Fortran 90 – User's Guide.* Incline Village, Nevada.

Lahey Computer Systems, Inc. (1994b). *Fortran 90 – Language Reference.* Incline Village, Nevada.

Lam, D. C. L., and Swayne, D. A. (1991). Integrating database, spreadsheet, graphics, GIS, statistical simulation models and expert systems: Experiences with the RAISON system on microcomputers. In: *Decision Support Systems in Water Resources Planning,* (ed. D. P. Loucks and J. R. Da Costa), NATO ASI Series, Vol. G 26. Springer, Berlin.

Landis, E. M. (1951). On functions representable as the difference of two convex functions. *Dokladi Akademii Nauk SSSR* 80, 9–11. (In Russian.)

Laporte, G. (1988). Location-routing problems. In: *Vehicle Routing: Methods and Studies*, (ed. B. I. Golden and A. A. Assad), pp. 163–197. Elsevier, New York.

Lasdon, L. S., and Waren, A. D. (1986). GRG2 User's Guide. School of Business Administration, University of Texas, Austin; and College of Business Administration, Ohio State University, Cleveland.

Lawler, E. L., Lenstra, J. K., Rinnooy Kan, A. H. G., and Schmoys, D. B. (1985). *The Travelling Salesman Problem: A Guided Tour of Combinatorial Optimization.* Wiley, Chichester.

Lawler, E. L. and Wood, D. E. (1969). Branch and bound method: A survey. *Operations Research* 14, 699–719.

Le Grand, S. M., and Merz, K. M., Jr. (1993). The application of the genetic algorithm to the minimization of potential energy functions. *Journal of Global Optimization* 3, 49–66.

Levy, A. V., and Gomez, S. (1980) The tunneling algorithm for the global optimization problem of constrained functions. Technical Report, Universidad Nacional Autonoma de Mexico.

Levy, A. V., and Gomez, S. (1985). The tunneling method applied to global optimization. In: *Numerical Optimization 1984*, (ed. P. T. Boggs), pp. 213–244. SIAM, Philadelphia.

Levy, A. V., and Montalvo, A. (1985), The tunneling algorithm for the global minimization of functions. *SIAM Journal of Scientific and Statistical Computation* 6, 15–29.

Lewandowski, A., and Wierzbicki, A. P. (1989). *Aspiration-Based Decision Support Systems.* Springer, Berlin.

Li, Z., Pardalos, P. M., and Levine, S. H. (1992). Space-covering approach and modified Frank–Wolfe algorithm for optimal nuclear reactor reload design. In: *Recent Advances in Global Optimization,* (ed. C. A. Floudas and P. M. Pardalos), pp. 593–615. Princeton University Press.

Lindley, D., and Singpurwalla, N. (1986). Reliability and fault tree analysis using expert opinion. *Journal of the American Statistical Association* 81, 87–90.

Lober, D. J. (1993). Beyond self-interest: A model of public attitudes towards water facility siting. *Journal of Environmental Planning and Management* 36, 345–363.

Lock, A. (1987). Integrating group judgment in subjective forecasts. In: *Judgement Forecasting,* (ed. G. Wright and P. Ayton). Wiley, New York.

Loehle, C. (1988). Robust parameter estimation for nonlinear models. *Ecological Modelling* 41, 41–54.

Lohani, B. N., and Thanh, N. C. (1979). Probabilistic water quality control policies. *Journal of Environmental Engineering Division, Proc. ASCE,* 105, 713–725.

Lootsma, F. A., Meisner, J., and Schellemanns, F. (1986). Multi-criteria decision analysis as an aid to the strategic planning of energy R&D. *European Journal of Operational Research* 25, 216–234.

Loucks, D. P., and da Costa, J. R., eds. (1991). *Decision Support Systems in Water Resources Planning.* NATO ASI Series Vol. G26. Springer, Berlin.

Loucks, D. P., Stedinger, J. R., and Haith, D. A. (1981). *Water Resources Systems Planning and Analysis.* Prentice-Hall, Englewood Cliffs, NJ.

Lycon, D. S. (1992). Sensitivity analysis of wastewater treatment models. ESIS Project Technical Note, School for Resource and Environmental Studies, Dalhousie University, Halifax, N.S.

Majthay, A., and Whinston, A. (1974). Quasiconcave minimization subject to linear constraints. *Discrete Mathematics* 9, 35–59.

Makarov, I. M., and Radashevich, Yu. B. (1981). Statistical estimation of the solution in constrained extremal problems. *Avtomatika i Telemekhanika*, No. 3, 41–48. (In Russian.)

Mandelbrot, B. B. (1983). *The Fractal Geometry of Nature.* (Updated and augmented edn.) W. H. Freeman and Co., New York.

Marks, D. H. (1975). Models in water resources. In: *A Guide to Models in Governmental Planning and Operations* (ed. S. I. Gass and R. L. Sisson), pp. 103–137. (Prepared for the United States Environmental Protection Agency, Washington, D.C.) Sauger Books, Potomac, Md.

Marquardt, D. W. (1963). An algorithm for least-squares estimation of nonlinear parameters. *SIAM Journal of Applied Mathematics* 11, 431–444.

Marti, K. (1980). Random search in optimization problems as a stochastic decision process (adaptive random search). *Methods of Operations Research* 86, 223–234.

Martos, B. (1971). Quadratic programming with a quasi-convex objective function. *Operations Research* 19, 87–97.

Martos, B. (1975). *Nonlinear Programming.* North-Holland, Amsterdam.

Martz, H., and Bryson, M. (1983). On combining data for estimating the frequency of low probability events. *Nuclear Science and Engineering* 83, 267–280.

Masliev, I., and Somlyódy, L. (1994). Uncertainty analysis and parameter estimation for a class of river dissolved oxygen models. Working Paper 94–9, International Institute for Applied Systems Analysis, Laxenberg, Austria.

Massart, D. L., Plastria, F., and Kaufman, L. (1983). Nonhierarchical clustering with MASLOC. *Pattern Recognition* 16, 507–516.

Massmann, J., and Freeze, R. A. (1987). Groundwater contamination from waste management sites: The interaction between risk-based engineering design and regulatory policy. Parts I–II. *Water Resources Research* 23, 351–367 and 368–380.

Mayne, D. Q., and Polak, E. (1984). Outer approximation algorithm for nondifferentiable optimization problems. *Journal of Optimization Theory and Applications* 42, 19–30.

McBean, E. A. (1987). Developing management positions for acidic emission reduction negotiations In: *Systems Analysis in Water Quality Management,* (ed. M. B. Beck), pp. 51–59. Pergamon Press, Oxford.

McBean, E. A., and Ponnambalam, K. (1989). Optimal emission reductions for varying deposition limits and meteorological conditions. (Presented at the Workshop on Integrated Models of Acidic Deposition, Toronto.)

McConway, K. (1981). Marginalization and linear opinion pools. *Journal of the American Statistical Association* 76, 410–411.

McCormick, G. P. (1976). Computability of global solutions to factorable nonconvex programs. Part I: Convex underestimating problems. *Mathematical Programming* 10, 147–175.

McCormick, G. P. (1983). *Nonlinear Programming: Theory, Algorithms and Applications.* Wiley, New York.

McCubbin Consultants, Inc. (1992). Best available technology for the Ontario pulp and paper industry. Report prepared for the Ontario Ministry of Environment, Toronto.

McDonald, C. M. and Floudas, Ch. A. (1994). Decomposition based and branch and bound global optimization approaches for the phase equilibrium problem. *Journal of Global Optimization* 5, 205–251.

McMurty, G. J., and Fu, K. S. (1966). A variable structure automaton used as a multimodal searching technique. *IEEE Transactions on Automatic Control* 11, 379–387.

Medova, E. (1981). A Bayesian procedure for resource evaluation of petroleum provinces in the early stages of exploration. Working Paper 81–66, IIASA, Laxenburg, Austria.

Meewella, C. C., and Mayne, D. Q. (1988). An algorithm for global optimization of Lipschitz-functions. *Journal of Optimization Theory and Applications* 57, 307–323.

Meewella, C. C., and Mayne, D. Q. (1989). Efficient domain partitioning algorithms for global optimization of rational and Lipschitzian functions. *Journal of Optimization Theory and Applications* 61, 247–270.

Meshcheryakov, A. A., and Rovkij, M. E. (1992). Automatic device for calibration and operative control of multi-channel measuring systems. *Pribory i Tekhnika Eksperimenta* 1, 224–225. (In Russian.)

Meyer, P. D., and Brill, E. D. (1988). A method for locating wells in a groundwater monitoring network under conditions of uncertainty. *Water Resources Research* 24, 1277–1282.

Michalewicz, Z. (1992). *Genetic Algorithms + Data Structure = Evolution Programs. [Analogies of Natural Processes, Evolution Principle → Survival of the Fittest],* (2nd edn.) Springer, Berlin.

Miranda, M. A., and Lopez-Barrio, C. A. (1993). Generation of optimized single distributions of weights for random built-in self-test. *Proceedings of the 24th IEEE International Test Conference* (Baltimore, Maryland), pp. 1023–1030. IEEE Service Center, Piscataway, NJ.

Mitten, L. G. (1970). Branch and bound method: General formulation and properties. *Operations Research* 18, 24–34.

Mladineo, R. H. (1986). An algorithm for finding the global maximum of a multimodal, multivariate function. *Mathematical Programming* 34, 188–200.

Mladineo, R. H. (1992a). Stochastic minimization of Lipschitz functions. In: *Recent Advances in Global Optimization,* (ed. C. A. Floudas and P. M. Pardalos), pp. 369–383. Princeton University Press.

Mladineo, R. H. (1992b). Convergence rates of a global optimization algorithm. *Mathematical Programming* 54, 223–232.

Mockus, J. (1967). *Multiextremal Problems in Design.* Nauka, Moscow. (In Russian.)

Mockus, J. (1972). On Bayesian methods of search for the extremum. *Automatics and Computers* 3, 53–62.

Mockus, J. (1989). *Bayesian Approach to Global Optimization.* Kluwer, Dordrecht.

Mockus, J. (1994). Application of Bayesian approach to numerical methods of global and stochastic optimization. *Journal of Global Optimization* 4, 347–365.

Mockus, J., Tiesis, V., and Žilinskas, A. (1978). The application of Bayesian methods for seeking the extremum. In: *Towards Global Optimisation,* Vol. 2, (ed. L. C. W. Dixon and G. P. Szegö), pp. 117–129. North-Holland, Amsterdam.

Moder, J. J., and Elmaghraby, S. E., eds. (1978). *Handbook of Operations Research.* Van Nostrand Reinhold, New York.

Moore, R. E. (1962). Interval arithmetic and automatic error analysis in digital computation. Ph.D. Dissertation, Stanford University.

Moore, R. E. (1966). *Interval Analysis.* Prentice-Hall, Englewood Cliffs, N.J.

Moore, R. E. (1979). *Methods and Applications of Interval Analysis.* SIAM, Philadelphia.

Moore, R. E., ed. (1988). *Reliability in Computing: The Role of Interval Methods.* Academic Press, Boston.

Moore, R., Hansen, E., and Leclerc, A. (1992). Rigorous methods for global optimization. In: *Recent Advances in Global Optimization,* (ed. C. A. Floudas and P. M. Pardalos), pp. 321–342. Princeton University Press.

Moore, R. E., and Ratschek, H. (1988). Inclusion functions and global optimization II. *Mathematical Programming* 41, 341–356.

Morris, J. M., and D'Amore, R. J. (1980). Aggregating and communicating uncertainty. PAR Report 79–71, Pattern Analysis and Recognition Corporation, Rome.

Morris, P. (1977). Combining expert judgments: A Bayesian approach. *Management Science* 23, 679–693.

Moshirvaziri, K. (1994). A generalization of the construction of test problems for nonconvex optimization. *Journal of Global Optimization* 5, 21–34.

Mosleh, A., and Apostolakis, G. (1986). The assessment of probability distributions from expert opinions. *Risk Analysis* 6, 447–461.

Motreanu, D., and Panagiotopoulos, P. D. (1995). Nonconvex energy functions, related eigenvalue hemivariational inequalities on the sphere and applications. *Journal of Global Optimization* 6, 163–177.

Mulvey, J., and Cowder, H. (1979). Cluster analysis: An application of Lagrangian relaxation. *Management Science* 25, 329–340.

Munn, R. E., ed. (1975). *Environmental Impact Assessment.* Wiley, New York.

Muroga, S. (1982). *VLSI System Design.* Wiley, New York.

Murtagh, B. A., and Saunders, M. A. (1983, 1987). MINOS 5.1 User's Guide. Report SOL 83-20R (revised Jan. 1987), Stanford University.

Murty, K. G. (1988). *Linear Complementarity, Linear and Nonlinear Programming.* Heidermann, Berlin.

Muu, L. D. (1993). Algorithm for solving convex programs with an additional convex-concave constraint. *Mathematical Programming* 61, 75–87.

Nefedov, V. N. (1987). Search for the global maximum of a function of several variables. *USSR Computational Mathematics and Mathematical Physics* 11, 38–54.

Neimark, J., and Strongin, R. G. (1969). Informational approach to the problem of search for the extremum of a function. *Engineering Cybernetics* 4, 17–26.

Neuman, S. P., Fogg, G. E., and Jacobson, E. A. (1980). A statistical approach to the inverse problem of aquifer hydrology: 2. Case study. *Water Resources Research* 16, 33–58.

Nijkamp, P. (1980). *Environmental Policy Analysis: Operational Methods and Models.* Wiley, New York.

Nitti, M. L., and Speranza, M. G. (1987). Heuristic control strategies in flexible flow lines. In: *Modern Production Management Systems,* pp. 97–105. North-Holland, Amsterdam.

Novotny, W., and Chesters, G. (1982). *Handbook of Nonpoint Pollution.* Van Nostrand Reinhold, New York.

Okamoto, M. (1958). Some inequalities relating to the partial sum of binomial probabilities. *Annals of the Institute of Statistical Mathematics* 10, 29–35.

Onstad, D. W. (1988). Population-dynamics theory: The roles of analytical and supercomputer models. *Ecological Modelling* 43, 111–124.

Ontario Waste Management Corporation (1989). *Industrial Waste Audit and Reduction Manual.* OWMC, Toronto, Ontario.

Orlob, G. T., ed. (1983). *Mathematical Modeling of Water Quality: Streams, Lakes and Reservoirs.* Wiley, New York.

Ortega, J., and Rheinboldt, W. C. (1970). *Iterative Solution of Nonlinear Equations in Several Variables.* Academic Press, New York.

Padgett, W. J., and Papadopoulos, A. S. (1979). Stochastic models for prediction of BOD and DO in streams. *Ecological Modelling* 6, 289–303.

Pang, J.-S. (1995). Complementarity problems. In: *Handbook of Global Optimization,* (ed. R. Horst and P. M. Pardalos), pp. 271-338. Kluwer, Dordrecht.

Papalambros, P. Y. (1993). Remarks on sufficiency of constraint-bound solutions in optimal design. *Journal of Mechanical Design, Transactions of the ASME* 115, 374–379.

Papalambros, P. Y., and Wilde, D. J. (1988). *Principles of Optimal Design: Modeling and Computation.* Cambridge University Press.

Pardalos, P. M. (1994). On the passage from local to global in optimization. In: *Mathematical Programming – State of the Art*, (ed. J. R. Birge and K. G. Murty), pp.102–131. The University of Michigan, Ann Arbor.

Pardalos, P. M. (1995). An open global optimization problem on the unit sphere. *Journal of Global Optimization* 6, 213.

Pardalos, P. M., Glick, J. H., and Rosen, J. B. (1987). Global minimization of indefinite quadratic problems. *Computing* 39, 281–291.

Pardalos, P. M., and Rosen, J. B. (1986). Methods for global concave minimization: A bibliographic survey. *SIAM Review* 28, 367–369.

Pardalos, P. M., and Rosen, J. B. (1987). *Constrained Global Optimization: Algorithms and Applications. Lecture Notes in Computer Science* 268. Springer, Berlin.

Pardalos, P. M., and Rosen, J. B. (1988). Global optimization approach to the linear complementarity problem. *SIAM Journal of Scientific and Statistical Computing* 9, 341–353.

Pardalos, P. M., and Vavasis, S. A. (1991). Quadratic programing with one negative eigenvalue is NP-hard. *Journal of Global Optimization* 1, 15–22.

Pardalos, P. M., and Xue, G. L., eds. (1994) Advances in computational biology, computational chemistry and protein folding. Preface. *Journal of Global Optimization* 4, No. 2. (Special issue.) iii.

Park, R. A., et al. (1974). A generalized model for simulating lake ecosystems. *Simulation* 23, 33–50.

Parvin, M., and Grammas, G. W. (1976). Optimization models for environmental pollution control: A synthesis. *Journal of Environmental Economics and Management* 3, 113–128.

Patel, N. R., and Smith, R. L. (19983). The asymptotic extreme value distribution of the sample minimum of a concave function under linear constraints. *Operations Research* 31, 789–794.

Patel, N. R., Smith, R. L., and Zabinsky, Z. B. (1988). Pure adaptive search in Monte Carlo optimization. *Mathematical Programming* 43, 317–328.

Patry, G. G., and Chapman, D. T., eds. (1989). *Dynamic Modelling and Expert Systems in Wastewater Engineering.* Lewis, Chelsea, Michigan.

Phong, T. Q., Tao, P. D., and An, L. T. H. (1995). A method for solving d.c. programming problems. Application to fuel mixture nonconvex optimization problem. *Journal of Global Optimization* 6, 87–105.

Pintér, J. (1978). On the maximal distance between two series of empirical distribution functions, with application to an inventory problem. *Methods of Operations Research* 29, 623–636.

Pintér, J. (1980a). On a stochastic model of reservoir system sizing. In: Proceedings of the 9th IFIP Conference on Optimization Techniques (Lecture Notes in Control and Information Sciences 23) (ed. K. Iracki, K. Malanowski, and S. Walukiewicz), pp. 546–558. Springer-Verlag, Berlin.

Pintér, J. (1980b). Stochastic models in regional water quality control. *Hidrológiai Közlöny* 60, 364–373. (In Hungarian.)

Pintér, J. (1980c). On a method of random search for unconstrained minimization. *Avtomatika i Telemekhanika*, No. 12, 76–85. (In Russian.)

Pintér, J. (1981). Stochastic procedures for solving optimization problems. *Alkalmazott Matematikai Lapok* 7, 217–252. (In Hungarian.)

Pintér, J. (1982). Hybrid procedures for solving non-smooth, stochastic constrained optimization problems. *Bulletin of the Moscow State University, Ser. 15. Computational Mathematics and Cybernetics*, No. 1, 39–49. (In Russian.)

Pintér, J. (1983a). An improved Chebyshev-inequality for function value estimates by Monte Carlo techniques. *Alkalmazott Matematikai Lapok* 9, 93–104. (In Hungarian.)

Pintér, J. (1983b). A unified approach to globally convergent one-dimensional optimization algorithms. Research Report IAMI 83–5, Istituto per le Applicazioni della Matematica e dell'Informatica, CNR, Milano.

Pintér, J. (1984a). Convergence properties of stochastic optimization procedures. *Optimization* 15, 405–427.

Pintér, J. (1984b). *Stochastic Optimization Methods.* (Lecture Notes, Eötvös University.) Tankönyvkiadó, Budapest. (In Hungarian.)

Pintér, J. (1985). A modified Bernstein-technique for estimating noise-perturbed function values. *Calcolo* 22, 241–247.

Pintér, J. (1986a). Globally convergent methods for n-dimensional multiextremal optimization. *Optimization* 17, 187–202.

Pintér, J. (1986b). Extended univariate algorithms for n-dimensional global optimization. *Computing* 36, 91–103.

Pintér, J. (1986c). Global optimization on convex sets. *Operations Research Spektrum* 8, 197–202.

Pintér, J. (1987a). A conceptual optimization framework for regional acidification control. *Systems Analysis, Modelling and Simulation* 4, 213–226.

Pintér, J. (1987b). Convergence qualification of partition algorithms in global optimization. Research Report 87–61, Department of Mathematics and Informatics, Delft University of Technology.

Pintér, J. (1988). Branch-and-bound methods for solving global optimization problems with Lipschitzian structure. *Optimization* 19, 101–110.

Pintér, J. (1989). Deterministic approximations of probability inequalities. *ZOR — Methods and Models of Operations Research, Series Theory* 33, 219–239.

Pintér, J. (1990a). On the convergence of adaptive partition algorithms in global optimization. *Optimization* 21, 231–235.

Pintér, J. (1990b). Solving nonlinear equation systems via global partition and search: Some experimental results. *Computing* 43, 309–323.

Pintér, J. (1990c). Globally optimized calibration of environmental models. *Annals of Operations Research* 25, 211–222.

Pintér, J. (1990d). Stochastic decision models for RA&M: A brief methodological overview. Research Report 90.068, Institute for Inland Water Management and Waste Water Treatment, Lelystad.

Pintér, J. (1990e). Model calibration: Problem statement, solution method and implementation manual. Research Report 90.024, Institute for Inland Water Management and Waste Water Treatment, Lelystad.

Pintér, J. (1991a). Stochastic modelling and optimization for environmental management. *Annals of Operations Research* 31, 527–544.

Pintér, J. (1991b). Global convergence revisited: Reply to A. Žilinskas. *Computing* 46, 87–91.

Pintér, J. (1991c). Simplicial partition strategies for Lipschitzian global optimization. Working Paper, Institute for Inland Water Management and Waste Water Treatment, Lelystad.

Pintér, J. (1992a). Convergence qualification of partition algorithms in global optimization. *Mathematical Programming* 56, 343–360.

Pintér, J. (1992b). Lipschitzian global optimization: Some prospective applications. In: *Recent Advances in Global Optimization,* (ed. C. A. Floudas and P. M. Pardalos), pp. 399–432. Princeton University Press.

Pintér, J. (1993). ESIS Project: Final Report. School for Resource and Environmental Studies, Dalhousie University, Halifax, N.S.

Pintér, J., Benedek, P., and Darázs, A. (1990). Risk management of accidental water pollution: An illustrative application. *Water Science and Technology* 22, 265–274.

Pintér, J., and Cooke, R. (1987). Combining expert opinions: An optimization framework. Research Report 87–84, Department of Mathematics and Informatics, Delft University of Technology.

Pintér, J., Fels, M., Lycon, D. S., Meeuwig, D. J., and Meeuwig, J. W. (1993). An intelligent decision support system for assisting industrial wastewater management. *Annals of Operations Research.* (To appear.)

Pintér, J., and Lycon, D. S. (1992a). Approximating TOX as a function of TSS and BOD. ESIS Project Technical Note, School for Resource and Environmental Studies, Dalhousie University, Halifax, N.S.

Pintér, J., and Lycon, D. S. (1992b). Statistical uncertainty analysis of wastewater treatment systems. ESIS Project Technical Note, School for Resource and Environmental Studies, Dalhousie University, Halifax, N.S.

Pintér, J., and Pesti, G. (1991). Set partition by globally optimized cluster seed points. *European Journal of Operational Research* 51, 127–135.

Pintér, J., and Somlyódy, L. (1986). A stochastic lake eutrophication management model. In: *Stochastic Optimization,* (ed. V. I. Arkin, A. N. Shiriaev, and R. Wets), pp. 501–512. *Lecture Notes on Control and Information Sciences* 81, Springer, Berlin.

Pintér, J., and Somlyódy, L. (1987). Optimization of regional water quality monitoring strategies. *Water Science and Technology* 19, 721–727.

Pintér, J., Szabó, J., and Somlyódy, L. (1986). Multiextremal optimization for calibrating water resources models. *Environmental Software* 1, 98–105.

Pintér, J., and Van der Molen, D. T. (1991). Calibration under different problem specifications: Application to the model SED. Research Report 91.049, Institute for Inland Water Management and Waste Water Treatment, Lelystad.

Piyavskii, S. A. (1967). An algorithm for finding the absolute minimum. Institute of Cybernetics (IK AN USSR), Kiev, pp. 13–24. (In Russian.)

Piyavskii, S. A. (1972). An algorithm for finding the absolute extremum of a function. *USSR Computational Mathematics and Mathematical Physics* 12, 57–67.

Polak, E. (1971). *Computational Methods in Optimization.* Academic Press, New York.

Polak, E., and Ribière, G. (1963). Note sur la convergence de méthodes de directions conjugées. *Revue Française d'Informatique et Recherche Opérationelle* 16, 35–43.

Poljak, B. T. (1978). Nonlinear programming methods in the presence of noise. *Mathematical Programming* 14, 87–97.

Porto, R. L., and Lopes, J. E. (1993). Global search techniques to calibrate rainfall-runoff models. In: *Proceedings of the Symposium on Engineering Hydrology* (San Francisco, July 25–30, 1993), pp. 1218–1222.

Powell, M. J. D. (1964). An efficient method for finding the minimum of a function of several variables without calculating derivatives. *Computer Journal* 7, 155–162.

Prékopa, A. (1973). Contributions to the theory of stochastic programming. *Mathematical Programming* 3, 202–221.

Prékopa, A., ed. (1980). *Studies in Applied Stochastic Programming.* Computer and Automation Institute, Hungarian Academy of Sciences, Budapest.

Prékopa, A. (1986). Boole-Bonferroni inequalities and linear programming. Research Report 4–86, RUTCOR, Rutgers University, New Brunswick, N.J.

Prékopa, A., and Wets, R. J.-B., eds. (1986). *Stochastic Programming. Mathematical Programming Study,* Vols. 27–28.

Press, W. H., Teukolsky, S. A., Vetterling, W. T., and Flannery, B. (1992). *Numerical Recipes in FORTRAN: The Art of Scientific Computing,* Cambridge University Press.

Rabinowitz, Ph., ed. (1970). *Numerical Methods for Nonlinear Algebraic Equations.* Gordon and Breach, London.

Raghavachari, M. (1969). On connections between zero-one integer programming and concave programming under linear constraints. *Operations Research* 17, 680–684.

Ralston, A., and Reilly, E. D., eds. (1993). *Encyclopedia of Computer Science.* (3rd edn.) Van Nostrand Reinhold, New York.

Rao, M. R. (1971). Cluster analysis and mathematical programming. *Journal of the American Statistical Association* 66, 622–626.

Rastrigin, L. A. (1974). *Systems of Extremal Control.* Zinatne, Riga. (In Russian.)

Ratschek, H. (1985). Inclusion functions and global optimization I. *Mathematical Programming* 33, 300–317.

Ratschek, H., and Rokne, J. G. (1987). Efficiency of a global optimization algorithm. *SIAM Journal of Numerical Analysis* 24, 1191–1201.

Ratschek, H., and Rokne, J. G. (1988). *New Computer Methods for Global Optimization.* Ellis Horwood, Chichester.

Ratschek, H., and Rokne, J. (1993), Experiments using interval analysis for solving a circuit design problem. *Journal of Global Optimization* 3, 501–518.

Rényi, A. (1968). *Probability Calculus.* Tankönyvkiadó, Budapest. (In Hungarian.)

Rich, L. G. (1973). *Environmental Systems Engineering.* McGraw-Hill, Tokyo.

Richardson, M. L., ed. (1988). *Risk Assessment of Chemicals in the Environment.* The Royal Society of Chemistry, London.

Rijkswaterstaat (1991). *ZWENDL User's Manual.* (In Dutch.) Rijkswaterstaat, The Hague.

Riley, M. R., and Stefan, H. G. (1988). MINLAKE: A dynamic lake water quality simulation model. *Ecological Modelling* 43, 155–182.

Rinnooy Kan, A. H. G., and Timmer, G. T. (1987a). Stochastic global optimization methods. Part I: Clustering methods. *Mathematical Programming* 39, 27–56.

Rinnooy Kan, A. H. G., and Timmer, G. T. (1987b). Stochastic global optimization methods. Part II: Multi-level methods. *Mathematical Programming* 39, 57–78.

Rinnooy Kan, A. H. G., and Timmer, G. T. (1989). Global optimization: A survey. In: *Handbooks of Operations Research,* Vol. 1. *Optimization,* (eds. G. L. Nemhauser, A. H. G. Rinnooy Kan, and M. J. Todd), pp. 631–662. North-Holland, Amsterdam.

Robinson, S. M., ed. (1980). *Analysis and Computation of Fixed Points.* Academic Press, New York.

Rockafellar, R. T. (1970). *Convex Analysis.* Princeton University Press.

Romeijn, H. E., and Smith, R. L. (1994). Simulated annealing for constrained global optimization. *Journal of Global Optimization* 5, 101–126.

Romero, C. (1986). A survey of generalized goal programming (1970–1982). *European Journal of Operational Research* 25, 183–191.

Rosen, J. B. (1983). Minimization of a linearly constrained concave function by partition of the feasible domain. *Mathematics of Operations Research* 8, 215–230.

Rosen, J. B., and Pardalos, P. M. (1986). Global minimization of large-scale constrained quadratic problems by separable programming. *Mathematical Programming* 34, 163–174.

Rous, P. J. (1993). Global approach to the search problem in surface crystallography by low-energy electron diffraction. *Surface Science* 296, 358–373.

Rousseeuw, P. (1987). Silhouettes: A graphical aid to the interpretation and validation of cluster analysis. *Journal of Computational and Applied Mathematics* 20, 53–65.

Rowe, W. (1977). *An Anatomy of Risk.* Wiley, New York.

Rubenstein, R. Y. (1981). *Simulation and the Monte Carlo Method.* Wiley, New York.

Russell, C. S. and Spofford, W. O. (1977). A regional environmental quality management model: An assessment. *Journal of Environmental Economics and Management* 4, 89–110.

Saari, D. G. (1993). *The Geometry of Voting.* Springer, Berlin.

Said, M., and Kerr, C. P. (1993). A numerical method for determining optimum synthesis gas injection rates along the length of a methanol reactor using a Rayleigh-Ritz technique. Paper presented at the Spring National Meeting of AIChE (Houston, Texas, March 28–31, 1993).

Savage, L. J. (1971). Elicitation of personal probabilities and expectations. *Journal of the American Statistical Association* 66, 783–801.

Sawaragi, Y., Nakayama, H., and Tanino, T. (1985). *Theory of Multiobjective Optimization.* Academic Press, New York.

Scavia, D. (1980). An ecological model of Lake Ontario. *Ecological Modelling* 8, 49–78.

Schaible, S. (1995). Fractional programming. In: *Handbook of Global Optimization,* (ed. R. Horst and P. M. Pardalos), pp. 495–608. Kluwer, Dordrecht.

Scheffer, M. (1990). Simple models as useful tools for ecologists. Ph.D. Dissertation. University of Utrecht.

Schoen, F. (1982). On a sequential search strategy in global optimization problems. *Calcolo* 19, 321–334.

Schoen, F. (1991). Stochastic techniques for global optimization: A survey of recent advances. *Journal of Global Optimization* 1, 207–228.

Schoen, F. (1993). A wide class of test functions for global optimization. *Journal of Global Optimization* 3, 133–137.

Scholten, H., de Hoop, B. J., and Herman, P. M. J. (1990). *SENECA 1.2 - A Simulation ENvironment for ECological Application.* Manual. Delta Institute for Hydrobiological Research, Yerseke.

Schrage, L. (1990). *User's Manual for LINDO.* Scientific Press, Palo Alto.

Schrage, L. (1991). *User's Manual for LINGO.* LINDO Systems Inc., Chicago.

Schweigman, C., Joosten, G., Klein Haneveld, W. K., Kijne, J. W., and Snijders, T. A. B. (1985). *O.R. Problems in Agriculture in Developing Countries.* Tanzania Publishing House, Dar es Salaam.

Seidel, W. (1991). An open optimization problem in statistical quality control. *Journal of Global Optimization* 1, 295–303.

Sergeyev, Ya. D. (1994). An information global optimization algorithm with local tuning. *SIAM Journal on Optimization.* (To appear.)

Sergeyev, Ya. D., and Grishagin, V. A. (1994a). A parallel method for finding the global minimum of univariate functions. *Journal of Optimization Theory and Applications* 80, 513–536.

Sergeyev, Ya. D., and Grishagin, V. A. (1994b). Sequential and parallel algorithms for global optimization. *Optimization Methods and Software* 3, 111–124.

Sergeyev, Ya. D., and Strongin, R. G. (1989). A global minimization algorithm with parallel iterations. *USSR Computational Mathematics and Mathematical Physics*

29, No. 3, 332–345. (In Russian.)

Shalloway, D. (1992). Packet annealing: A deterministic method for global optimization. Application to molecular conformation. In: *Recent Advances in Global Optimization*, (ed. C. A. Floudas and P. M. Pardalos), pp. 433–478. Princeton University Press.

Shaw, R.W. (1986). A proposed strategy for reducing sulfate deposition in North America. II: Methodology for minimizing costs. *Atmospheric Environmen* 20, 201–206.

Shen, Z., and Zhu, Y. (1987). An interval version of Shubert's iterative method for the localization of the global maximum. *Computing* 38, 275–280.

Shepilov, M. A. (1987). Determination of the roots and the global extremum of a Lipschitz function. *Cybernetics* 23, 233–238.

Sherman, B. (1957). Percentiles of the ω_n statistics. *Annals of Mathematical Statistics* 28, 259–261.

Shubert, B. O. (1972). A sequential method seeking the global maximum of a function. *SIAM Journal on Numerical Analysis* 9, 379-388.

Simmons, D. M. (1975). *Nonlinear Programming for Operations Research.* Prentice-Hall, Englewood Cliffs, N.J.

Singer, I. (1980). Minimization of continuous convex functionals on complements of convex subsets of locally convex spaces. *Optimization* 11, 221–234.

Singpurwalla, N. D. (1988). An interactive P.C. based procedure for reliability assessment incorporating expert opinion and survival data. *Journal of the American Statistical Association* 83, 44–51.

Smith, R. L. (1987). Estimating tails of probability distributions. *Annals of Statistics* 15, 1174–1207.

Snyder, W. (1978). Decision making with risk and uncertainty: The case of horse racing. *American Journal of Psychology* 91, 201–209.

Solis, F. J., and Wets, R. (1981). Minimization by random search techniques. *Mathematics of Operations Research* 6, 19–30.

Somlyódy, L. (1980). Preliminary study on wind induced interaction between water and sediment for Lake Balaton. In: *Proceedings of the Second Hungarian Academy of Sciences/International Institute for Applied Systems Analysis Task Force Meeting on Lake Balaton Modeling,* (ed. G. van Straten et al.), Vol. 1, pp. 26–50. MTA VEAB Veszprém, Hungary.

Somlyódy, L. (1982a). Water quality modelling: Comparison of transport-oriented and ecology-oriented approaches. *Ecological Modelling* 17, 183–207.

Somlyódy, L. (1982b). Modeling a complex environmental system: The Lake Balaton case study. *Mathematical Modelling* 3, 481–502.

Somlyódy, L. (1983). Lake eutrophication management models. In: *Eutrophication of Shallow Lakes: Modelling and Management. The Lake Balaton Case Study,* (ed. L. Somlyódy, S. Herodek, and J. Fischer), pp. 207–250. Collaborative Paper 83–S3, IIASA, Laxenburg.

Somlyódy, L., and Pintér, J. (1984). Water quality management models. Research Report 7614/3/38. Water Resources Research Centre, Budapest. (In Hungarian.)

Somlyódy, L., and Szilágyi, F. (1984). Influence of sediment resuspension on the light conditions in the Keszthely basin, Lake Balaton. Research Report 7614/3/56. Water Resources Research Centre, Budapest. (In Hungarian.)

Somlyódy, L., and van Straten, G., eds. (1986). *Modeling and Managing Shallow Lake Eutrophication.* Springer, Berlin.

Somlyódy, L., and Varis, O. (1992). Water quality modeling of rivers and lakes. Working Paper 92–041, International Institute for Applied Systems Analysis, Laxenburg, Austria.

Somlyódy, L., and Wets, R. (1986). Stochastic optimization models for lake eutrophication management. *Operations Research* 36, 669–681.

Steuer, R. E. (1986). *Multiple Criteria Optimization: Theory, Computation and Application.* Wiley, New York.

Stewart, W. R., and Golden, B. L. (1983). Stochastic vehicle routing: A comprehensive approach. *European Journal of Operational Research* 14, 371–385.

Stoer, J., and Witzgall, C. (1970). *Convexity and Optimization in Finite Dimensions.* Springer, New York.

Stokoe, P. K., Lane, P. A., Côté, R. P., and Wright, J. A. (1989). Evaluation of holistic marine ecosystem modelling as a potential tool in environmental impact assessment. Manuscript. Bedford Institute of Oceanography, Dartmouth, N.S.

Stortelder, W. J. H. (1995). Parameter estimation in dynamic systems. Working Paper, Department of Numerical Mathematics, CWI, Amsterdam. (Submitted.)

Streeter, H. W., and Phelps, E. B. (1925). A study of the pollution and natural purification of the Ohio River. *Public Health Bulletin* 146, U.S. Public Health Service, Washington, D.C.

Strigul, O. I. (1987). Search for a global extremum in a certain subclass of functions with the Lipschitz condition. *Cybernetics* 21, 812–818.

Strongin, R. G. (1973). On the convergence of an algorithm for finding a global extremum. *Engineering Cybernetics* 11, 549–555.

Strongin, R. G. (1978). *Numerical Methods for Multiextremal Problems.* Nauka, Moscow. (In Russian.)

Strongin, R. G. (1984). Numerical methods for multiextremal mathematical programming problems with nonconvex constraints. In: *Lecture Notes in Economics and Mathematical Systems* 255, (ed. V. F. Demyanov and D. Pallaschke), pp. 278–282. Springer, Berlin.

Strongin, R. G. (1992). Algorithms for multiextremal mathematical programming problems employing the set of joint space-filling curves. *Journal of Global Optimization* 2, 357–378.

Strongin, R. G., and Sergeyev, Ya. D. (1992). Global multidimensional optimization on parallel computer. *Parallel Computing* 18, 1259–1273.

Sturua, E. G., and Zavriev, S. K. (1991). A trajectory algorithm based on the gradient method. I: The search on the quasi-optimal trajectories. *Journal of*

Global Optimization 1, 375–388.

Sukharev, A. G. (1971). On optimal strategies for searching the extremum. *USSR Computational Mathematics and Mathematical Physics* 11, 910–924.

Sukharev, A. G. (1985). The equality of errors in classes of passive and sequential algorithms. *USSR Computational Mathematics and Mathematical Physics* 25, 193–195.

Suki, B., Csendes, T., Daróczy, B. (1990). Mechanical impedance of the canine diaphragm: I. Theoretical model and parameter estimation. *Medical and Biological Engineering and Computing* 28, 367–373.

Syaiful, S., Wilhelm, A. M., and Delmas, H. (1983). Bubble column and three phase fluidized beds: A comparison of axial disperson and gas-liquid mass transfer by dynamic absorption. *Chemical Engineering and Processing* 32, 149–154.

Szöllősi-Nagy, A. (1982). On the discretization of the continuous linear cascade by means of the state-space analysis. *Journal of Hydrology* 58, 223–236.

Taha, H. A. (1973). Concave minimization over a convex polyhedron. *Naval Research Logistics Quarterly* 20, 533–548.

Taha, H. A. (1975). *Integer Programming Theory, Applications and Computations.* Academic Press, New York.

Talman, A. J. J., and Van der Laan, G. (1987). *The Computation and Modelling of Economic Equilibria.* North-Holland, Amsterdam.

Tang, Z. B. (1993). Partitioned random search to optimization. *Proceedings of the 1993 American Control Conference* (San Francisco, California), pp. 2330–2334. IEEE Service Center, Piscataway, N.J.

Tebbutt, T. H. Y. (1992). *Principles of Water Quality Control.* (4th edn.) Pergamon, Oxford.

Thach, P. T. (1988). The design centering problem as a D.C. programming problem. *Mathematical Programming* 41, 229–248.

Thach, P. T. (1993). D.C. sets, D.C. functions and nonlinear equations. *Mathematical Programming* 58, 415–428.

Thach, P. T., and Konno, H. (1993). Generalized Dantzig-Wolfe decomposition principle for a class of nonconvex programming problems. *Mathematical Programming* 62, 239–260.

Thach, P. T., and Tuy, H. (1987). Global optimization under Lipschitzian constraints. *Japan Journal of Applied Mathematics* 4, 205–217.

Thach, P. T., and Tuy, H. (1990). The relief indicator method for constrained global optimization. *Naval Research Logistics* 37, 473–497.

Thoai, N. V. (1988). A modified version of Tuy's method for solving D.C. programming problems. *Optimization* 19, 665–674.

Thoai, N. V., and Tuy, H. (1980). Convergent algorithms for minimizing a concave function. *Mathematics of Operations Research* 5, 556–566.

Thomann, R. V., and Mueller, J. A. (1987). *Principles of Surface Water Quality Modeling and Control.* Harper and Row, New York.

Timmer, G. T. (1984). Global optimization: A stochastic approach. Ph.D. Dissertation. Erasmus University, Rotterdam.

Timonov, L. N. (1977). An algorithm for search of a global extremum. *Engineering Cybernetics* 15, 38–44.

Tind, J. (1990). Decomposition in global optimization. Publication No. 90/7, Department of Operations Analysis, Aarhus University, Aarhus.

Törn, A. A. (1978). A search clustering approach to global optimization. In: *Towards Global Optimisation,* Vol. 2, (ed. L. C. W. Dixon and G. P. Szegö), pp. 49–62. North-Holland, Amsterdam.

Törn, A. A., and Viitanen, S. (1992). Topographical global optimization. In: *Recent Advances in Global Optimization,* (ed. C. A. Floudas and P. M. Pardalos), pp. 384–398. Princeton University Press.

Törn, A. A., and Viitanen, S. (1994). Topographical global optimization using presampled points. *Journal of Global Optimization* 5, 267–276.

Törn, A. A., and Žilinskas, A. (1989). *Global Optimization. Lecture Notes in Computer Science* 350. Springer, Berlin.

Treccani, G. (1975). On the convergence of Branin's method: A counter example. In: *Towards Global Optimisation,* Vol. 1, (ed. L. C. W. Dixon and G. P. Szegö), pp. 107–116. North-Holland, Amsterdam.

Triantaphyllou, E. (1994). Inference of a minimum size Boolean function from examples by using a new effcient branch-and-bound algorithm. *Journal of Global Optimization* 5, 69–94.

Tung, Y.-K., and Hathhorn, W. E. (1988). Probability distribution for critical DO location in streams. *Ecological Modelling* 42, 61–74.

Tuy, H. (1964). Concave programming under linear constraints. *Soviet Mathematics Doklady* 5, 1437–1440.

Tuy, H. (1987a). Global minimization of the difference of two convex functions. *Mathematical Programming Study* 30, 150–182.

Tuy, H. (1987b). Convex programs with an additional reverse convex constraint. *Journal of Optimization Theory and Applications* 52, 463–485.

Tuy, H. (1995). D.C. optimization: Theory, methods and algorithms. In: *Handbook of Global Optimization,* (ed. R. Horst and P. M. Pardalos), pp. 149–216. Kluwer, Dordrecht.

Tuy, H., and Al-Khayyal, F. A. (1992). Global optimization of a nonconvex single facility location problem by sequential unconstrained convex minimization. *Journal of Global Optimization* 2, 61–71.

Tuy, H., and Horst, R. (1988). Convergence and restart in branch-and-bound algorithms for global optimization. Application to concave minimization and D.C. optimization problems. *Mathematical Programming* 41, 161–183.

Tuy, H., Thieu, T. V., and Thoai, N. Q. (1985). A conical algorithm for globally minimizing a concave function over a closed convex set. *Mathematics of Operations Research* 10, 489–515.

Twilt, F. (1981). Newton flows for meromorphic functions. Ph.D. Thesis, University of Technology, Twente.

Ueing, U. (1972). A combinatorial method to compute the global solution of certain non-convex optimization problems. In: *Numerical Methods for Nonlinear Optimization,* (ed. F. A. Lootsma), pp. 223–230. Academic Press, New York.

Ulanowicz, R. E. (1988). On the importance of higher-level models in ecology. *Ecological Modelling* 43, 45–56.

United States Environmental Protection Agency (1974). *A Guide to Models in Governmental Planning and Operations.* Washington, D.C.

United States Environmental Protection Agency (1980). Acid rain. Research Report EPA–600/9–79–036. Research Triangle Park, North Carolina.

United States Environmental Protection Agency (1988). *Waste Minimization Opportunity Assessment Manual.* Research Report EPA–625/7–88/003, Cincinnati.

Van der Molen, D. T. (1991). A simple, dynamic model for the simulation of the release of phosphorous from sediments in shallow, eutrophic systems. *Water Research* 25, 737–744.

Van der Molen, D. T., and Pintér, J. (1993). Environmental model calibration under different problem specifications: An application to the model SED. *Ecological Modelling* 68, 1–19.

Van Laarhoven, P. J. M., and Aarts, E. H. L. (1987). *Simulated Annealing: Theory and Applications.* Kluwer, Dordrecht.

Van Noortwijk, J. M., Dekker, R., Cooke, R. M., and Mazzucchi, T. A. (1992). Expert judgment in maintenance optimization. *IEEE Transactions on Reliability* 41, 427–432.

Van Raaphorst, W., Kloosterhuis, H. T., Cramer, A., and Bakker, K. J. M. (1990). Nutrient early diagenesis in the sandy sediments of the Dogger Bank area, North Sea: Pore water results. *Netherlands Journal of Sea Research* 26, 25–52.

Vavasis, S.A. (1995) Complexity issues in global optimization: A survey. In: *Handbook of Global Optimization,* (ed. R. Horst and P. M. Pardalos), pp. 27–41. Kluwer, Dordrecht.

Vaysbord, E. M., and Yudin, D. B. (1968). Multiextremal stochastic approximation. *Engineering Cybernetics* 5, No. 1, 1–10.

Veinott, A. F. (1967). The supporting hyperplane method for unimodal programming. *Operations Research* 15, 147–152.

Vicente, L. N., and Calamai, P. H. (1994). Bilevel and multilevel programming: A bibliography review. *Journal of Global Optimization* 5, 291–306.

Vilkov, A. V., Zhidkov, N. P. and Schedrin, B. M. (1975). A method for finding the global minimum of a function of one variable. *USSR Computational Mathematics and Mathematical Physics* 15, 1040-1042.

Vincke, Ph. (1986). Analysis of multicriteria decision aid in Europe. *European Journal of Operational Research* 25, 160–168.

Vinod, H. D. (1969). Integer programming and the theory of grouping. *Journal of the American Statistical Association* 64, 506–519.

Vizvári, B., and Yilmaz, F. (1994), An ordering (enumerative) algorithm for nonlinear 0–1 programming. *Journal of Global Optimization* 5, 277–290.

Voogt, L., van Zetten, J., Bak, C., and Pintér, J. (1991). Calibration of the one-dimensional flow model ZWENDL in the Noordelijk Deltabekken region: Some illustrative results. Research Report 91.028, Institute for Inland Water Management and Waste Water Treatment, Lelystad.

Vysotskaya, I. N., and Strongin, R. G. (1983). A method for solving nonlinear equations using prior probabilistic root estimates. *USSR Computational Mathematics and Mathematical Physics* 23, 3–12. (In Russian.)

Wacker, H.-J., ed. (1978). *Continuation Methods.* Academic Press, New York.

Wagner, B. J., and Gorelick, S. M. (1987). Optimal groundwater quality management under parameter uncertainty. *Water Resources Research* 23, 1162–1174.

Wallace, S. W. (1988). A two-stage stochastic facility location problem with time-dependent supply. In: *Numerical Techniques for Stochastic Optimization,* (ed. Yu. M. Ermoliev and R. J.-B. Wets), pp. 489–513. Springer, Berlin.

Wang, T., Xing, X., and Zhuang, X. (1993). Correspondenceless 3-D surface reconstruction using global optimization. *Acta Automatica Sinica* 19, 497–499.

Warwick, J. J., and Cale, W. G. (1988). Estimating model reliability using data with uncertainty. *Ecological Modelling* 41, 169–181.

Wasilkowski, G. W. (1992). On average complexity of global optimization problems. *Mathematical Programming* 57, 313–324.

Wegscheider, W., Lankmayr, E. P., and Budna, K. W. (1982). A chromatographic response function for automated optimization of separations. *Chromatographia* 15, 498-504.

Wegscheider, W., Lankmayr, E. P., and Otto, M. (1983). Relationships between chromatographic response functions and performance characteristics. *Analytica Chimica Acta* 150, 87–103.

Welch, J. W. (1983). Algorithmic complexity: Three NP-hard problems in computational statistics. *Journal of Statistical Computation and Simulation* 15, 17–25.

Wets, R. (1983). Stochastic programming: Solution techniques and approximation schemes. In: *Mathematical Programming: The State of the Art,* (ed. A. Bachem, M. Grötschel, and B. Korte). Springer, Berlin.

Wets, R. (1994). Challenges in stochastic programming. Working Paper 94–32. International Institute for Applied Systems Analysis, Laxenburg, Austria.

White, J. A., Manis, P. B., and Young, E. D. (1992). The parameter identification problem for the somatic shunt model. *Biological Cybernetics* 66, 307–318.

Wilde, D. J. (1978). *Globally Optimal Design.* Wiley Interscience, New York.

Wilde, D. J. (1993). Convexity analysis in detecting a steel plant hidden global optimum. *Journal of Global Optimization* 3, 117–131.

Wingo, D. R. (1984). Fitting three-parameter lognormal models by numerical global optimization: An improved algorithm. *Computational Statistics and Data Analysis* 2, 13–25.

Winkler, R., and Makridakis, S. (1983). The combination of forecasts. *Journal of the Royal Statistical Society* A146, 150–157.

Winston, W. L. (1994). *Operations Research: Applications and Algorithms* (3rd edn.). Wadsworth, Belmont.

Wolfram, S. (1991) *Mathematica – A System for Doing Mathematics by Computer.* (2nd edn.) Addison-Wesley, Redwood City, California.

Wood, G. R. (1985). Multidimensional bisection and global minimization. Working Paper, University of Canterbury, Christchurch, New Zealand.

Wood, G. R. (1991). Multidimensional bisection applied to global optimization. *Computers and Mathematics with Applications* 21, 161–172.

Wood, G. R. (1992). The bisection method in higher dimensions. *Mathematical Programming* 55, 319–337.

World Health Organization (1985). *European Regional Programme on Chemical Safety: Risk Management in Chemical Safety.* Extended report (Proc. WHO Work Group, Ulm, November 8–10, 1984).

Yeh, W. W.-G. (1986). Review of parameter identification procedures in groundwater hydrology: The inverse problem. *Water Resources Research* 22, 95–108.

Yudin, D. B. (1979). *Mathematical Methods of Control under Uncertainty.* Sovietskoye Radio, Moscow. (In Russian.)

Zabinsky, Z. B., Graesser, D. L., Tuttle, M. E., and Kim, G.-I. (1992). Global optimization of composite laminates using improved hit and run. In: *Recent Advances in Global Optimization,* (ed. C. A. Floudas and P. M. Pardalos), pp. 343–368. Princeton University Press.

Zabinsky, Z. B., and Smith, R. L. (1992). Pure adaptive search in global optimization. *Mathematical Programming* 53, 323–338.

Zanakis, S. H., and Evans, J. R. (1981). Heuristic optimization: why, when and how to use it. *Interfaces* 11, No. 5, 84–91.

Zanakis, S. H., and Kyparisis, J. (1986). A review of maximum likelihood estimation methods for the three-parameter Weibull distribution. *Journal of Statistical Computation and Simulation* 25, 53–73.

Zangwill, W. I. (1967). Nonlinear programming via penalty functions. *Management Science* 13, 344–358.

Zenios, S. A. (1989). Parallel numerical optimization: Current status and an annotated bibliography. *ORSA Journal on Computing* 1, 20–43.

Zhang, B., Wood, G. R., and Baritompa, W. P. (1993). Multidimensional bisection: The performance and the context. *Journal of Global Optimization* 3, 337–358.

Zhidkov, N. P., and Schedrin, B. M. (1968). A certain method of search for the minimum of a function of several variables. In: *Computing Methods and Programming* 10, pp. 203–210. Moscow University Press. (In Russian.)

Zhigljavsky, A. A. (1991). *Theory of Global Random Search,* (ed. J. Pintér). Kluwer, Dordrecht.

Zhigljavsky, A. A. (1993). Semiparametric statistical inference in global random search. *Acta Applicanda Mathematicae* 33, 69–88.

Zhigljavsky, A. A., and Žilinskas, A. (1991). *Methods for Searching the Global Extremum.* Nauka, Moscow. (In Russian.)

Zielinski, R. (1981). A statistical estimate of the structure of multiextremal problems. *Mathematical Programming* 21, 348–356.

Zielinski, R., and Neumann, P. (1983). *Stochastische Verfahren zur Suche nach dem Minimum einer Funktion.* Akademie–Verlag, Berlin.

Žilinskas, A. (1981). Two algorithms for one-dimensional multimodal minimization. *Optimization* 12, 53–63.

Žilinskas, A. (1982). Axiomatic approach to statistical models and their use in multimodal optimization theory. *Mathematical Programming* 22, 104–116.

Žilinskas, A. (1986). *Global Optimization.* Mokslas, Vilnius. (In Russian.)

Žilinskas, A. (1992). A review of statistical methods for global optimization. *Journal of Global Optimization* 2, 145–153.

Index

Nonconvex Optimization and Its Applications

1. D.-Z. Du and J. Sun (eds.): *Advances in Optimization and Approximation.* 1994. ISBN 0-7923-2785-3
2. R. Horst and P.M. Pardalos (eds.): *Handbook of Global Optimization.* 1995 ISBN 0-7923-3120-6
3. R. Horst, P.M. Pardalos and N.V. Thoai: *Introduction to Global Optimization* 1995 ISBN 0-7923-3556-2; Pb 0-7923-3557-0
4. D.-Z. Du and P.M. Pardalos (eds.): *Minimax and Applications.* 1995 ISBN 0-7923-3615-1
5. P.M. Pardalos, Y. Siskos and C. Zopounidis (eds.): *Advances in Multicriteria Analysis.* 1995 ISBN 0-7923-3671-2
6. J.D. Pintér: *Global Optimization in Action.* Continuous and Lipschitz Optimization: Algorithms, Implementations and Applications. 1996 ISBN 0-7923-3757-3

KLUWER ACADEMIC PUBLISHERS – DORDRECHT / BOSTON / LONDON